The Economy of Nature

The Economy of Nature

of Nature SECOND EDITION

A TEXTBOOK IN BASIC ECOLOGY

Robert E. Ricklefs
UNIVERSITY OF PENNSYLVANIA

Chiron Press, INCORPORATED/NEW YORK AND CONCORD

THE ECONOMY OF NATURE *Second Edition*

Sales: Chiron Press, Incorporated
            Publishers Storage & Shipping Corp.
            231 Industrial Park
            Fitchburg, Massachusetts 01420

Editorial: Chiron Press, Incorporated
            24 West 96th Street
            New York, New York 10025

Library of Congress Catalogue Card Number 82-83643

ISBN 0-913462-09-8

First Printing, March 1983
Second Printing, August 1983
Third Printing, December 1983
Fourth Printing, July 1985

International Edition ISBN 0-913462-10-1

First Printing, July 1985

Drawings by John Woolsey
Cover painting by Nara Vesely

Contents

Preface

In the seven years since the first edition of *The Economy of Nature* appeared, there have been exciting developments in the discipline of ecology. They include the broader application of evolutionary principles to the study of ecological problems, the statistical appraisal of patterns in biological communities, and the investigation of traditional population problems at the level of organism physiology and biochemistry. In addition, ecologists have greatly increased the use of experimentation to study populations, communities, and ecosystems.

The second edition of *The Economy of Nature* reflects these developments. It also incorporates a more quantitative and theoretical treatment of population processes in the chapters on population regulation, competition, and predation. The book has four more chapters than the first edition, the increase resulting from new material and from reorganization of subject matter. A chapter on natural selection gives more background for evolutionary principles referred to throughout the second edition. The chapters on the relations of organisms to their physical and chemical environments have been increased in number and content. The entire book has been brought up to date. The section on suggested readings has been expanded, and there is an appendix on the International System of Units of Measurement with useful conversion factors.

These extensive changes have not, however, altered the character of *The Economy of Nature.* Like the first edition, the second is meant to be a broad, integrated exposition of the principles of ecology in a book of moderate length, written for introductory courses in ecology and environmental biology. The response to the first edition was extremely favorable, and colleges and universities in the United States, Canada, and the United Kingdom adopted the book widely. I am grateful to many students, teachers, and colleagues who took the time to write to me with comments, suggestions, and encouragement. I hope that you who are now about to read this new edition will derive from it pleasure, stimulation, and knowledge.

Philadelphia   R.E.R.

Preface to the First Edition

The recent proliferation of books in ecology still leaves a gap between long texts that are more or less comprehensive and short books, usually paperbacks, that are inadequate even for beginning courses lasting a quarter or a semester. I have written *The Economy of Nature* to provide a broad integrated treatment of ecological principles in a book of moderate length.

I have emphasized the dynamics of populations, communities, and ecosystems while keeping sight of the organism as the basic unit of the biological community. I have tried to balance ideas and principles on the one hand with examples of structure and functioning of natural systems on the other. I believe such balance makes clear the complementary roles of theory and observation in the development of science. I have also tried to convey the diversity of biological communities and the remarkable manifestation of basic principles under different environmental conditions.

I have deliberately avoided the problem of man's ecological crisis. One could not hope to do justice to a topic of such importance and complexity in a chapter or two tacked on to the end of a text whose subject is the basic principles of ecology. Where that problem can be understood or solved, students will readily discern applications of the principles discovered in the study of natural systems. I have omitted descriptions of mathematical and statistical techniques in ecology because they do not so much aid our understanding of principles as they provide useful tools for the professional ecologist.

The Economy of Nature is meant to be a basic exposition of ecology, not a source book for advanced students and professional ecologists. I have accordingly omitted literature references and, wherever possible, scientific names in the text so as to give force to the narrative itself. References, selected readings, and source books are listed by chapter at the back of the book.

Philadelphia   R. E. R.

The Economy of Nature

1 | Introduction

ECOLOGY IS THE STUDY of plants and animals, both as individuals and together in populations and biological communities, in relation to their environments. The word ecology is derived from the Greek *oikos* meaning house, man's immediate surroundings. The origin of the word in the middle of the last century is obscure, but its general usage can be traced to the definition given by the influential German biologist, Ernst Haeckel, in 1870. "By ecology," he wrote, "we mean the body of knowledge concerning the economy of nature—the investigation of the total relations of the animal both to its organic and to its inorganic environment; including above all, its friendly and inimical relation with those animals and plants with which it comes directly or indirectly into contact—in a word, ecology is the study of all the complex interrelations referred to by Darwin as the conditions of the struggle for existence."

The period in which Haeckel and Darwin worked was a period of exploration. Naturalists were just beginning to discover the bewildering variety of plants and animals and their peculiar ways of life. Charles Darwin's theory of evolution by natural selection had placed the organism in the context of the environment: form and behavior were adapted to the particular environment in which the organism dwelled. Ecology first flourished as the study of natural histories of organisms, the life stories of animals and plants: where and when they were found, what they ate, what ate them, and how they responded to changes in their surroundings. This narrow view of ecology gave way, towards the end of the nineteenth century, to a broader perception of the interrelationships of all plants and animals. Whereas *autecology* related the organism to its surroundings, *synecology*, as this broader view has been called, recognized that assemblages of plants and animals had characteristic properties of structure and function shaped by the environment.

Synecology, and its inherent treatment of all animals and plants together as a biological unit, the *community*, achieved its extreme expression in conceiving the community as a superorganism. The parallels between community and organism are obvious. Each is made up of distinctive subunits. The organism has its liver, muscle, and heart; the community its green

1

plants, predators, and decomposers. A forest community grows on cleared land, progressing through field and shrub stages of succession to its mature form, just as the organism progresses through its developmental stages.

Ecologists have always viewed organisms and communities in the context of their physical environments. Evolutionary adaptations and developmental responses enable plants and animals to respond to variation in the environment. Biological structure and function are molded by their physical surroundings. As recently as the 1940s, ecologists began to realize that the biological community and its environment could be considered together as a single unit. The physical and biological worlds form a larger system, called the *ecosystem*, within which the material substances of life are continually passed back and forth between the earth, air, and water on one hand and plants and animals on the other.

A gradually enlarging view has not been the only contributor to the maturity of ecology. Since the first part of this century, ecology has been a meeting ground for ideas from genetics, physiology, mathematics, agriculture, and animal husbandry. Indeed, for many years the infusion of ideas and approaches from other areas of endeavor had diverted the central movement of ecology to many seemingly divergent paths. One led to population biology, another to physiological ecology, a third to the study of community energetics, and so on. We now seem to be on the threshold of a period of coalescence, a recognition of the ties between the separate disciplines of ecology and an achievement of unity.

The Organization of Ecology

I cannot imagine the form of the ultimate unification of ecology as a science, nor, indeed, be certain whether it shall ever take a form as well ordered as the structure given by theorems to mathematics and by laws to physics. There is a principle that prevents a scientific discipline from gelling prematurely into a mold shaped by misconception and ignorance. I shall, however, try to provide a basic organization, or plan, of ecology. I do not mean to describe all the conceptual, often semantic, pigeonholes into which ecologists stuff their observations of nature. To keep these distinctions to a minimum is to hasten progress toward understanding nature which, except for use by pigeons, has no pigeonholes. Organizations, plans, schemes of classification help us along the way. We take what help we can find, and this book is in part a catalogue, an encyclopedia, of devices, useful now, but to be put aside as our understanding increases.

Ecology is a two-dimensional construction of horizontal layers, representing a hierarchy of biological organization from the individual through the population and community to the ecosystem, and of vertical sections cutting through all layers, representing form, function, development, regulation, and adaptation. If we follow the community layer across its sections, we find a form section with the numbers and relative abundances of species; a function section with the interaction between predator and prey populations and the mutually depressing influences of competitors; a development section with plant succession, as in the reversion of cleared land to forest; a regulation section with the elusive property of inherent community stability hiding in a corner; and an adaptation section with the evolution of antipredator adaptations. If we take the particular stack of sections representing function, for example, we find energy flow and nutrient cycles in the ecosystem layer; predator-prey interactions and competition between species in the community layer; birth, death, immigration, and emigration in the population layer; physiology and behavior of individuals in the organism layer.

Each layer of ecological organization has unique properties of structure and function. Each section in each layer of the construction represents a unique constellation of observed phenomena, perceived patterns, and abstracted concepts. Yet all are presumably governed by fundamental natural laws. Such laws are the quest of ecology the science.

The Study of Ecology

Because each individual organism both influences and is influenced by its environment, ecology can be fully appreciated only by studying the roles of organisms in their natural surroundings. This is one way in which ecology differs from other biological subjects, such as physiology and cell biology, in which structure and function are often studied by isolating a system from its surroundings and carefully characterizing it under controlled conditions. Ecology also differs from other aspects of biology in being concerned with the diversity of life's interrelationships with the environment. Whereas physiologists wish to learn the common properties of function shared by organisms, ecologists are more concerned with how environment and organism interact to create the diversity of habitats and biological communities that occupy different regions of the earth. To understand natural diversity, ecologists must learn both the common principles of function shared by all ecological systems and the manner in which the systems respond to the conditions of the environment. These principles and their expression in natural diversity are the subject of this book.

An ecosystem by definition comprises the organisms—plants, animals, and microbes—and the physical and chemical factors that interact to determine the structure and function of the biological community. When they study ecosystems, ecologists strive towards two goals. The first is to describe the structure and function of the system much as a mechanic would describe the parts of an engine and how they move with respect to each other. The second is to understand how the system developed and is regulated and maintained. Here the analogy with the engine breaks down. An engine is manufactured according to a set of plans and, once built, continues to function in much the same manner until it falls into disrepair or wears out. The ecological system conforms to no such set of plans. Its apparent design results from the ways in which its self-directed parts fit together and interact with one another. Replacement parts for the engine are manufactured according to plans and permit the engine to continue to function in its customary manner. Individuals that constitute the biological community die and are replaced by others. Hence the structure of the community is in a continual state of flux. If the community is to remain unchanged, each individual that dies must be replaced by another of the same kind. Because no engineer oversees this job, the regulation of the ecosystem must come from within the system itself: ecological systems are self-regulated. For the ecologist, to understand the mechanisms of regulation is to understand the dynamic aspects of the ecosystem and how it responds to different conditions in the environment.

As in other branches of science, ecologists measure structure and function by the size or amount of each component of the system—the standing crop—and the rate and pathways of fluxes in the system. Standing crop is the quantitative description of the system at a particular instant—a snapshot of the system frozen in time. Quantities of things may be measured in a variety of units, such as individuals in populations, food energy locked up in the organic chemicals of plants and animals, calcium or phosphorus per unit of area of the habitat, and so on. To understand the dynamics of the system, one must also know the speed at which the amounts are changing and the direction in which they move. Changes, or fluxes of energy and materials, might be measured by the rate of turnover of individuals in populations, the rate of assimilation of energy from the sun by green plants and its transfer along the food chain, or the rate of uptake of calcium from the soil. Ecosystems can be described in these terms, even when we deal with such complicated ecological behavior as the response of ecosystems to natural perturbations (drought, fire, storms) or to such human disturbances as pollution, pesticides, and introduced diseases, predators, and competitors.

Description has two functions. The first is to assemble a set of organized facts about nature—a phenomenology—that may suggest new ideas about the development and regulation of ecological systems. The observation that most groups of organisms are represented by a greater number of species in the tropics than in temperate and arctic regions led to a number of theories of species diversity based on the obvious changes in climate between the equator and the poles. Observations of how forests became reestablished following the abandonment of agriculture on cleared fields suggested theories concerning competition of plant populations for space and renewable resources. The second function of description is its practical application to the management of ecological systems for the benefit of mankind. The accumulation of facts about how ecological systems behave provides practical knowledge useful in their management and manipulation. For example, many years of observation have revealed a relationship between fall and winter populations of ducks on hunting grounds in the United States and rainfall on breeding grounds in the Canadian prairies. This observed relationship allows game managers to adjust hunting limits according to past weather conditions.

Description of ecological systems comes about in a variety of ways. The simplest methods of ecologists are observation and measurement of the state of natural systems. One may count the number of individuals in populations, measure the kinds and amounts of plant life in a forest, measure the amounts of nutrients or energy contained in the organisms in a pond, and so on. Descriptions of the response of systems to perturbations may derive from natural variation in the environment or they may derive from experimentation. Experiments give the investigator a measure of control over nature. Waiting for natural perturbations may be unsuccessful, as well as frustrating and expensive. To examine the effect on understory vegetation of light admitted to the forest floor, one need not wait for a tree to be toppled by a hurricane. A chainsaw can create the required situation quite nicely. The effects of certain human disturbances caused, for example, by fertilization of streams or application of pesticides may be investigated on small and relatively harmless scales by experimentation before there are untoward results from large-scale programs with potentially disastrous consequences. Experiments also enable one to isolate the effects of two or more factors that may vary simultaneously in natural systems. Heat and drought often occur together, but experimental irrigation or other provision of water may enable the investigator to expose the independent effect of heat on the system. In a purely descriptive study, one's objective is to find generalities about relationships. Patterns may be revealed that allow one to deal with nature in a rational way.

Differing totally from the goal of description is the goal of understanding how a system is regulated and maintained. One may determine by observation that the productivity of duck populations is related to water levels on the breeding grounds. It may be more difficult to learn how water level exerts its influence. It might do so by affecting the amount of food available to ducklings, by altering vegetation so as to make nests more vulnerable to predators, or by reducing suitable habitat for feather molt in late summer. It may be even more difficult to understand why each adult in the population is replaced by a single adult on average and the population neither grows nor declines over long periods, or to determine the maximum hunting harvest that the population can sustain, without driving the population to extinction while trying to find the proper level of hunting pressure by trial and error.

Research on the dynamics of ecological systems, although stimulated by and ultimately based upon the phenomenology arising from experience and observation, relies more on experimentation and mathematical modeling for its methods. Experiments are designed to test ideas about how a system functions. Theories and hypotheses about how systems work may allow one to predict the response of a system to a particular manipulation. If the experiment fails to produce the predicted result, something is wrong with the hypothesis; hence our understanding of the system is deficient. If the experiment produces the predicted result, the hypothesis could be correct, although we could not rule out the possibility that other hypotheses might explain the same facts equally well.

Knowledge of population interactions led ecologists to suspect in the 1950s and 1960s that selective hunting by predators could regulate the diversity of prey species if the predators preferred to eat highly competitive species which themselves excluded weaker competitors from a locality. This theory predicted that if such predators were removed, superior competitors thereby released from predator pressure would increase in number and crowd out others. To test this prediction, predators were experimentally removed from a number of habitats (rocky shores, for example) and the response of prey populations was monitored. Indeed, in the absence of predators (starfish, for example) some prey populations (mussels, for example) became very dense and others (barnacles, for example) were crowded out. Had the experiment produced no change in prey populations, one might have ruled out predators as having a major role in regulating prey diversity. As it turned out, the experiments revealed that predators greatly influence populations of their prey, which was not surprising, and did so selectively. But even though predators figure prominently, one must stop short of concluding that they alone regulate prey diversity or are solely responsible for variations in diversity among communities. After all, predators are but one part of the system, and interactions between prey populations undoubtedly are influenced by other factors as well. By analogy, if one takes an engine running at 3000 rpm and removes half of its spark plugs, the engine is bound to slow down or quit altogether. One would not, however, conclude from this experiment that the speed of an engine is normally regulated by adding and removing spark plugs; the amount of gas admitted to the cylinders can produce the same effect. Lacking this knowledge, we might surmise the primary role of fuel flow in regulating engine speed from observations on the state of the carburetor in engines running at different speeds. Thus we might surmise the role of predation in regulating the diversity of the community by observing the activities of predators in communities having different numbers of prey species.

As we learn more about the dynamics of a system, it may become possible to construct mathematical models of the system that mimic its behavior. Such models have two functions. First, they summarize our understanding. To the extent that the behavior of the model resembles that of the natural system, the model is consistent with reality and our understanding is likely correct. Discrepancies between natural systems and their models can be used to assess precision and degree of our understanding. A model of predators and their prey that predicted an increase in the prey population when the predator population also increased would be inconsistent with most observations and probably would reflect a fundamental misunderstanding of what predators do to their prey.

As in testing other predictions, concordance between model and system does not necesssarily validate the model. For example, during the 1920s, population biologists wrote simple equations for the interactions between predator and prey populations that predicted oscillations in both, with the ups and downs of the predators lagging slightly behind those of the prey. The theory and the observations available at that time were consistent, for such cycles were revealed by data on many predator-prey systems, most notably that of the hare and lynx in northern Canada. But more recent models have shown that cycles may be caused by a variety of factors affecting the interaction between predators and their prey other than those described by the simpler equations of early models. Furthermore, detailed reanalysis of the data recently has revealed finer points of disagreement with the original models. Distinguishing between competing models often requires refined matching of models to natural systems, sometimes after further observation or experiment. Combined with these latter approaches, modeling serves a useful role in scientific enquiry.

A second function of a model may be to predict the behavior of a system when exposed to novel influences, or to predict some behavior about which one may not generalize. Used in this way, models have important applications to conservation, wildlife management, and pollution control. By understanding precisely how water levels in prairie ponds affect breeding productivity and survival of ducks, it may be possible to predict the numbers of individuals in hunted populations without having to make direct counts in the autumn. In this particular case, small errors in measuring the variables in the model probably would make the predictions less reliable than direct counts of the population. In principle, however, models may have widespread application, and their utility increases as our knowledge of ecological systems increases.

As a scientific endeavor, ecology is still in its youth. Much ecological work during the present century has been descriptive, but our knowledge of natural systems remains woefully incomplete. Because the world is so diverse and complex, we may never describe completely even the simplest systems. Even so, for many of the questions they ask, ecologists have achieved the level of phenomenology necessary for experimenting and modeling to be applied fruitfully. In some specialties, particularly population biology, experiments and models have substantially increased understanding of the dynamics of natural systems. Attempts to extend these models to predict the structure of biological communities sometimes expose tremendous gaps in our understanding, but they are a beginning to the study of structure, function, and regulation of complex assemblages of species.

Each of the major areas of ecology is at a different stage of intellectual progress. Our ability to capitalize on knowledge with practical applications varies, too. These differences will become apparent to the reader as we deal with physical processes, organism function, population dynamics, evolutionary response to the environment, community organization, and ecosystem structure and function. The study of physical influences includes physical and chemical processes in the soil and water, and seeks to understand how nutrients required by life are made available to and transported through the ecosystem. The study of organism function seeks to understand how individuals modify their morphology, behavior, and physiology to respond to changes in their environments. Such studies include, for example, observations and experiments on heat stress and water conservation in desert animals and on changes in the growth form of plants in response to light levels. Population ecologists are concerned with the interactions between organism and environment as they are expressed by population statistics of fecundity and mortality. Population ecology is an active area for modelers who wish to project changes in populations through time resulting from particular patterns of birth and death rates. More complicated models involving two or more populations hold much promise for understanding the regulation of biological communities. When population processes express genetic differences among individuals in the population, evolutionary change can result. Ecologists are interested in both the process of evolution and the evolutionary value of adaptations in particular environments. Studies of ecosystem structure and function summarize, in terms of energy and nutrient cycles, all interactions between the physical and biological parts of the system. Inasmuch as ecosystem properties are predictable from underlying processes, the ecoystem is the ultimate level upon which ecological theory and understanding should be tested.

2 | Life and the Physical Environment

WE OFTEN CONTRAST the living and the nonliving as opposites: biological versus physical and chemical, organic versus inorganic, biotic versus abiotic, animate versus inanimate, active versus passive. While these two great realms of the natural world are almost always readily distinguishable and separable, they do not exist one apart from the other. The dependence of life upon the physical world is obvious. The impact of living beings on the physical world is more subtle, but this impact is equally important to the continued existence of life on earth. Soils, the atmosphere, lakes and oceans, and many sediments turned to stone by geological forces owe their characteristics in part to the activities of plants and animals.

The Uniqueness of Life

All forms of life have many properties in common that set organisms apart from stones and other inanimate objects. But although living beings are distinct from inanimate objects, life also must be viewed as an elaboration of the physical world, not its alternative. Organisms function within constraints set by physical laws. They are like internal combustion engines transforming energy to perform work. The earth is also a giant heat machine, utilizing the energy in sunlight to drive the winds and ocean currents. But here lies the difference between physical and biological systems. In physical systems, energy transformations act to even out differences in energy level throughout the system, always following the path of least resistance. In biological systems, whether energy transformation is directed toward pursuing prey, producing seeds, keeping warm, or maintaining such basic body functions as breathing, blood circulation, and salt balance, it is used purposefully by the organism to maintain itself *out* of equilibrium with the physical forces of gravity, heat flow, diffusion, and chemical reaction. In a

9

sense, this is the secret of life. A boulder rolling down a steep slope releases energy during its descent, but no useful work is performed. The source of energy, gravity in this case, is external, and as soon as the boulder comes to rest in the valley below, it is once more brought into equilibrium with the forces in its physical environment.

A bird in flight must constantly expend energy to maintain itself aloft against the pull of gravity. The bird's source of energy is internal, being the food that it has assimilated into its body, and the work performed serves a purpose useful to the bird in its pursuit of prey, escape from predators, or migration. To be able to act against external physical forces is the one common property of all living forms, the source of animation that distinguishes the living from the nonliving. Bird flight may be a supreme expression of animation, but plants just as surely perform work to counter physical forces when they absorb soil minerals into their roots or synthesize the highly complex carbohydrates and proteins that make up their structure.

Physical forces in the environment could not be held at bay without the expenditure of energy to perform work. The ultimate source of energy for life in the physical world is light from the sun. Plants have special pigments, among them chlorophyll, that absorb light and capture its energy. That energy is then converted to food energy during the manufacture of sugars from simple inorganic compounds—carbon dioxide and water. The energy-trapping process is called *photosynthesis*, literally, putting together with light. Energy locked up in the chemical bonds of sugars, and thence proteins and fats, is used by plants, by animals that eat plants, and by those that eat other animals that eat plants, and so on, to perform the work required of an animate existence.

The Interdependence of Life and the Physical World

Life is totally dependent on the physical world. Organisms receive their nourishment from it and their distributions are limited by tolerance of its conditions. The heat and dryness of deserts prevent the occurrence of most life forms, just as the bitter cold of polar regions prevents the establishment of all but the most hardy organisms. The form and function of plants and animals also must respect the physical world. The viscosity and density of water require that fish be streamlined according to stringent hydrodynamic rules if they are to be swift. The concentration of oxygen in the atmosphere, at 21 per cent, places upper bounds on the metabolic rates of organisms. Similarly, the ability of plants and animals to dissipate body heat—accomplished by the purely physical means of evaporative cooling, conduction, and radiation of heat from the body surface to the surroundings—limits their rate of activity and their safe exposure to direct sunlight.

The activities of organisms also affect the physical world, sometimes in a profound manner. The oxygen that we take for granted with every breath was produced largely by the photosynthetic activities of green plants. Before green plants evolved in primitive seas, the atmosphere of the earth was composed mostly of methane (CH_4), ammonia (NH_3), water vapor (H_2O), and hydrogen (H_2). As early aquatic plants began to utilize sunlight as a source of energy, they began to liberate oxygen, some of which escaped from the oceans and accumulated in the atmosphere. Over the past two billion years, the span of life on earth, most of the hydrogen once contained in the earth's primitive atmosphere has escaped into outer space. Plants have assimilated the carbon contained in atmospheric methane and the nitrogen in ammonia, and their place in the atmosphere has been partly taken by oxygen released during photosynthesis.

Plants play an equally influential role in the development of soil properties. Plant roots find their way into tiny crevices and pulverize rock as they grow and expand. Bacteria and fungi hasten the weathering of rock by chemical means. Fungi secrete acids to dissolve minerals out of unaltered rock, thereby weakening the crystalline structure of the rock and speeding its decomposition. Rotting plant detritus also releases acids that do the work of chemical decay, while fragments of detritus alter the physical structure of the soil. Animals, by burrowing, trampling, and defecating, play their part in the development of soil.

The role of plants and animals in maintaining soil characteristics is shown most dramatically when communities are disturbed. The development of the Dust Bowl in the midwestern part of the United States during the 1930s provides a vivid example. The Dust Bowl area is normally dry and windy, but the root systems of the natural vegetation, mostly perennial grasses, are extensive enough to hold the soil in place. When the prairies were plowed for agriculture, the perennial grasses were replaced by annual crops, with less extensive root systems. A series of dry years reduced crop growth and turned the soil surface into fine dust. The result, shown in Figure 2-1, is legendary.

Plants also influence movement of water. Rain does not accumulate where it falls. If it did, New York State would be under 200 feet of water within a lifetime. Some water flows over the soil surface or through the underlying earth to enter rivers, lakes, and, eventually, the ocean. The remainder escapes by evaporation from the ground surface and vegetation. The leaf area of an eastern deciduous forest is, on the average, about four times the area of the ground surface; that is, there are about four acres of leaf surface per acre of forest floor. Plants are the major pathway of evaporation. When a forest is cut, most of the water that normally would have evaporated from the leaves flows instead into rivers. The consequences of clear-cutting without provision for extensive replanting are flooding, increased erosion and the silt deposition that accompanies it, and removal of mineral nutrients from the denuded soil.

Evaporation of water from plant leaves tends to retain water in a locality, for much of the water vapor quickly condenses and falls as rain nearby. In some areas, particularly in the tropics, the presence of forest vegetation

FIGURE 2-1 The Dust Bowl area of the midwestern United States. Wind erosion begins when soils are plowed but there is insufficient rain for crop growth. Above, a winter-wheat crop failure in Finney County, Kansas, has resulted in soil blowing (March 1954). Below, wind-blown soil particles—a dust storm— approaches Springfield, Colorado, in May 1937 during the height of the Dust Bowl tragedy. Dust storms completely destroyed some farming areas.

increases local precipitation. Extensive clearing of forests for agriculture has caused significant drying trends in some local climates.

The Ecosystem Concept

The interdependence of the physical and biological realms is the basis of the ecosystem concept in ecology. In spite of the ecosystem's being the largest and, in many ways, the most fundamental unit of ecology, the term itself was not used until 1935, when it was coined by the English botanist A. G. Tansley. The ecosystem, he wrote, includes ". . . not only the organism-complex, but also the whole complex of physical factors forming what we call the environment of the biome—the habitat factors in the widest sense. Though the organism may claim our primary interest, when we are trying to think fundamentally we cannot separate them from their special environment, with which they form a physical system."

The biotic and abiotic parts of the ecosystem are linked by a constant exchange of material through cycles of nutrients driven by energy from the sun. The basic pattern of energy and material flux in the ecosystem is shown in Figure 2-2. Plants manufacture organic compounds, utilizing energy obtained from sunlight and nutrients from soil and water. The plants use these compounds as a source of building materials for their tissues and as

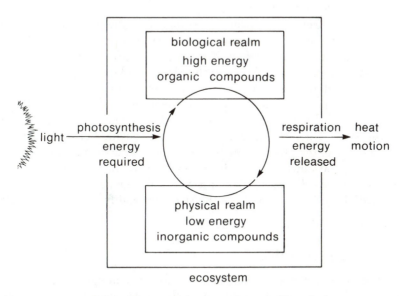

FIGURE 2-2 Schematic diagram of the flow of energy through the ecosystem and the cycling of chemical nutrients within the ecosystem. The biological and physical realms, represented by organic and inorganic compounds, together comprise the ecosystem.

a source of energy for their maintenance functions. To release stored chemical energy, plants break apart organic compounds into their original inorganic constituents—carbon dioxide, water, nitrates, phosphates, and so on—thus completing the nutrient cycle.

Plants manufacture their own "food" from raw materials. Hence they are referred to as *autotrophs*, literally self-nourishers. Animals obtain their energy in ready-made food by eating plants or other animals. Animals are therefore referred to as *heterotrophs*, meaning nourished from others. The specialization of living forms as food-producers and food-consumers creates an energetic structure, called the *trophic structure*, in biological communities, through which energy flows and nutrients cycle. The food chain from grass to caterpillar to sparrow to snake to hawk depicts the path of organic materials and the energy and nutrient minerals they contain. Each link in the food chain, each trophic level in the community, dissipates most of the food energy it consumes as heat, motion, and, in the case of luminescent organisms, light. None of these energy forms is useful to other organisms. Hence, with each step in the food chain, the total amount of usable energy that passes through to the next higher trophic level becomes smaller. It is no wonder, then, that all the grass in Africa heaped into a big pile would dwarf the mound of grasshoppers, gazelles, zebras, wildebeests, rhinoceroses, and all other animals that eat grass. So much would the piles of plants and herbivores overwhelm us that we would probably not even notice the pitiful heap of lions and hyenas nearby.

Ecosystem structure and function summarize the activities of all the organisms in the biological community—their interactions with the physical environment and with each other. We must never lose sight of the fact that the lives of organisms are played out for their own benefit, and not for the purpose of fulfilling some role in the ecosystem as an actor fills a role in a drama. The properties of the ecosystem result from the self-serving activities of the plants and animals it contains, and this is where understanding of the ecosystem structure and function must be sought. Before considering the properties of the physical environment and overall organic productivity, energy flow, and nutrient cycles in ecosystems, we ought to stop briefly to consider the lives of a few kinds of organisms, intriguing by themselves and indicative of the forces that shape the structure and functioning of ecosystems.

The Giant Red Velvet Mite

The Mohave Desert is a forsaken land of searing summer heat, bitter winter cold, and year-long drought. Little rain falls because clouds are intercepted by the mountains that rise between the desert and the coast of southern California. Except for struggling desert plants, little life is in evidence during most of the year. But the desert's stillness is occasionally broken during the milder days of the winter by swarms of insects and other crea-

FIGURE 2-3   Adult giant red velvet mite, on the surface of the ground (left) and in a vertical burrow (right, a cutaway view of the underside of the mite).

tures that mysteriously appear and disappear on or above the surface. One of the more conspicuous of these creatures is the giant red velvet mite (Figure 2-3) whose scientific name, *Dinothrombium pandorae*, paints a vivid picture of this close relative of spiders. The generic name, *Dinothrombium*, is derived from the Greek *deinos*, meaning terrible, and *thrombos*, a lump or clot. This particular species is named after the mythological woman Pandora, who carried a box containing all the ills humans might suffer. When curiosity compelled Pandora to open the box, the ills escaped and have plagued the human race ever since.

Several years ago, biologists Lloyd Tevis and Irwin Newell became interested in the behavior of the mite and began a study of its activity in relation to the physical conditions of its environment. They found that the mites spend most of the year in their burrows, which are dug only where the grains of sand measure less than one-half millimeter in diameter. The particular environmental conditions that are favorable for the emergence of the giant red velvet mite occur infrequently in the Mohave Desert. During four years of observation, adult mites appeared above ground only ten times, always during the months of December, January, or February, when temperatures reach a level low enough for the mite to tolerate the desert's surface. On the basis of their observations, Tevis and Newell could predict that emergence would occur on the first sunny day after a rain of more than three-tenths of an inch, provided that air temperatures were moderate. An individual mite appeared only once each year.

On the day of a major emergence, the mites come out of their burrows between 9:00 and 10:00 A.M., and by late morning many hundreds of mites scurry across the desert sands in all directions. At midday, between 11:30 and 12:30, the mites dig back into the sand where they wait until the following year before emerging again. The mites do not leave their burrows to terrorize unsuspecting desert travellers; during their two- or three-hour stay above ground each year, mites perform two important functions: feeding and mating. On the same day the mites emerge, large swarms of termites appear, flying over the desert sands, their emergence presumably triggered by the same physical conditions that cause the mites to leave their burrows. It is upon these termites that the mites feed. Flying termites cannot, of course, be caught by the flightless mites; each mite must locate its prey after the termites drop to the ground and shed their wings but before they burrow into the sand. All this happens very quickly, giving the mites about an hour to find their prey.

Because the mites are solitary, they must mate as well as feed during their brief sojourn above ground each year. The courtship pattern of the giant red velvet mite resembles that of spiders and their relatives. The males walk nervously around and over a feeding female, tapping and stroking her, and cover the sand around her with loosely spun webs. Males court *feeding* females for two reasons. First, the females have ravenous appetites (they haven't eaten for a year, after all) and are probably just as likely to devour a male mite as a termite. Second, and probably more important, females can produce eggs only after they have had a meal. Thus, by mating with a feeding female, the male guarantees that his reproductive efforts will not be wasted.

At about midday, after feeding and mating have taken place, the mites congregate in troughs on the windward sides of sand dunes, where surface temperatures and the size of the sand particles are just right, and burrow into the sand. The mites continue to dig their new burrows on the first day until the coolness of the late winter afternoon slows their activity. Burrowing continues on subsequent days, when the sand becomes warm enough, until the burrows are complete. During the rest of the year, the adult mite spends its time moving up and down in its burrow to follow the movement of the suitable temperature zone as the surface of the sand is heated and cooled each day.

Eggs are laid early in spring. They soon hatch, and young mites crawl to the surface of the desert to search for a host to attach themselves to, usually a grasshopper. While they are growing, the young mites remain attached to their host, obtaining their nourishment from its body fluids. When they are full grown, the mites drop off their unwilling hosts, and seek suitable spots to dig their own burrows in the sand, thus renewing the life cycle of the giant red velvet mite.

The Oropendola and the Cowbird

The tropical surroundings of the chestnut-headed oropendola in Panama are a far cry from the rigorous environment of the giant red velvet mite. In the tropics, air temperature varies little during the year and abundant rainfall maintains the lush tropical vegetation. Life abounds, and diverse animals and plants are intricately interwoven into a rich fabric of biological interactions. In the harsh environments of the desert and polar regions such interactions are noticeably simplified. In the tropics we are not so aware of adaptations to the physical environment, rather we are impressed by adaptations to biological environments.

The situation we are going to examine involves two birds—the chestnut-headed oropendola and its brood parasite the giant cowbird (Figure 2-4). As we shall see, two insects—a fly and a wasp—also play an integral part in the interaction between the two birds. The biological sleuthing that uncovered this story was performed by Neal Smith, a staff biologist at the Smithsonian Tropical Research Institute in the Panama Canal Zone. Smith's insight, inventiveness, and perseverance proved a good match for the complexity of the oropendola–cowbird relationship.

Brood parasites (cowbirds and cuckoos are familiar ones) have been known for a long time. They are so named because the female lays her eggs in the nest of another species, the host, and the young are raised by the foster parents. Naturalists used to believe that the presence of a brood parasite in a nest always reduced the survival of host young because brood parasite young compete for food brought in by the host parents. As one would expect, many potential hosts can detect the presence of parasite eggs in their nest and eject them. To counter this defense, many brood parasite species have evolved an elaborate egg mimicry to fool the host species into accepting the alien egg as one of its own.

Smith was aware that nesting colonies of the chestnut-headed oropendola, usually consisting of ten to a hundred nests, were often parasitized by the giant cowbird. He observed a curious phenomenon. Examining the eggshells that had been thrown out of the nests by the females after the young had hatched, Smith noted that under some colonies the eggshells of the cowbirds were distinctly different from those of the oropendolas, whereas, in other colonies, eggshells of the two species so closely resembled each other that they could be distinguished only on the basis of shell thickness. Thus it appeared that in some colonies the cowbirds had evolved to mimic the eggs of their host while in others they had not. These promising observations were to form the basis of a detailed study of brood parasitism.

Smith's first problem was to devise methods of studying bird nests suspended from the outlying limbs of large trees, 20 to 60 feet above his head. One generally avoids climbing trees in the tropics because of all the other creatures that might be climbing the same tree. And even if one could reach the top of a tree without an unpleasant encounter, climbing out on the small branches from which the nests are treacherously suspended would be challenging even for a circus acrobat. To remove the nests, examine and manipulate their contents, and then replace the nests in their

FIGURE 2-4   Female chestnut-headed oropendola at her nest (above) and giant cowbird (below). The oropendola egg (left) is compared to a mimetic cowbird egg (center) and a nonmimetic cowbird egg (right).

original position seemed impossible. The nest is constructed in the form of a long, pensile bag, made of interwoven grasses and small vines. But it should be feasible, Smith reasoned, to cut or rip the nests down with the aid of long poles and then reattach them on the original site. Most ecologists would have difficulty imagining themselves standing atop a 15-foot ladder, balancing 48 feet of extendable aluminum poles with instruments at the tip controlled from below by long ropes. But with this apparatus, Smith performed the delicate task of lowering a nest of fragile eggs 50 feet to the ground, then examining the contents and replacing it—all of this carried out at night, under the duress of incessant mosquito attack, so as not to provoke the female oropendolas to abandon their nests.

Using a formidable, Rube Goldberg array of pincers and flaps connected by lever and pulley, Smith was able to replace the nests in their original positions with a sticky variety of contact tape or contact glue. In the most highly developed version of nest replacement, he stapled rat snaptraps to the end of the nest. These clung tenaciously to any tree limb they were pressed against.

Having mastered the technique of working with nests, Smith set out to determine whether the presence or absence of egg mimicry in the cowbirds elicited different oropendola behavior toward foreign eggs. A number of objects—including mimicking cowbird eggs, nonmimicking cowbird eggs, other kinds of eggs, and a variety of other objects only remotely, if at all, resembling eggs—were put into nests in both kinds of oropendola colonies, those that tolerated nonmimicking eggs and those that did not. Smith found, as he had originally suspected, that in oropendola colonies where the cowbird eggs closely mimicked those of their hosts, the oropendolas removed virtually everything from their nests except their own eggs and very closely matching cowbird eggs. These discriminating oropendolas tried, often in vain, to discover and eject the cowbird eggs. On the other hand, in oropendola colonies where cowbirds were poor egg mimics, the oropendolas were willing to accept all sorts of foreign objects. Smith had thus identified two types of oropendola colonies, one in which the oropendolas discriminated against cowbirds and tried to eject everything from their nests but their own eggs, the other in which oropendolas were nondiscriminators and accepted notably different eggs and other materials as well.

In what other ways did the two colonies differ? In the nondiscriminator colonies, were the cowbirds actually beneficial to the oropendolas? Why else would their eggs be tolerated by the oropendolas? While making an extensive survey of the oropendolas in the Panama Canal Zone and nearby Panama, Smith found that in nondiscriminator colonies, young oropendolas were often infested with the larvae of a species of bot fly. The parasites sometimes killed the nestling oropendolas and frequently so weakened them that their chances of surviving to adulthood were slim. In discriminator colonies these parasites were rarely present. Here was a major difference between the two types of colonies. Was it possible that the role of the cowbird in the colonies was linked to bot fly parasitism?

Smith examined oropendola young in nondiscriminator colonies (susceptible to bot fly parasitism) and discovered that the incidence of bot flies was higher in nests that did not contain cowbirds than in nests which did, as shown by the following data:

	Number of nestling oropendolas in nests	
	With cowbirds	Without cowbirds
With bot fly parasites	57	382
With no parasites	619	42

Further observations plainly showed that the nestling cowbirds would snap at anything small, including adult bot flies, that moved within the nest and, furthermore, that they would remove bot fly larvae from the skin of the nestling oropendolas. This behavior on the part of the young cowbird benefitted the oropendola and accounted for the acceptance of the brood parasites in nondiscriminator colonies.

Cowbird young are well suited for grooming their nest mates. They hatch five to seven days before the oropendola young and develop precociously. Their eyes are open within 48 hours after hatching, whereas the eyes of oropendola nestlings open six to nine days after hatching. Also, the cowbird young are born with a thick covering of down, absent in the oropendola young, which presumably deters bot flies from laying their eggs on the skin of the young cowbirds. By the time the oropendolas hatch, the cowbirds are sufficiently developed to groom the oropendola young. In discriminator colonies, which are not troubled by bot fly parasitism, cowbird young perform no such useful function for the oropendolas, and because they compete with the oropendola young for food, they are detrimental to the productivity of the colony.

The role of the cowbird in this story is now evident, but we have not yet determined why bot flies are present in some colonies and absent from others. Smith noted that all the discriminator colonies, and none of the nondiscriminator colonies, were built near the nests of wasps or bees, which swarm in large numbers around their nests and virtually fill the air throughout the oropendola colony. Wasps and bees presumably prevent the bot flies from entering the colonies. Occasionally Smith found the detached wings of bot flies beneath the nests of discriminator colonies. By hanging up rolls of fly paper in the two types of colonies, Smith found that adult bot flies rarely enter the area around discriminator nests. But the protection is not perfect. Nests on the periphery of the discriminator colonies, and thus at some distance from the wasp and bee nests, are occasionally parasitized by bot flies. Because the protective influence of the wasp extends over a limited distance, discriminator colonies tend to be much more compact than nondiscriminator colonies, with nests placed close together around the wasp nests at the center.

For the oropendola, the presence or absence of wasps in the vicinity of the colony completely alters the role of the cowbird as a factor of the environment. Accordingly, the behavior of the oropendolas towards the cowbirds and their eggs also varies between the two types of colonies, and in turn affects the environment that molds the behavior of the cowbird. In discriminator colonies, adult oropendolas not only eject the eggs of cowbirds from their nests if they can distinguish them, but they also chase adult cowbirds out of the colony. The oropendolas in nondiscriminating colonies are indifferent to cowbirds. Behavior of the cowbird in the two types of colonies reflects this difference. In discriminator colonies, female cowbirds are cautious and always enter the colony singly. Behavioral adaptation has gone so far that the cowbirds mimic the behavior of the discriminator oropendolas; they often gather the stems of small vines and act as if they are beginning to build a nest, a most uncharacteristic behavior of brood parasites. On the other hand, cowbirds that parasitize nondiscriminator colonies are often gregarious and enter the colonies in small groups. They behave aggressively towards the oropendolas and sometimes even chase them from their nests.

It is not sufficient for a successful egg mimic merely to produce an egg that is indistinguishable from host eggs. It must also take care to lay the egg at the proper time. A female oropendola normally lays her two eggs on consecutive days. She is very sensitive to the appearance of new eggs in the nest before or after her laying period, and will frequently desert her nest if it contains more than three eggs. Even when confronted by perfect egg mimicry, a discriminating oropendola can be fooled only if a single egg is laid by the cowbird soon after the first oropendola egg has been laid. If the cowbird lays its egg a day too early or too late, it reveals its presence, and the oropendola will abandon the nest and start another. In nondiscriminator colonies, however, neither the number of cowbird eggs nor their appearance matters to the oropendola. Commonly, a female cowbird lays two to five eggs over several days in the nest of a single oropendola. Smith refers to these birds as dumpers.

The interaction between the oropendola and the cowbird shows how the

TABLE 2-1   Biological attributes of discriminator and nondiscriminator colonies of the chestnut-headed oropendola.

	COLONY TYPE	
	Discriminator	Nondiscriminator
Wasp nests	Present	Absent
Possibility of bot fly parasitism	Slight or absent	Heavy
Effect of cowbird on oropendola	Disadvantageous	Advantageous
Foreign objects in nest	Rejected	Accepted
Cowbird eggs	Mimetic	Nonmimetic
Cowbird eggs per nest	One	Several
Cowbird behavior in colony	Timid	Aggressive
Colony structure	Compact	Open
Nesting season	Late	Early

environment determines the adaptations of the organism and how two or more kinds of organisms can be major factors in each other's environments (Table 2-1). The oropendola, the cowbird, the bot fly, and the wasp are all mutually important to each other and affect the evolution of one another. This relationship differs sharply from the relationship of an organism to the physical environment, which neither evolves nor responds by adaptation to changes in the biotic environment. Whereas the physical environment is passive, the biotic environment is responsive. It continually readjusts to evolutionary changes in any one of its living components. We might expect evolution in a biotically dominated environment to differ from evolution in a physically dominated environment. Different populations, like the oropendola and the cowbird, may coevolve with respect to each other into a mutually beneficial relationship that could not be achieved between a population and its physical environment.

3 | Natural Selection

 HE ADAPTATIONS of the velvet mite and oropendola are well suited to their particular environments. Every detail of their morphology, physiology, and behavior seems capable of meeting the challenge of their surroundings. The close correspondence between organism and environment is no accident. Only those individuals that are well suited to the environment survive and produce offspring. The inherited traits which they pass on to their progeny are preserved. Unsuccessful individuals do not survive and reproduce, hence their less suitable traits are eliminated from the population. This process, which Charles Darwin called *natural selection*, allows the population to respond to its environment over periods of many generations and slowly refines the adaptations of organisms to fit the requirements of the environment.

Natural selection can be outlined in general terms and by specific example as follows:

a. The reproductive potential of populations is great, but

b. populations tend to remain constant in size, because

c. populations suffer high mortality.

d. Individuals vary within populations, leading to

e. differential survival of individuals.

f. Traits of individuals are inherited by their offspring.

g. The composition of the population changes by the elimination of unfit individuals.

a. Rabbits should cover the earth, but

b. they don't, because

c. many are caught by predators.

d. Some rabbits run faster than others,

e. and escape from predators.

f. So do their young.

g. Populations of rabbits, as a whole, tend to run faster than their predecessors.

As it is described above, natural selection occurs because of three properties of organisms and their relationship to the environment: (1) genetic variability, (2) inheritance of traits, and (3) influence of the environment on survival and reproduction. Design is not inherent to the process of natural selection. The environment itself is the template for the design we see in organisms. Selection merely acts as an agent for realizing that pattern. Whether a rabbit runs slowly or rapidly is inconsequential to natural selection; only the influence of its swiftness on the number of offspring that it leaves is important.

Inheritance and genetic variation within populations are facts of genetics obvious to anyone who has noticed both the variability of physique, facial appearance, eye color, and hair among humans and the tendency of related individuals to share many of these traits. The relationship between particular genetic traits, on one hand, and survival and reproduction, on the other, is less obvious. By our own technological devices we have surrounded ourselves with an environment largely of our own making, one that seems to have little influence on physique, eye color, blood type, baldness, and other genetic traits. To be sure, some inherited diseases—such as retinoblastoma, achondroplastic dwarfism, aniridia, and sickle-cell anemia—have a profound effect on survival. These are deleterious, often fatal traits. They occur rarely and produce major disruption of body function. But they hardly seem representative of the traits that might have fostered eagles from primitive single-celled organisms balanced uncertainly on the fence between the coming kingdoms of plants and animals a billion years ago.

Plant and animal breeders have known for centuries that by carefully selecting breeding lines for a desired trait, the appearance of a population could be altered toward a desired end: longer wool, increased egg and milk production, sweeter fruit. Selection was practiced on domesticated plants and animals long before its role in shaping biological communities was appreciated.

The demonstration of natural selection acting to produce evolutionary changes in natural populations did not come until more than half a century after Charles Darwin originally proposed the mechanism. The first evidence that a change in the environment could select a new trait in a population came from early programs designed to control insect pests. Agricultural researchers had found that cyanide gas could be used to control populations of scale insects on citrus crops in southern California. As early as 1914, however, populations in some groves had become tolerant of the gas, and fumigation was no longer effective. Laboratory experiments demonstrated not only that tolerance was inherited, but also that some individuals naturally resisted cyanide poisoning in areas where the fumigant had never been used. This resistance was caused by a mutation—an error in the genetic code—which appeared rarely and sporadically in individual scale insects. In the absence of cyanide, the mutation for cyanide resistance would be of no value to the organism and could be detrimental if it altered normal body functions. Where the gas treatment was used to control scale insects, it resulted in survival of individuals carrying the resistant trait, while killing all others. The more recent development of a wide variety of insecticides has further revealed the presence of resistant traits among individuals of numerous pest species. More than performing their intended function, pesticides have often selected resistant populations that are all the more difficult to control.

Industrial Melanism

The English have always been avid butterfly and moth collectors, and such enthusiasts look carefully for rare variant forms. Early in the last century, occasional dark (or melanistic) specimens of the common peppered moth (*Biston betularia*) were collected. Over the succeeding hundred years, the dark form, referred to as *carbonaria*, became increasingly common in some industrial areas, until at present it makes up nearly 100 per cent of some populations. The phenomenon aroused considerable interest among geneticists, who showed by cross-mating light and dark forms of the moth that melanism is a simple inherited trait.

In the early 1950s, H. B. D. Kettlewell, an English physician who had been practicing medicine for fifteen years and who was also an amateur butterfly and moth collector, changed the pattern of his life to pursue the study of industrial melanism. Several facts about melanism were already known before Kettlewell began his studies: (1) the melanistic trait is an inherited characteristic, hence the widespread occurrence of melanism had been the result of genetic changes in populations; (2) the earliest records of *carbonaria* were from forests near heavily industrialized regions of England; (3) the dark form occurs most frequently in populations near modern industrial centers (Figure 3-1); (4) where there is relatively little industrialization, the light form of the moth still prevails. It was also known that dark forms have similarly appeared in many other moths and other insects. Melanism is not unique to the peppered moth.

Kettlewell knew that the peppered moth inhabits dense woods and rests on tree trunks during the day. He reasoned that where melanistic individuals had become common, the environment must have been altered in some way to give the dark form a greater survival advantage than the light form. Could natural selection have led to the replacement of the "typical" light form by *carbonaria*? To test this hypothesis, Kettlewell had to find some measure of fitness other than the relative evolutionary success of the two forms.

To determine whether *carbonaria* had a greater fitness than typical peppered moths in areas where melanism occurred, Kettlewell chose the mark-release-recapture method. Large numbers of individuals of both forms were to be marked with cellulose paint and then released in a suitable forest. The area would be thoroughly trapped for moths, and the number

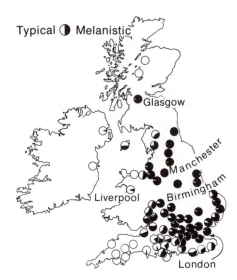

Typical ◑ Melanistic

FIGURE 3-1 The frequency of melanistic individuals in populations of the peppered moth (*Biston betularia*) in various localities in the British Isles. The map is based on more than 20,000 records, from 83 centers, collected during 1952 to 1956.

of marked individuals of each type that were recaptured would be noted. Any difference in the survival of the two forms would appear in the percentages of the light and dark forms recaptured. Kettlewell did not design his experiment to detect differential survival of the larvae of the two forms of *Biston*. Nor could it detect differences in the fecundity of adult moths. Thus, if the results of the experiment turned out negative—that is, if the two forms had similar survival rates as adults—Kettlewell could not have flatly rejected the original hypothesis that the two forms had different fitnesses. Even if predators were detecting and killing equal numbers of light and dark moths, a dark moth might bear more eggs than a light one, or its offspring may have better survived the caterpillar stage. Kettlewell was banking heavily on measurements of adult survival to reveal differences in fitness between the light and dark forms.

Because large numbers of individuals had to be released to guarantee even a small number of recaptures, Kettlewell collected and raised more than 3,000 caterpillars to provide adult moths for his experiments. Two patches of woods, one in a polluted area and one in an unpolluted area, were chosen for the experiment. Adult moths of both forms were marked with a dot of cellulose paint and then released. The mark was placed on the underside of the wing so that it would not call the attention of predators to a moth resting on a tree trunk. Moths were recaptured by attracting them to a mercury vapor lamp in the center of the woods and to caged virgin females around the edge of the woods. (Only males could be used in the study because females are attracted neither to lights nor to virgin females.)

In one experiment, Kettlewell released 201 typicals and 601 *carbonaria* in a polluted wood near Birmingham. The results were as follows:

Birmingham (industrial area)	Typicals	Melanics
Number of moths released	201	601
Number of moths recaptured	34	205
Per cent recaptured	16.0	34.1

These figures indicate that the dark form survived better than the light.

Although consistent with Kettlewell's original hypothesis, these results could be interpreted otherwise: as differential attraction of the two forms to the traps used or as differential dispersion of the two forms away from the point of release. Variables besides differential mortality had to be accounted for.

To test the hypothesis of natural selection unequivocally, Kettlewell ran a control experiment in an unpolluted forest near Dorset with the following results: of 496 marked typicals released, 62 (12.5 per cent) were recaptured; of 473 marked melanics, 30 (6.3 per cent) were recaptured. Thus, in the unpolluted forest, light adults had a higher survival rate than dark adults. If typicals and melanics were differently attracted to light traps or dispersed from a release point at different rates, the level of pollution in the forest should not have influenced the result of the experiment. Only differential fitness could account for a reversal in the relative rate of recapture of typical and melanistic forms in polluted and unpolluted forests. The differences in relative survival of the two forms in the different environments confirmed Kettlewell's hypothesis and established that natural selection was responsible for the high frequency of *carbonaria* in industrial areas.

Having demonstrated that natural selection had been responsible for the replacement of typical by melanistic forms of the peppered moth in industrial regions, Kettlewell then sought to determine the specific agent of selection. He reasoned that in industrial areas pollution had darkened the trunks of trees so much that typicals stood out against them and were easily found by predators. Any aberrant dark forms would be better camouflaged against the darkened tree trunks, and their coloration would confer survival value to them. Eventually, differential survival of dark and light forms would lead to changes in their relative frequency in a population. To test this hypothesis, Kettlewell had to determine whether tree trunks are, in fact, darker in areas where the melanic form was prevalent, and then to demonstrate that camouflage is important to the survival of the moths.

A clean handkerchief rubbed against a tree trunk can satisfy even the most doubting critic that the trunks of trees in polluted areas are darker than those in nonpolluted areas. And as one might have expected, the *carbonaria* stands out against tree trunks in unpolluted woods, whereas typicals are more conspicuous in polluted settings (Figure 3-2).

Kettlewell was certain that conspicuousness of light-colored forms resting on darkened backgrounds must have greatly increased predation on them by visually oriented animals. The dark forms must have a similar

FIGURE 3-2 Typical and melanistic forms of the peppered moth at rest on a lichen-covered tree trunk in unpolluted countryside (left) and on a soot-covered tree trunk near Birmingham (from the experiments of H. B. D. Kettlewell).

disadvantage in unpolluted woods. To test this, Kettlewell placed equal numbers of the light and dark forms on tree trunks in polluted and unpolluted woods and watched them carefully at some distance from behind a blind. (A blind is a tentlike structure intended to conceal an observer from his subjects, more appropriately called a "hide" by the English.) He quickly discovered that several species of birds regularly searched the tree trunks for moths and other insects and that these birds more readily found the moth that contrasted with its background than the moth that blended with the bark. Kettlewell tabulated the following instances of predation:

| | Individuals taken by birds | |
	Typicals	Melanics
Unpolluted woods	26	164
Polluted woods	43	15

These data are fully consistent with the results of the mark-release-recapture experiments. Together they clearly demonstrate the operation of natural selection that, over a long period, resulted in genetic changes in populations of the peppered moth in polluted areas. Many decades were required for the replacement of one form by the other. The agents of selection were insectivorous birds whose ability to find the moths depended on the coloration of the moth with respect to its background. The evolution of industrial melanism shows clearly how the interaction between the organism and its environment determines the organism's fitness.

The theory of natural selection enables us to predict genetic changes in a population from known changes in the environment. If pollution were to be controlled in industrialized areas, and if this allowed forests to revert to their natural state, we would predict that the frequency of the light form of the peppered moth would begin to increase. In fact, smoke-control programs were started in Manchester in 1952. Collections of the peppered moth over the last twenty years in the Manchester area do show a statistically significant increase in the proportion of the light form in the population.

Human Disturbance and Evolutionary Change

The rapid evolutionary change in coloration of peppered moths was stimulated by the effects of industrial pollutants on the appearance of tree trunks. Man has a propensity for creating rapid change in the environment. As a result, many plants and animals have undergone remarkable evolutionary adjustments of their adaptations in response to new conditions. A few examples are worth noting briefly.

The house sparrow, a common bird of cities and farms, was introduced to the east coast of the United States from England and Germany in 1852. The species rapidly spread throughout North America from southern Canada to southern Mexico and can be found in regions as varied as desert, prairie, and coniferous forest. In little more than a century, local populations of sparrows have diverged in body size and coloration in response to local variation in the environment over their extensive range. Northern populations are now conspicuously larger than southern populations; birds from hot, arid localities are paler in color than birds from cool, humid localities. Moreover, geographical variation in the house sparrow parallels similar variation in native species such as the song sparrow. Thus, when accidentally introduced to a variety of new habitats, the sparrow populations quickly responded to the new conditions through evolutionary change.

In many mining operations, ore of too low grade to be refined is often dumped at the surface surrounding the mine shaft. If the ore contains toxic concentrations of lead, zinc, or copper, the presence of these metals in the soil that forms on the mine dump grossly affects the local vegetation. On many long-abandoned mine dumps, species from the surrounding pastures have colonized the toxic soils and appear to grow perfectly well. But plants of the same species growing in unaffected pastures cannot tolerate high metal concentrations and die if they are transplanted into mine dump soil. Most individuals of these species are clearly not inherently tolerant of high concentrations of heavy metals, but toxic elements have strongly selected those few individuals with mutations that enabled them to tolerate the mine dump soils.

Plants sampled along a transect in North Wales, Great Britain, that passes from soil contaminated by a zinc mine to natural pasture reveal an abrupt boundary at which several adaptations change. Zinc tolerance is much greater in plants raised from seed collected on the mine dump soils, as we would have expected, but differences in height, degree of self-fertility, and other characteristics are also evident. The evolution of self-fertility is thought to reduce pollination of mine dump plants by plants on natural pastures where zinc tolerance has not evolved. Self-pollination would thereby help to preserve favorable combinations of genes in the progeny of mine dump plants by keeping nontolerant genes out of the population.

Rapid evolution of plant morphology and physiology has appeared in a variety of situations. Plants directly at the base of old, galvanized (zinc-plated) fences have been shown to tolerate high zinc concentrations in the soil, while plants of the same species growing six feet from the fence died when grown in soil containing zinc. Lawn mowers and grazing pressure by cattle have exerted a strong selection for short, rapidly growing varieties of many pasture and lawn plants. Rapid growth improves a plant's prospects for producing seed before becoming fodder.

At times, evolutionary responses have directly confounded attempts by man to control populations of injurious pests. The evolution of resistance to cyanide by the red scale insect was followed by the appearance of tolerance in several other species of scale. The evolution of resistant strains greatly diminished the value of cyanide gas as an insecticide. Recently developed chemicals, such as DDT, have also stimulated the evolution of resistance in flies and mosquitos where the insecticides have been used repeatedly. Strains of bacteria resistant to antibiotics have similarly reduced the effectiveness of many disease control programs.

For several decades, plant geneticists have been waging an evolutionary battle with wheat rust, a variety of fungus that infects wheat crops and greatly reduces production of grain. Each time a new wheat strain has been developed to resist infection by known types of wheat rust, some new variety of rust invariably has appeared with pathogenic effects on the wheat. The genetic characteristics that distinguish new infective strains of rust were undoubtedly present as rare mutations in the rust population. The new strain of wheat created favorable conditions for the proliferation of the infective rust strain, and it flourished.

Artificial Selection

Man has consciously created by "artificial" selection a variety of domesticated animals and plants tailored to his particular needs. Cows have been selected to give more milk, chickens to lay more eggs, sheep to yield more wool, cereal crops to produce more grain—all these changes directed toward some goal.

Man has accomplished desired evolutionary changes by providing particular environments for the animals and plants he raises. These new environments are designed to increase the fitness of desired traits so that those traits are incorporated into the population by selection. If, by controlled breeding programs, man greatly increases the relative fecundity of cows with high milk production by preventing inferior producers from reproducing, genetic factors for high milk production will be passed on to future generations, while genetic factors for low production are eliminated. In the environment of the milk barn, low production drastically reduces fecundity.

The environments of tropical forests and the arctic tundra have characteristic patterns that select for widely differing adaptations among the organisms that live in these environments. If man consciously were to change, say, the environmental temperature for a population, he would not change the process of natural selection, but merely the adaptations that it produced. The following detailed example demonstrates this point.

Fruit flies of the genus *Drosophila* are frequently used for experiments on selection because they are easily raised in the laboratory, they exhibit many variable morphological traits that are amenable to selection, and they have a short life span, about two weeks, so that selection can be applied to many generations of flies within time limits determined by the patience of the investigators. The experiments we shall consider were performed with *D. melanogaster* by K. Mather and B. J. Harrison in the Department of Genetics at the University of Birmingham, England, between 1942 and 1946. A single trait was chosen for selection: the number of bristles on the ventral surface of the fourth and fifth abdominal segments. These bristles normally number about forty per individual, with some variation in wild populations of flies; the investigators were interested in attempting to both increase and decrease the number of bristles.

The procedure followed to select for an increased or decreased number of bristles is illustrated in Figure 3-3. To begin the experiment, two pairs of flies were allowed to mate and lay eggs in a small culture bottle, with a layer of food on the bottom. The adults were removed from the bottle just before the first of their progeny began to emerge as adults from the pupal stage of development (about two weeks). As the progeny emerged, they were removed from the culture bottles and the sexes were kept separate to assure the virginity of the females. Twenty offspring of each sex were chosen at random and the number of bristles on each were counted. For the high line, Mather and Harrison selected the two individuals of each sex with the largest number of bristles and mated these to produce the next generation. The four flies were first placed together in a small vial for one to three days to mate and they were then transferred to larger culture bottles where they laid their eggs, starting the next generation. The investigators followed a comparable procedure for the low line.

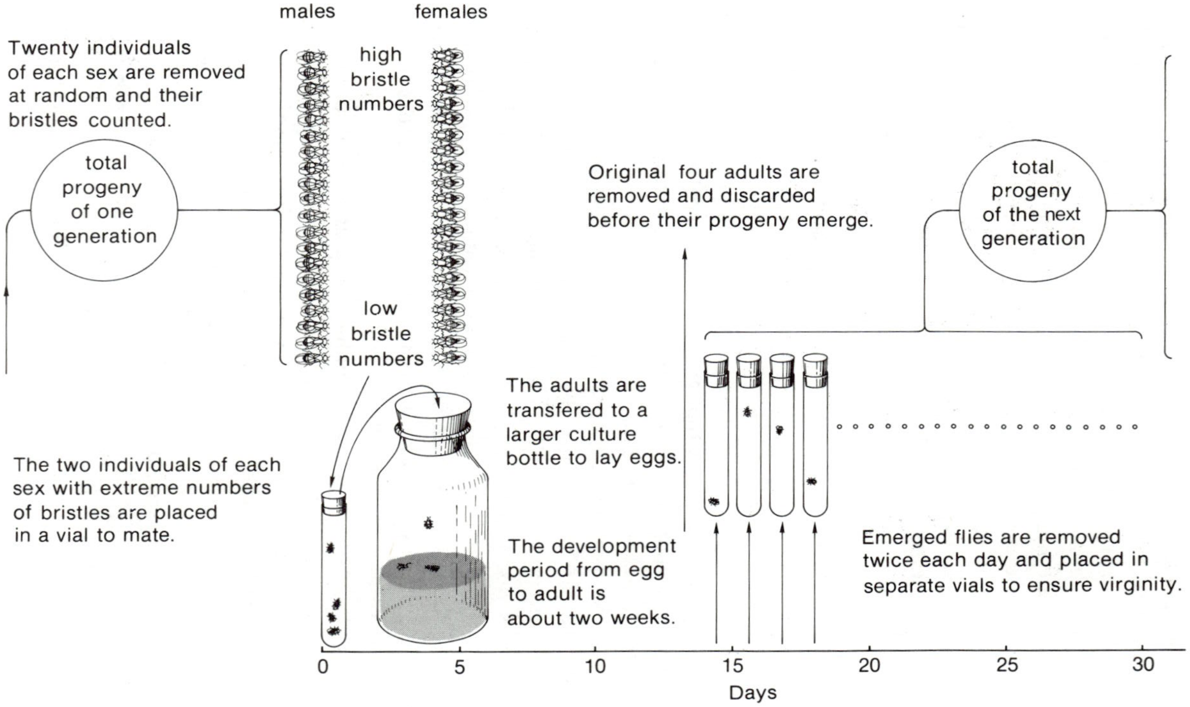

males
females
Twenty individuals of each sex are removed at random and their bristles counted.
total progeny of one generation
high bristle numbers
low bristle numbers
The two individuals of each sex with extreme numbers of bristles are placed in a vial to mate.
The adults are transfered to a larger culture bottle to lay eggs.
The development period from egg to adult is about two weeks.
Original four adults are removed and discarded before their progeny emerge.
total progeny of the next generation
Emerged flies are removed twice each day and placed in separate vials to ensure virginity.
0
5
10
15
20
25
30
Days

FIGURE 3-3 Schematic diagram of Mather and Harrison's method for artificial selection of high and low numbers of bristles on the fourth and fifth abdominal segments of the fruit fly, *Drosophila melanogaster*. The details for the high- and low-selected lines (high and low numbers of bristles) are identical. Each generation of selection required about thirty days. The initial population for the first generation of selection was formed by interbreeding two wild strains, Oregon and Samarkand.

In each successive generation only one-tenth of the flies in each group of progeny, those with the extreme numbers of bristles, were selected and allowed to reproduce. This selection regime produced dramatic results (Figure 3-4). In twenty generations the number of bristles per individual in the high line increased from 36 at the beginning of the experiments to almost 56. Selection in the low line produced a somewhat slower, but nonetheless significant decrease in the number of bristles.

Clearly, Mather and Harrison were able to change the genetic composition of populations of fruit flies according to their own design. They did not, however, alter the process of selection itself, because the genetic changes they produced were brought about through the differential reproduction of individuals with different genetic constitutions, just as Darwin envisioned the mechanism of natural selection. In the high line, for example, only those individuals with the highest number of bristles were allowed to mate and reproduce. Thus the genetic factors that tend to produce high numbers of bristles were transmitted into future populations at a greater rate than those that result in lower bristle numbers. Only the environment, not the mechanism of selection, had been changed.

By applying artificial selection, man alters the pattern of the environment in such a way that selection produces desired changes in populations. In the peppered moth, certain patterns in the forest—the background color of the bark and the hunting characteristics of visually oriented bird predators—provide the evolutionary "reasons" for the adaptations of wing color in the moths. In experiments with artificial selection, such as Mather and Harrison's, man merely changes the pattern of the environment, and

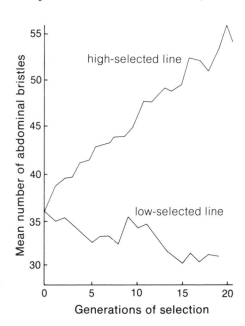

FIGURE 3-4   The results of artificial selection for high and low numbers of abdominal bristles in the fruit fly, *Drosophila melanogaster*.

through the process of natural selection, the design of the organism is changed accordingly.

Life is inseparable from its surroundings. Plants and animals continually exchange energy and materials with the environment. Their adaptations of form and function reflect the evolutionary history of their relationships to the environment. The next five chapters are devoted to the environment itself. At first we shall look at the needs of animals and plants to determine what aspects of their physical surroundings deserve our closest attention. Then we shall look at these properties, as they differ in aquatic and terrestrial environments, as they develop in the course of soil formation, and as they vary in time and space. Finally, we shall survey the variety of biological communities supported by the diverse physical environments at the earth's surface.

4 | Properties and Requirements of Life

LIFE IS AN EXTENSION of the physical world. Biological systems have unique properties, but they nonetheless must obey the constraints imposed by physical and chemical properties of the environment and of organisms themselves. Life is intimately bound to the physical world for three essential needs: water, as a medium for life processes; energy, to fire the engines of living systems; and chemical nutrients (carbon, nitrogen, phosphorus, and others), which are the building blocks of life.

But in spite of the utter dependence of life on the physical environment, the essence of living systems is that they maintain themselves out of equilibrium with the physical world. Although living systems need much of what surrounds them, they must also maintain their distinctiveness. Terrestrial animals obtain oxygen from the atmosphere, but they must guard against the loss of water from their bodies at the same time. Marine fish drink water and pump salts out of their bodies through their gills and kidneys to maintain the proper salt concentration of their blood, which is only half that of seawater. This standoff posture is expensive. Its cost may be measured by the energy transformations of living systems, just as one measures the cost of transportation by fuel used per mile or the cost of electric heating by kilowatt-hours. Animals and plants are open, dynamic systems, continually exchanging with the environment to maintain their activity, continually using the fruits of their activities to maintain their integrity.

Living systems have always had to come to terms with their surroundings. The physiologist Lawrence J. Henderson remarked in his book, *The Fitness of the Environment*, published in 1913, that "Darwinian fitness is compounded of a mutual relationship between the organism and the environment. Of this, fitness of environment is quite as essential a component as the fitness which arises in the process of organic evolution; and in fundamental characteristics the actual environment is the fittest possible abode of life." Although we understand that the physical world does not adapt to the presence of life, it is no accident that life, as we know it, is both constrained by the physical world and takes advantages of its opportunities.

Life processes take place in an aqueous medium. All organisms are composed mostly of water, whether they dwell in the oceans, lakes, and rivers, or on the land. Because the physical and chemical properties of water are well suited to the requirements of life, it is no accident that life is a water-based phenomenon. Water is generally abundant on the earth's surface, and within the temperature range normally encountered, it is present in a liquid state. This property is fundamental to the occurrence of life, for liquids are dense and, at the same time, fluid. If life processes are to work, various compounds must be brought into close proximity to react with each other. Gases are too diffuse and have too little substance to support such processes; solids are too rigid and constraining. Water also has many thermal and solvent properties that are favorable to life. A large amount of heat energy must be added to or removed from water to change its temperature, and water, therefore, does not change temperature rapidly. Because of this property, and the fact that heat travels rapidly through water, the temperatures of organisms and aquatic environments tend to be relatively constant and homogeneous. In addition, the formidable capacity of water to dissolve inorganic compounds makes these materials accessible to living systems.

The immense solvent properties of water are based on the strong attraction of water molecules for other compounds. Molecules are composed of electrically charged atoms (electrically charged atoms, or parts of molecules, are referred to as *ions*). Common table salt, sodium chloride (NaCl), is made up of a positively charged sodium atom (Na^+) and a negatively charged chlorine atom (Cl^-). When salt is placed in water, the attraction of the water molecules for the charged sodium and chlorine atoms is so great, compared with the bonds that hold the molecule together, that the salt molecule readily dissociates into its component atoms. Thus, as the salt molecules dissolve in water, they dissociate and become closely held by hydrogen bonds to water molecules. The dissociation of sodium chloride into its component ions may be written

$$NaCl \rightleftharpoons Na^+ + Cl^-$$

or as

$$NaCl + H_2O \rightleftharpoons H_2ONa^+ + Cl^-$$

if the role of the water molecules as a solvent is to be portrayed. The arrows indicate that even in solution, ions continually rejoin as well as dissociate. It is the equilibrium between these processes that determines the solubility of the substance. The ability of a substance to dissolve in a liquid depends, in part, on the strength of attraction of the solvent molecules compared with the intramolecular bonds of the substance to be dissolved. For biological processes to occur, it is important that substances be dissolved in water because the dissociated ions of the substance are available to react with other ions in the solution.

Temperature and Life

Life processes, as we know them, are restricted to the temperatures at which water is liquid: 0 C to 100 C. Relatively few organisms can survive above 45 C. A few kinds of bacteria occur in hot springs close to the boiling point of water, and blue-green algae are found in water as hot as 75 C. Temperatures greater than 50 C are relatively rare—being found only in hot springs and at the soil surface in hot deserts—but temperatures below the freezing point of water occur commonly over large portions of the earth's surface. If living cells are frozen, the crystal structure of ice prevents the occurrence of most life processes and damages delicate cell structures, rapidly leading to death. A large number of species successfully cope with freezing temperatures in their environments either by maintaining their body temperatures above the freezing point of water or by activating one of a number of mechanisms to resist freezing or to tolerate its effects.

Many organisms reduce the freezing points of their body fluids by adding large quantities of glycerol or another similar organic compound to their blood. These substances act like antifreeze and allow, for example, antarctic fish to remain active in sea water that is colder than the normal freezing point of blood for fish in temperate or tropical seas. Other species are known to be able to become supercooled. That is, their body fluids can fall below the freezing point without ice crystals being formed. Ice crystals generally form around some object, called a seed, which can be a small ice crystal or some other particulate matter. In the absence of crystal seeds, water or blood may be cooled more than 30 C below its freezing point.

Within the range of temperatures from the freezing point of water to the highest temperatures normally encountered on the earth's surface, temperature has two opposing effects on the life processes. First, heat increases the kinetic energy of molecules and thereby accelerates chemical reactions; the rate of biological processes commonly increases between two and four times for each 10 C rise in temperature. Second, the specific biological compounds that catalyze biological reactions (enzymes) become unstable and do not function properly at high temperatures. The combination of these two factors results in an optimum temperature range for the occurrence of biological systems. Enzymes are usually adapted to function best within the particular temperature range that corresponds to the normal body temperature range of the organism. No enzyme is fully active over a broad range of temperatures. Enzymes in organisms that inhabit cold environments usually function better in cold temperatures than the enzymes of organisms from warmer climates.

Inorganic Nutrients

Organisms require a wide variety of chemical elements to form their structure and maintain their proper function. The elements required in greatest amount, after hydrogen, oxygen, and carbon (which are assimilated by plants during photosynthesis), are nitrogen, phosphorus, sulfur, potassium, calcium, magnesium, and iron. Their primary functions are summarized in Table 4-1. Many other nutrients are known to be required in smaller quantity.

Mineral nutrients are acquired by plants in the form of dissolved ions. Plants obtain nitrogen in the form of ammonia ion (NH_4^+) or nitrate ion (NO_3^-), phosphorus in the form of phosphate ion (PO_4^{3-}), calcium and potassium in the form of their simple (elemental) ions (Ca^{++}, K^+), and so on. The solubility of these substances, which determines their availability, varies with temperature, acidity, and the presence of other ions.

All natural waters contain some dissolved substances. Although rain water is nearly pure, it invariably acquires some dissolved minerals from dust particles and droplets of ocean spray in the atmosphere. Most lakes and rivers contain 0.01 to 0.02 per cent dissolved minerals, roughly 1/200 to 1/400 the average salt-concentration of the oceans (3.5 per cent), where salts and other minerals have accumulated over the millennia. In hot climates,

TABLE 4-1   Major nutrients required by living organisms, and their functions.

Element*	Function
Nitrogen (N)	Structural component of proteins and nucleic acids
Phosphorus (P)	Structural component of nucleic acids, phospholipids, and bone
Potassium (K)	Major solute in animal cells
Sulfur (S)	Structural component of many proteins
Calcium (Ca)	Regulator of cell permeability; structural component of bone and material between woody plant cells
Magnesium (Mg)	Structural component of chlorophyll; involved in function of many enzymes
Iron (Fe)	Structural component of hemoglobin and many enzymes
Sodium (Na)	Major solute in extracellular fluids of animals

* Chemical symbol in parentheses

TABLE 4-2 Percentage composition of dissolved minerals in rivers (fresh water), in sea water, and in the blood plasma and cells of frogs.

Mineral ion	Delaware River*	Rio Grande River*	Sea water	Frog plasma	Frog cells
Sodium	6.7	14.8	30.4	35.4	1.3
Potassium	1.5	0.9	1.1	1.3	77.7
Calcium	17.5	13.7	1.2	1.2	3.1
Magnesium	4.8	3.0	3.7	0.4	5.3
Chlorine	4.2	21.7	55.2	39.0	0.8
Sulfate	17.5	30.1	7.7	—	—
Carbonate	33.0	11.6	0.4	22.7	11.7

* The percentages of the negatively charged ions (Cl^-, $SO_4^=$, and $CO_3^=$) exceed those of the positively charged ions because, ion for ion, anions are much the heavier. The numbers of positive and negative ions are approximately equal.

where dissolved substances are concentrated by the evaporation of water more rapidly than rainfall dilutes them, lakes without natural drainage outlets—the Dead Sea and the Great Salt Lake of Utah are examples—may contain up to ten per cent dissolved substances.

Dissolved minerals in fresh and salt water differ in composition as well as in quantity (Table 4-2). Sea water abounds in sodium and chlorine, with respectable amounts of magnesium and sulfate. Fresh water contains a more even distribution of diverse ions, but calcium is usually the most abundant cation (positively charged ion) and carbonate and sulfate the most abundant anions (negatively charged ions). The difference in composition between fresh and salt water is a result of the different solubilities of different substances. Calcium carbonate is relatively insoluble (0.0014 per cent in pure water) and precipitates to form limestone sediments before attaining a high concentration in the oceans. At the other extreme, the solubility of sodium chloride (35.7 g/100 g water, or 26.3 per cent) far exceeds its concentration in sea water; nearly all the sodium chloride that has been washed into ocean basins remains dissolved. The solubility of magnesium sulfate (26.0 g/100 g water) is intermediate.

The pervasive tension between biological systems and the physical world extends to minerals in the environment. On one hand, organisms must obtain minerals from the soil, water, or their food. On the other, they must maintain concentrations of some minerals in their bodies at different levels from those that occur in the environment. For terrestrial and fresh-water organisms this means conserving ions within the body fluids and preventing their being washed out of the body either through the skin or in urine. For many salt-water organisms, abundant ions in the surrounding milieu must be kept out. The blood plasma of the frog, and most vertebrates whether terrestrial or aquatic, resembles sea water in its ionic composition, although it is only one-third as concentrated (Table 4-2). The intracellular medium contains abundant potassium, but plasma ions are kept out to a large extent.

Salt Balance

Left to their own devices, ions diffuse across cell membranes from regions of high to low concentration, thereby tending to equalize their concentrations. Water also moves across membranes (*osmosis*) toward regions of high ion concentrations, tending to dilute concentrations of dissolved minerals. Maintaining an ionic imbalance between the internal medium of the organism and the surrounding environment (*osmoregulation*) against the physical forces of diffusion and osmosis requires the expenditure of energy by organs specialized for salt retention or excretion.

In terrestrial and fresh-water organisms, ion retention is critical. In fresh-water fish, for example, water continually enters the body by osmosis through the mouth and gills, which are the most permeable tissues exposed to the surroundings. (The skin is relatively impermeable.) To counter this influx, water is continually eliminated in the urine, but if dissolved ions were not selectively retained, the fish would soon be a lifeless bag of water. Retention is effected by the kidneys, where salts are actively removed from the urine and fed back into the bloodstream. In addition, the gills are capable of selectively absorbing dissolved ions from the surrounding water and secreting them into the bloodstream. Terrestrial animals acquire minerals in the water they drink and the food they eat; plants absorb ions dissolved in soil water.

It would be quite futile for a salt-water fish, with blood containing half the concentration of mineral ions as sea water, to try to dilute the oceans with water from its body. Keeping ions out of the body is as big a problem for salt-water fish as the retention of ions is for fresh-water species. In contrast to their fresh-water relatives, the gills and kidneys of salt-water fish actively *excrete* ions to counter the tendency of ions to diffuse into the body from the surrounding water. Marine fish also drink sea water to replenish water lost in urine and by osmosis across the surfaces of the gills and, to a lesser extent, the skin. The sharks and rays have achieved a rather elegant solution to the problem of osmotic (water) balance. Urea—$CO(NH_2)_2$—a normal nitrogenous waste product of metabolism in vertebrates, is retained in the bloodstream, instead of being excreted in the urine, to raise the ionic concentration of the blood to the level of sea water without having to increase the concentration of sodium chloride. Although sharks and rays must regulate the diffusion of specific ions in and out of the body, the high level of urea in the blood effectively cancels the tendency of water to leave the body by osmosis. And inasmuch as sharks do not have to drink water to replace osmotic losses of water, they also do not ingest large quantities of salt. It is interesting to note that fresh-water sharks and rays do not accumulate urea in their blood.

Light Energy

Light is the primary source of energy for the ecosystem. Green plants absorb light and assimilate its energy into manufactured organic compounds by *photosynthesis*. The resulting chemical energy in turn is used by plants, and by the animals that consume plants, as a source of energy for other biological processes. The basic reaction in photosynthesis is as follows:

carbon dioxide + water + light energy →

glucose (high in chemical energy) + oxygen

Light energy is absorbed by organic pigments, such as chlorophyll and various types of carotenes. This energy is then used, by way of a long series of steps, to combine carbon dioxide and hydrogen obtained from water molecules into glucose ($C_6H_{12}O_6$). The excess oxygen produced by photosynthesis is released into the atmosphere or into the water surrounding the plant. Glucose may be split and its atoms rearranged and combined to form other organic compounds. Other elements (nitrogen, sulphur, phosphorus, and so on) may be added along the way to form proteins, complex sugars, fats, nucleic acids, and pigments.

The absorption of light by plant pigments is the first step in the production of all organic compounds. But not all the light striking the earth's surface is useful in photosynthesis. Rainbows and prisms show that light is made up of a spectrum of different wavelengths that we perceive as different colors. Wavelengths of light are expressed in terms of nanometers (nm), which are one billionth of a meter (or 10^{-9} meters). The visible spectrum lies between 400 and 700 nm. The energy content of light varies with wavelength and hence with color; short-wavelength blue light has a higher energy level than longer-wavelength red light.

The light that reaches the earth from the sun actually extends in quality far beyond the visible range: through the ultraviolet region toward the short-wavelength, high-energy X rays at one end of the spectrum, and through the infrared region to such extremely long-wavelength, low-energy radiation as radio waves at the other end of the spectrum. As light passes through the upper atmosphere of the earth, most of its ultraviolet components are absorbed, primarily by a form of oxygen known as ozone, which occurs in the upper atmosphere. Because of its high energy level, ultraviolet light can damage exposed cells and tissues. (Sunburn is a symptom of overexposure to ultraviolet radiation.)

Vision and the photochemical conversion of light energy to chemical energy by plants occur primarily within that portion of the solar spectrum containing the greatest amount of energy. Absorption of radiant energy depends on the nature of the absorbing substance. Water has relatively little capacity to absorb light whose characteristic wavelengths fall in the visible region of the spectrum of energies and, as a result, its appearance is "colorless." Dyes and pigments are strong light absorbers of some wavelengths in the visible region and reflect or transmit light of definite colors that become identifying characteristics. Plant leaves contain several kinds of pigments, particularly chlorophylls (green) and carotenoids (yellow), that absorb light and harness its energy. Carotenoids, which give carrots their orange color, absorb primarily blue and green light (Figure 4-1) and reflect light in the yellow and orange portions. Chlorophyll absorbs light in the red and violet portions of the spectrum, and reflects green and blue. When chlorophylls and carotenoids occur together in a leaf, green light is absorbed least, hence the color of the leaf.

Oxygen and Biological Activity

The chemical energy in organic compounds is made available to living systems primarily by the converse process to photosynthesis, called *respiration*. In the respiratory metabolism of a simple sugar like glucose, the sugar molecule is oxidized to produce carbon dioxide and water (the raw materials for photosynthesis) and release energy to drive other biological processes. The overall equation for respiration is

$$\text{glucose} + \text{oxygen} \rightarrow \text{carbon dioxide} + \text{water} + \text{energy}$$

Because oxygen plays such an important role in making energy available, its occurrence in the environment can limit the level of metabolic activity of the organism. The manner in which different kinds of organisms procure oxygen demonstrates the importance of the physical environment to the design of organisms (Table 4-3).

Most small aquatic organisms obtain oxygen by diffusion from the surrounding environment into their tissues. Carbon dioxide produced by respiration also escapes the body by the same route. If the concentration of oxygen in an organism's tissues is lower than that of the surrounding me-

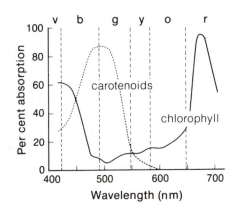

FIGURE 4-1 Absorption of light of different wavelengths (the absorption spectrum) by two groups of plant pigments—chlorophyll and carotenoids—which capture light energy for photosynthesis. The colors of the spectrum are violet (v), blue (b), green (g), yellow (y), orange (o), and red (r).

TABLE 4-3 Summary of some problems in the evolution of high rates of oxygen delivery to the tissues of large organisms.

Problem	Solution	Biological occurrence
Not critical in small or inactive organisms	Oxygen is obtained by simple diffusion through cells	Protozoa, sponges, coelenterates
Diffusion distance from surface to body core is too great in large organisms	Circulatory system to pump fluids from surface to core	Widespread: pumping by body muscles in roundworms; open system without capillaries in arthropods and many molluscs; closed capillary systems in vertebrates
The solubility of oxygen in water limits oxygen transport by circulating fluids	Incorporation of oxygen-binding proteins (*e.g.*, hemoglobin) in blood	Hemoglobin is widespread in vertebrates, but sporadic in lower groups where other pigments may be found; insects notably lack blood pigments because air is carried directly to cells by a tracheal system
High concentrations of protein increase osmotic level of blood	Respiratory proteins are tightly packed in red blood cells	All vertebrates, some molluscs and echinoderms

dium, oxygen will diffuse into the body. And because the oxygen is used in respiratory metabolism, its concentration in the body is kept low.

Diffusion can satisfy the oxygen needs of very small aquatic organisms (and the leaves of terrestrial plants), but in large organisms, the distance between the external environment and the center of the body is too great for diffusion to ensure a rapid supply of oxygen. In fact, diffusion is ineffective at distances greater than about one millimeter. One solution to this problem has been the evolution of circulatory systems. Oxygen that diffuses across the surface of the organism, or across the surfaces of specialized structures with large areas in direct contact with the environment (lungs and gills) is carried to other parts of the organism by circulating body fluids.

The movement of aqueous fluids through a system of vessels greatly aids the distribution of oxygen and other materials throughout the body, but water itself often cannot carry enough dissolved oxygen to support a high rate of activity. The solubility of oxygen in water (up to 1 per cent by volume or about fourteen parts per million by weight) just cannot supply sufficient oxygen to active tissues.

To increase the oxygen-carrying capacity of their blood, most groups of animals have complex protein molecules, such as hemoglobin, to which oxygen molecules readily attach and thereby are taken out of solution. When oxygen becomes attached to hemoglobin—four molecules of oxygen for each one of hemoglobin—a complex known as oxyhemoglobin is formed. The binding process must be reversible so that oxygen can be released to the tissues. While blood plasma itself carries only limited oxygen in solution, up to fifty times more is transported in the bloodstream, bound to oxygen-carrying molecules. Hemoglobin is most effective in binding oxygen when its molecules are present in very high concentration, close to the point of crystallization. If such high concentrations of hemoglobin occurred in the plasma, the osmotic level of the blood would become too great for proper physiological function, particularly for proper salt balance, so hemoglobin is concentrated inside red blood cells (erythrocytes), each of which may contain upward of a quarter billion hemoglobin molecules. A further advantage to packaging hemoglobin in red blood cells may be that the close association of hemoglobin molecules may alter their interaction among each other and with the surrounding blood environment in such a way as to facilitate oxygen binding and release. An analogous arrangement is found in several worms and snails, which have hemoglobin but lack special blood cells. In these species, hemoglobin is aggregated into large groups of, perhaps, twenty to fifty molecules.

Another recurrent adaptation to enhance the uptake of dissolved oxygen from water is countercurrent circulation, a particular arrangement of the structure of gills whereby water and blood flow in opposite directions (Figure 4-2). In a countercurrent system, as blood picks up oxygen from the water flowing past, it comes into contact with water having progressively greater oxygen concentration. This is possible because the water has flowed past a progressively shorter distance of the gill lamella (Figure 4-3). With this arrangement, the oxygen concentration of the blood plasma can approach very nearly the concentration in the surrounding water. If blood and

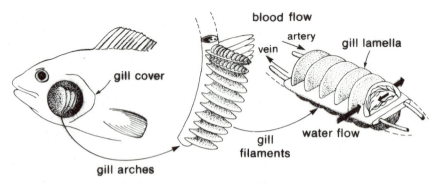

FIGURE 4-2 A fish's gill consists of several gill arches each of which carries two rows of filaments. The filaments bear thin lamellae oriented in the direction of the flow of water through the gill. Within the lamellae, blood flows in a countercurrent manner in opposite direction to the movement of water past the surface.

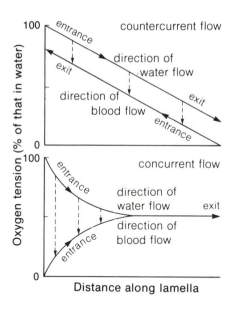

FIGURE 4-3 A schematic comparison of changes in oxygen tension of blood and water in concurrent and countercurrent systems. In the latter, a constant gradient is maintained across which oxygen can diffuse, resulting in a higher oxygen tension in the blood leaving the system, approaching that of the incoming water.

water were to flow together through the gill, an equilibrium oxygen concentration would be established with equal and intermediate levels in the blood and water. The countercurrent system keeps the blood and water out of equilibrium and maintains a constant gradient across which oxygen can flow.

The rate of oxygen uptake depends, over and above the oxygen gradient between the gill and the surrounding water, on the area of the gills, or other respiratory surface. Active fish with high oxygen requirements have large gills compared with their more sedentary relatives. Pound for pound of body weight, the gills of the predatory mackerel have fifty times the surface area of the gills of the sedentary, bottom-dwelling goosefish. The total amount of hemoglobin in the bloodstream is also subject to adjustment. For example, the goosefish has a total oxygen capacity in its blood of 5 per cent by volume; the mackerel, 16 per cent. This difference reflects the relative hemoglobin concentration in the blood and parallels the relative sizes of the gills in the two species.

The basic requirements of animals and plants are independent of the environment they inhabit. All forms of life need water, solutes, energy, and nutrients. But the way in which they procure these and their manner of tolerating the physical conditions of the environment vary—nowhere so widely as between aquatic and terrestrial habitats, as we shall see in the next chapter.

5 | Aquatic and Terrestrial Environments

Life arose in the sea. Conditions in shallow coastal waters were ideal for the development and diversification of the first plants and animals. Temperature and salinity varied little; sunlight, dissolved gases, and minerals were abundant. Water itself is buoyant and supports both delicate structures and massive bodies with equal ease.

The difficulty of the first step in colonizing the land can be measured by the gap of several hundred million years between the time modern life began to flourish in the sea and the appearance of life on land. Yet in spite of the harshness of terrestrial environments, life has generally attained a higher degree of organic diversity and productivity on the land.

Perhaps we should not distinguish environments as being primarily aquatic or terrestrial, for the sea is as surely underlain by land as the terrestrial environment is drenched in an ocean of air. To appreciate fully the distinction between aquatic and terrestrial environments, we should contrast the properties of water and air rather than those of water and earth. The qualities of water that overwhelmingly determine the form and functioning of aquatic organisms are its density (about 800 times that of air) and its ability to dissolve gases and minerals. Water provides a complete medium for life; most marine organisms are independent of the land beneath them except for those that use it as a site for attachment in shallow water or a place to burrow. In contrast, terrestrial life is narrowly confined to the interface between the atmosphere and the land, each of which makes essential contributions to the environment of life. Air provides oxygen for respiration and carbon dioxide for photosynthesis, while soil is the source of water and minerals. Air also offers less resistance to motion than water, and thus constrains movement less.

46

Buoyancy and Viscosity of Water and Air

Because water is dense, it provides considerable support for organisms that after all are themselves mostly water. But organisms also contain bone, proteins, dissolved salts, and other materials that are more dense than salt or fresh water, and thus generally tend to sink. To counter this, aquatic plants and animals have a variety of adaptations either to reduce their density or to retard their rate of sinking. Such adaptations are crucial to the tiny plants and animals of the plankton that cannot move actively. Many fish have swim bladders, small enclosures within the body that are filled with gas to equalize the density of their bodies and the surrounding water. Large kelps, a type of seaweed found in shallow waters, have analogous gas-filled organs. The kelps are attached to the bottom by holdfasts, and gas-filled bulbs float their leaves to the sunlit surface waters. Many microscopic unicellular plants (*phytoplankton*) that float in great numbers in the surface waters of lakes and oceans contain droplets of oil that are less dense than water and compensate for the natural tendency of cells to sink. Fish and other large marine organisms also make use of lipids to provide buoyancy. Most fats and oils have a density of about 0.90 to 0.93 g cc^{-1} (90 to 93 per cent of the density of water) and thus tend to float. Aquatic organisms also lighten their bodies by reducing skeleton, musculature, and perhaps even the salt concentration of the body fluids. It has been argued that the low osmotic concentration of the blood plasma of aquatic vertebrates (about one-third to one-half that of sea water) is an adaptation to reduce density. Accumulation of less dense oils and fats and reduction of more dense body components are particularly important to organisms that inhabit very deep water where, under high pressure, the density of gases in the swim bladder increases to nearly the density of water and hence provides little buoyancy.

The high viscosity of water lends a hand to some organisms that would otherwise sink more rapidly, but hampers the movement of others. Tiny marine animals often have long, filamentous appendages that retard sinking, just as a parachute slows the fall of a body through air (Figure 5-1). The wings of maple seeds, the spider's silk thread, and the tufts on dandelion and milkweed seeds provide a similar function and increase the dispersal range of land species. But to reduce the drag encountered in moving through a medium as viscous as water, fast-moving animals must assume streamlined shapes. Mackerel and other schooling fish of the open ocean closely approach the hydrodynamicist's body of ideal proportions (Figure 5-2). Of course, air offers far less resistance to movement, having less than 1/50 the viscosity of water.

Because water is more buoyant than air, gravity does not limit the maximum size of organisms to the extent that it does on land. Blue whales may attain more than 100 feet in length and can weigh more than 100 tons, dwarfing the largest land animals. (Large elephants weigh only seven tons.)

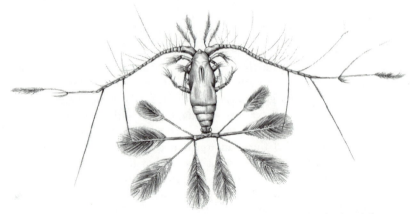

FIGURE 5-1   Filamentous and feathery projections from the body of the tropical marine planktonic crustacean, *Calocalanus pavo* (overall length about 1.2 millimeters).

That water provides excellent support against gravity is illustrated by the skeletons of sharks, which are composed of flexible cartilage, and therefore would offer little support on land. Even the air-breathing whales suffocate quickly when they are accidently stranded on a beach because their great weight deflates their lungs. By contrast, terrestrial organisms have rigid structures to keep their bodies upright against the pull of gravity. The bony

FIGURE 5-2   The streamlined shapes of young mackerel reduce the drag of water on the body and allow the fish to swim rapidly with minimum energy expenditure.

internal skeletons of vertebrates, the chitinous exoskeleton of insects, the cellulose walls of plant cells all provide support. Rigid structures occur in aquatic organisms more for protection (the shells of molluscs) or to provide rigid attachment sites for muscles (the shells of crabs and bony skeletons of fish) than to support the weight of the body.

Light

Light strikes the surface of the oceans and the land with equal intensity. On land, most light is absorbed or reflected by the leaves of plants. But unlike air, water absorbs or scatters light strongly enough to severely limit the depth of the sunlit zone of the sea. The transparency of a glass of water is deceptive. In pure sea water, the energy content in the visible part of the spectrum diminishes to 50 per cent of its surface value within a depth of 10 meters, and to less than 7 per cent within a depth of 100 meters. Furthermore, different wavelengths are affected differently by water. Longer wavelengths are absorbed more rapidly, virtually all infrared radiation within the topmost meter of water. Short light waves (violet and blue) tend to be scattered by water molecules and thus do not penetrate deeply. A consequence of the absorption and scattering of light by water is that with greater depth, green light tends to predominate (Figure 5-3). The photo-

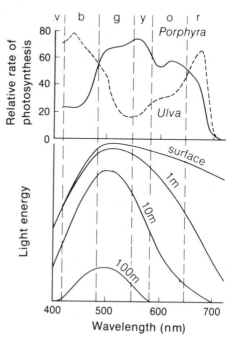

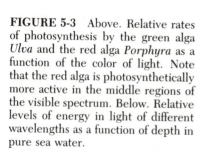

FIGURE 5-3 Above. Relative rates of photosynthesis by the green alga *Ulva* and the red alga *Porphyra* as a function of the color of light. Note that the red alga is photosynthetically more active in the middle regions of the visible spectrum. Below. Relative levels of energy in light of different wavelengths as a function of depth in pure sea water.

synthetic pigments of plants are adapted to this spectral shift. Plants near the surface of the oceans, such as the green alga *Ulva* (sea lettuce), have pigments resembling those of terrestrial plants and mostly absorb light of blue and red color (compare Figures 5-3 and 4-1). But the deepwater red alga *Porphyra* has additional pigments that enable it to utilize green light more effectively in photosynthesis.

Because photosynthesis requires light, the depth to which plants are found in the oceans is limited by the penetration of light to a fairly narrow zone close to the surface, in which photosynthesis exceeds plant respiration, called the *euphotic* zone. The lower limit of the euphotic zone, where photosynthesis just balances the rate of respiration, is called the *compensation point*. If algae in the phytoplankton sink below the compensation point or are carried below it by currents, and are not soon returned to the surface by upwelling, they will die.

In some exceptionally clear marine and lake waters, particularly in tropical seas, the compensation point may be a hundred meters below the surface, but this is a rare condition. In productive waters with dense phytoplankton, or in turbid waters with suspended silt particles, the euphotic zone may be as shallow as one meter. In some polluted rivers, little light penetrates beyond a few centimeters.

Because plants require light, large benthic algae (forms attached to the bottom) occur only near the edges of continents where the depth of the water does not exceed a hundred meters. In the vast open reaches of the ocean, as well as in shallower coastal waters, one-celled floating plants compose the phytoplankton of the euphotic zone. The small floating animals (*zooplankton*) that prey upon the phytoplankton are also found primarily in this region where their food is most abundant. But animal life is not restricted to the upper layers of water. Even the deepest parts of the ocean, under several miles of water, harbor a diverse fauna, supported by the constant rain of dead organisms that sink from the sunlit regions above.

Oxygen and Carbon Dioxide

Nearly all organisms, including green plants, require oxygen for respiration. Although oxygen abounds in the atmosphere, comprising about one-fifth its volume, oxygen does not dissolve readily in water. The solubility of oxygen is affected by both temperature and salinity, but even at its maximum possible solubility, at 0 C in fresh water, the concentration of oxygen is about ten milliliters per liter (10 ml l^{-1}), which is 1 per cent by volume, or one-twentieth that of air. Such concentrations of oxygen are never reached in natural bodies of water, where values probably range from a maximum of about 6 ml l^{-1} to zero, that is, completely *anaerobic* (no-oxygen) conditions.

To remove oxygen from the environment, aquatic organisms must bring water into contact with their respiratory structures. Inasmuch as the high viscosity of water makes it difficult to move water about, this property also restricts the availability of oxygen to aquatic organisms. Terrestrial animals can move air in and out of their lungs rapidly, compared to the rate at which a fish or clam can pump water past its gills. The physiologist Knut Schmidt-Nielsen has estimated that with a favorable concentration of oxygen (7 ml l^{-1}), an aquatic organism must pump 100,000 grams of water (about 26 gallons) past its gills to obtain one gram of oxygen. Air breathers would have to inhale only five grams of air (about one gallon) to obtain the same amount of oxygen. Clearly, oxygen uptake in the aquatic environment requires considerable expenditure of energy to move water, if it is to be done rapidly. It is not surprising that in most water breathers, water flows in one direction through their respiratory organs. This arrangement eliminates the need to use energy to periodically stop the flow of a large mass of water and accelerate it in the opposite direction. Such breathing in and out presents a negligible cost to terrestrial organisms because of the small quantities of air involved.

Although oxygen is evenly distributed in the atmosphere, the concentration of dissolved oxygen in water varies considerably. Oxygen is generally most abundant near the air-water interface, and its concentration decreases with depth. Still water usually contains less oxygen than rapidly flowing water, where mixing of water and air occurs in stream riffles, waterfalls, and waves. Photosynthesis by aquatic plants can also be an important source of dissolved oxygen. But the consumption of oxygen by animals and microorganisms in poorly mixed waters has a more dramatic effect on oxygen tension. Whereas photosynthesis occurs mostly in sunlit surface waters that are normally well aerated, animal and microbial respiration occurs at all depths and is often most intense in sediments at the bottom, where respiration tends to deplete the oxygen, leading to anaerobic conditions and a slowing or stopping of life processes. It was such anaerobic conditions in stagnant swamps and deep ocean basins that allowed organic sediments to escape microbial decomposition and become what are now oil and coal deposits.

In addition to oxygen for respiration, plants require carbon dioxide for photosynthesis. Carbon dioxide is one of the lesser gases of the atmosphere, accounting for only 0.03 per cent of the volume of air (0.3 ml l^{-1}). But the solubility of carbon dioxide in water is roughly thirty times that of oxygen, and under ideal conditions water contains about 0.3 ml of dissolved carbon dioxide per liter, the same concentration as in the atmosphere. In addition, carbon dioxide and water readily form bicarbonate ion (HCO_3^-), which is extremely soluble in water (69 g of $NaHCO_3$ per liter of water, for example). Sea water normally contains the equivalent of 34 to 56 ml carbon dioxide per liter in the form of bicarbonate. Because bicarbonate and dissolved carbon dioxide exist in equilibrium

$$H_2O + CO_2 \rightleftharpoons HCO_3^- + H^+$$

dissolved carbon dioxide utilized by plants is readily replenished from bicarbonate ion. And for this reason, shortages of carbon dioxide rarely occur.

Water Loss by Terrestrial Organisms

Terrestrial environments pose a severe problem to life: the conservation of water within the body. This problem has had an overriding influence on adaptations of form and function. The outer coverings of most truly terrestrial organisms—the chitin of arthropods (insects, spiders, and others), the skin of reptiles, birds, and mammals, the bark and cuticle of flowering plants and conifers—are nearly impermeable to water. Moreover, the respiratory organs, whose surfaces must be kept moist to effect gas exchange, have been relocated from external positions (as in the gills of fish and amphibians) to more protected internal positions: the lungs of vertebrates and the tracheae (air passages) of insects. Gas exchange in terrestrial plants is limited to small openings (*stomata*) distributed over the surface of the leaf (Figure 5-4). These morphological adaptations considerably reduce the loss of water through evaporation. Terrestrial organisms lacking well-developed water-conserving adaptations—earthworms, for example—are restricted to moist soil within which air is saturated with water vapor.

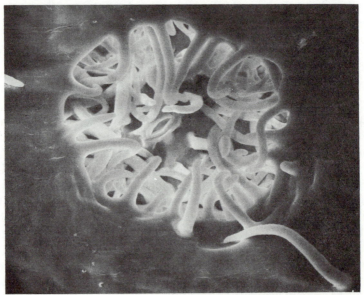

FIGURE 5-4 The leaf of oleander, a drought-resistant plant. The stomata lie deep within hair-filled pits on the leaf's undersurface. The hairs reduce air movement and trap moisture, thereby reducing water loss from the leaf. The scanning electron micrograph shows a hair-filled pit from underneath the leaf surface magnified about 500 times.

FIGURE 5-5   Merriam's kangaroo rat is well-adapted to hot deserts in the south-western United States. Its kidneys concentrate urine, thereby conserving water, and it ventures out of its cool burrows to feed only at night when heat stress on the surface is least.

Where fresh water is scarce, animals cannot drink to replenish water lost by evaporation from the lungs and in the excrement and urine. To achieve water balance under these conditions, organisms must restrict avenues of water loss if they are to survive. Some desert mammals conserve water so efficiently that their need for water may be satisfied by the water produced by respiratory metabolism. Among mammals, the kangaroo rat is well adapted for life in a nearly waterless environment (Figure 5-5). In the large intestine of the kangaroo rat's digestive tract, water is resorbed from waste material to such an extent that the feces are nearly dry. Much of the water that evaporates from the lungs is recovered by condensation in enlarged nasal passages. The tissues within the nose are cooler than the lung surfaces, and water exhaled in the warm air from the lungs is condensed there. This simple mechanism takes advantage of the high heat of vaporization of water and works as follows: When the kangaroo rat inhales dry air, moisture in its nasal passages evaporates, cooling the nose and saturating the inhaled air with water. When moist air is exhaled from the lungs, much of its water is condensed and retained in the nasal passages. By alternating condensation with evaporation during breathing, the kangaroo rat minimizes its respiratory water loss.

Water loss and salt balance are as intimately linked as are the physical processes of diffusion and osmosis. Animals that eat meat or other animal food obtain salts in their food in excess of their requirements. Where water is abundant, organisms merely have to drink large quantities of water to flush out salts that would otherwise tend to accumulate in the body. Where water is scarce, however, organisms must produce a concentrated urine to conserve water. Animals in moist and arid habitats excrete similar amounts of salt, but desert animals cannot afford the copious loss of water that would result from a dilute urine. And so, as one would expect, desert animals have champion kidneys. For example, whereas humans can concentrate salt ions in their urine to about four times the level in blood plasma, the kangaroo rat's kidney produces urine with a salt concentration as high as eighteen times that of the blood.

Carnivores consume excess nitrogen, as well as excess salts, in their food. This nitrogen, ingested in the form of proteins, must be eliminated from the body when proteins are metabolized. Animals lack the biochemical mechanisms possessed by some microorganisms for producing nitrogen gas and, consequently, they cannot dispose of nitrogen as a gas that would escape from the blood through the lungs. Many inorganic forms of nitrogen, nitrate for one, are highly poisonous and cannot be produced in quantity without toxic effects. To solve the problem of nitrogen excretion, most aquatic organisms produce the simple metabolic by-product, ammonia (NH_4^+). Although it is mildly poisonous to tissues, aquatic organisms can rapidly eliminate ammonia in a copious, dilute urine before it reaches a dangerous concentration. Because terrestrial animals cannot afford to lose so much water for the sake of nitrogen excretion, they produce a less toxic metabolic by-product of protein metabolism, which can be concentrated in the blood and urine without dangerous effects. In mammals, this waste product is urea—$CO(NH_2)_2$—which, because it dissolves in water, requires some urinary water loss, the amount depending on the concentrating power of the kidneys. Birds and reptiles have carried adaptation to terrestrial life one step further by producing uric acid—$C_5H_4N_4O_3$—as a nitrogenous waste product. Uric acid has the distinct advantage in desert environments of being crystallized out of solution and thereby greatly concentrated in the urine. This water-conserving adaptation allows birds and reptiles to be active during the heat of the day in the desert when small mammals are forced to retreat to their underground burrows.

Water loss is a greater problem for terrestrial plants than for terrestrial animals for, although most plants do not use water to excrete salts and nitrogenous wastes, they lose much more water through normal gas exchange. The reason for this difference is quite simple. Animals breathe to obtain oxygen, which constitutes 20 per cent of air. Plants require carbon dioxide, which constitutes only 0.03 per cent of air. Therefore, to obtain one milliliter of carbon dioxide, plants must expose themselves to almost 700 times as much air, and to 700 times the opportunity to lose water vapor to the atmosphere, as an animal does in procuring the same volume of oxygen. No wonder plants have roots! Their demand for water requires a continuous supply from the soil.

Drought-adapted plants do have numerous adaptations that reduce water loss. These involve modifications to reduce transpiration across plant surfaces, to reduce heat loads (rate of evaporation increases with temperature), and to tolerate higher temperatures and thereby avoid having to evaporate water to reduce temperature during hot spells. When plants absorb sunlight, they heat up. Overheating can be avoided by increasing the surface area for heat dissipation and protecting the plant surface from direct sunlight with dense hairs and spines. Spines also produce a still, boundary layer of air that traps moisture and reduces evaporation from the plant surface. Transpiration is further reduced in some plants by means of a thick, waxy cuticle that is impervious to water and by recessing the stomata (openings in the leaf for gas exchange) in deep pits (Figure 5-4). In many species, biochemical pathways of photosynthesis are adapted either for rapid uptake of carbon dioxide, whereby the period required for a given amount of gas exchange is reduced, or for storage of carbon dioxide in an altered chemical form so that gas exchange can occur at night when temperatures are lower and moisture stress is reduced.

And so we see some of the major differences between aquatic and terrestrial habitats and the unique problems each poses for organisms. The characteristics of aquatic and terrestrial environments are determined primarily by the density and thermal properties of water and air, and the relative availability of oxygen, water, and minerals. One characteristic aspect of almost all terrestrial environments, absent from aquatic habitats, is soil. As we shall see in the following chapter, soil is a complex combination of weathered rock and decaying organic detritus, within which much of the energy transformation and mineral exchange in terrestrial ecosystems takes place. Soil does not form in aquatic habitats because wave action prevents the accumulation of loose sediments in many areas, and aquatic plants do not have leathery leaves and woody branches which, when they decompose, contribute greatly to the structure of terrestrial soils. Probably the closest thing to soil in the sea are mucky, organic sediments in ocean basins. These sediments are habitats for many burrowing detritus feeders, as are terrestrial soils. When churned up by water movement, they also contribute dissolved minerals to open waters. But the analogy stops there. Unlike aquatic plants, the trees, shrubs, and herbs of terrestrial habitats obtain all their minerals from the soil and, in turn, contribute to its structure and composition. Also, the properties of soils are determined partly by the underlying bedrock, whereas marine sediments merely accumulate on top of the ocean floor. Finally, aquatic sediments are fully saturated with water at all times and do not develop the zones characteristic of terrestrial soils.

In the next chapter, we shall take a close look at the process of soil formation and the environmental factors which determine the structure and composition of soils in different regions.

6 | Soil Development

WE TAKE THE DIRT under our feet for granted, unwisely it would seem, because most of the vital mineral exchange between the biosphere and the inorganic world occurs in the soil. Plants obtain water and nutrients from the soil. When they die, they return to the soil where they are decomposed and their mineral nutrients released. Organisms responsible for decomposition—the myriad bacteria and fungi, the minute arthropods and worms, the termites and millipedes—abound in the surface layers of the soil where dead organic matter is freshest. Their activities contribute to the development of soil properties from above, while physical and chemical decomposition of the bedrock contribute to the soil from below.

As with climate, soil formation is determined by physical and chemical processes whose results are as varied as the conditions under which they occur. Soil characteristics vary greatly over the world and both influence and reflect the distribution of vegetation types. Five factors are largely responsible for variation in soils: climate, parent material, vegetation, local topography, and, to some extent, age. In general, the decomposition and weathering of parent rock and the addition of organic material to the soil proceed most rapidly in warm, wet climates. As a result, the influence of parent rock on the structure and composition of soil *decreases* with increasing rainfall, temperature, and age.

Once formed, soils remain in a dynamic state. Some minerals are removed by ground water; others blow in as dust or are released by the decomposition of underlying rock layers. Although soil is in a constant state of flux, soils of most regions attain characteristic steady-state properties. In dry areas, rainfall is so sparse that chemical weathering of bedrock or other parent material is slow, and plant production is so low that little organic detritus is added to the soil. Soils of arid regions are typically shallow, and bedrock lies close to the surface (Figure 6-1). Weathering may extend to the depth of only one foot; such soils can be shallower or even absent where

FIGURE 6-1 Profile of a poorly developed soil in Logan County, Kansas, illustrating shallow soil depth and absence of soil zonation.

erosion removes weathered rock and organic detritus as rapidly as they are formed. The faces of cliffs and rocks in the upper regions of intertidal zones at the edge of the sea are extreme examples of sites where soil formation is prevented by erosion. Soil development is also stopped short on alluvial deposits, where the weathering process does not have a chance to work owing to the fresh layers of silt deposited each year by floodwaters. At the other extreme, parent material is most highly weathered in parts of the humid tropics, where chemical alteration of the parent material may extend to depths of 200 feet or more. Most temperate soils are intermediate in depth, usually extending to a few feet.

Soil Horizons

Where a recent roadcut or excavation exposes the soil in cross section, one is often struck by the presence of distinct layers, called *horizons* (Figure 6-2). Soil horizons have been described with complex and sometimes conflicting terminology by soil classifiers. A generalized, and somewhat simplified, soil profile has four major divisions: O, A, B, and C horizons, with two subdivisions of the A horizon. Arrayed in order descending from the surface of the soil, the horizons and their predominant characteristics are:

O   primarily dead organic litter. Most soil organisms are found in this layer.

A_1 a layer rich in humus, consisting of partly decomposed organic material mixed with mineral soil.

A_2 a region of extensive *leaching* (or *eluviation*) of minerals from the soil. Because minerals are dissolved by water (mobilized) in this layer, plant roots are concentrated here where the minerals are most readily available.

B   a region of little organic material whose chemical composition closely resembles that of the underlying rock. Clay minerals and oxides* of aluminum and iron leached out of the overlying A_2 horizon are sometimes deposited here (*illuviation*).

C   primarily weakly weathered material, which closely resembles the parent rock. Calcium and magnesium carbonates accumulate in this layer, often forming hard, impenetrable layers within the C horizon.

The soil horizons demonstrate the decreasing influence of climate and the increasing influence of bedrock with increasing depth. Soil formation is greatly complicated, however, by the movement of mineral elements upward and downward through the soil profile. Before considering these processes in detail, we shall examine the initial weathering of the bedrock and how it influences soil characteristics.

Weathering

Decomposition of rock at the bottom of a soil profile, or on a newly exposed rock surface, is fostered by the action of both physical and chemical agents. Repeated freezing and thawing of water in crevices breaks up rock into smaller pieces and exposes new surfaces to chemical action. Initial chemical alteration of the rock occurs when water dissolves some of the more soluble minerals, particularly sodium chloride (NaCl) and calcium sulfate ($CaSO_4$), and leaches them from the soil profile. Other minerals, particularly the oxides of titanium, aluminum, iron, and silicon, do not dissolve readily and are thus resistant to leaching under most conditions.

The weathering of granite exemplifies some basic processes of soil formation. Granite is an igneous rock formed when the less dense molten material deep within the earth rose to the surface, cooled, and crystallized. Granite consists chiefly of three minerals: feldspar, mica, and quartz. Feldspar, which consists of aluminosilicates of potassium ($K_2O \cdot Al_2O_3 \cdot 6\ SiO_2$), weathers rapidly owing to the removal of potassium (K) in the presence of

---

* Oxides are compounds consisting of oxygen and one or more other elements.

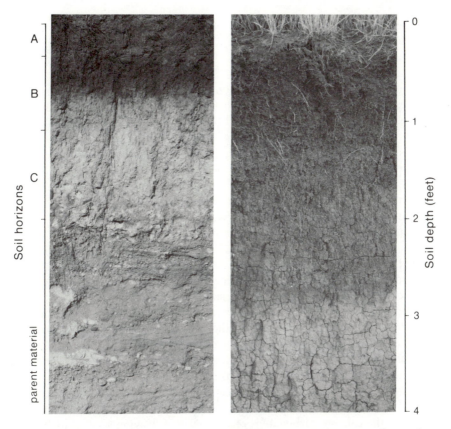

FIGURE 6-2 Soil profiles from the central United States illustrating distinct horizons. The profile at left, from eastern Colorado, is weathered to a depth of about two feet where the subsoil contacts the original parent material, consisting of loosely aggregated, calcium-rich, wind-deposited sediments (loess). The A_1 and A_2 horizons are not clearly distinguished except that the latter is somewhat lighter-colored. The B horizon contains a dark band of redeposited organic materials which were leached from the uppermost layers of the soil. The C horizon is light-colored and has been leached of much of its calcium. Some of the calcium has been redeposited at the base of the C horizon and at greater depths in the parent material. The profile at right is that of a typical prairie soil from Nebraska. Rainfall is sufficient to leach readily soluble ions completely from the soil. Hence there are no B layers of redeposition, as in the drier Colorado soil at left, and the profile is more homogeneous. The A horizon is weakly subdivided into a darker upper layer and lighter lower layer. The weathered soil lies upon a parent material composed of loess, the wind-blown remnants of glacial activity. The depth scale, in feet, at right, applies to both profiles; the soil horizons, at left, apply only to the left-hand profile.

carbonic acid. (Carbonic acid [H_2CO_3], formed when carbon dioxide dissolves in water, is always present in rain water.) The remainder of the feldspar mineral is reorganized with water to form one of several types of silicate clays, such as kaolinite ($Al_2O_3 \cdot 2\ SiO_2 \cdot 2\ H_2O$), depending on weathering conditions. As a general class of materials, clays perform an extremely important function in the soil, that of providing sites for ion exchange between the soil and plants. We shall look in detail at the role of clay particles later.

The mica grains in granite are composed of aluminosilicates of potassium, magnesium (Mg), and iron (Fe). When granite weathers, potassium and magnesium are removed rapidly, and the remaining oxides of iron, aluminum, and silicon form clay particles. Quartz, a form of silica (SiO_2), is relatively insoluble in acidic water and, therefore, remains more or less unaltered in the soil as sand grains. Changes in chemical composition of granite as it weathers from rock to soil are summarized in Figure 6-3. Calcium, magnesium, sodium, and potassium disappear quickly, while aluminum, silicon, and iron remain.

Removal of minerals from weathered granite rock varies greatly with climate. Similar parent materials in localities with progressively higher temperature and greater rainfall (for example, Massachusetts, Virginia, and Guyana) exhibit greater loss of total rock volume and increased removal of specific elements, particularly silicon and potassium (Figure 6-4).

The decomposition of granite illustrates some of the principal chemical influences on soil formation, but weathering follows quite different courses on different types of bedrock. Pure quartz sand (silica) and pure limestone (calcium carbonate) do not produce clay readily because they lack iron and aluminum oxides; soil formation thus proceeds slowly unless other materials are mixed into the bedrock. Limestone frequently has a high percentage of clay particles originally derived from eroded soils. When such limestones weather, the calcium carbonate is readily leached, leaving a soil of high clay content. In general, the composition of the bedrock and its initial weathering determine the relative amounts of clay and sand in derived soils. These qualities in turn influence the availability of mineral ions in the soil and the capacity of the soil to hold water.

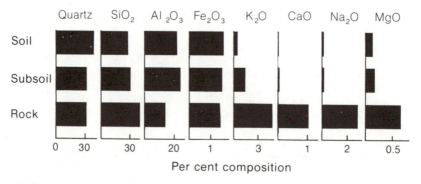

FIGURE 6-3 Percentage composition of soil layers and parent rock (granite) for each mineral in a soil profile from Guyana.

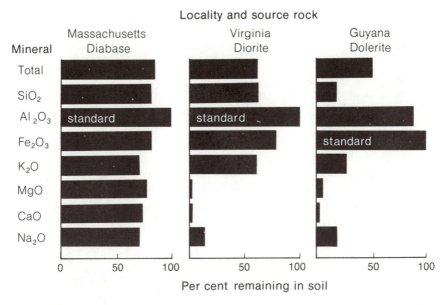

FIGURE 6-4 Differential removal of minerals from granitic rocks as a result of weathering in Massachusetts, Virginia, and Guyana. Values are compared to either aluminum or iron oxides, which are assumed to be the most stable components of the mineral soil.

Cation-exchange Capacity

Plants can obtain minerals from the soil in the form of dissolved ions, which are electrically charged atoms or compounds, whose solubility is determined by their electrostatic attraction to water molecules. Because ions dissolve in water, they would quickly be washed out of the soil if they were not strongly attracted to stable soil particles. Clay and humus particles, separately or associated in a complex, are large enough to form a stable component of the soil. These particles and complexes, referred to as *micelles*, actively play a role in the flux of mineral ions in the ecosystem. The surface of each particle has numerous negative electrical charges that attract positive ions (*cations*), such as calcium, magnesium, and potassium, and so retain them in the soil (Figure 6-5). The quantity of these negatively charged attachment sites in the soil is the *cation-exchange capacity* of the soil.

The role of the clay and humus particles in soil chemistry is not, however, so simple. The bonds between the mineral ions and the micelle are relatively weak, so they constantly break and re-form. When a potassium ion (K^+) dissociates from a micelle, its place may be taken by any other ion that is close by. Some ions cling more strongly to micelles than others. In order of decreasing tenacity, the common ions are hydrogen (H^+), calcium (Ca^{++}), magnesium (Mg^{++}), potassium (K^+), and sodium (Na^+). Hydrogen

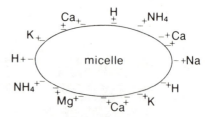

FIGURE 6-5 Schematic representation of a clay or humus particle (micelle) with hydrogen ions and mineral ions attracted by negative charges at its surface.

ions thus tend to displace calcium and all other ions on the micelle. If ions were not added to or removed from the soil, the relative proportions of mineral ions associated with clay-humus particles would reach a steady state. But carbonic acid in rain water and organic acids produced by the decomposition of organic detritus continually add hydrogen ions to the upper layers of the soil; the hydrogen ions readily displace other minerals, which are then washed out of the soil and into ground water. The influx of hydrogen ions in water percolating through the soil is largely responsible for the mobility of ions in the soil and the differentiation of layers in the soil profile, as we shall see below.

Soil Water

In addition to determining the clay content of the soil, the mineral composition of the bedrock determines the size and abundance of sand grains and silt particles, collectively known as the *skeleton* of the soil. The materials of the soil skeleton are chemically inert, but they influence the physical structure of the soil and its water-holding capacity.

Water is sticky. The capacity of water molecules to cling to each other and to surfaces they touch underlies the familiar phenomena of surface tension and the rise of water against gravity in capillary tubes. Water clings tightly to surfaces of the soil skeleton. Because total surface area of particles in the soil increases as particle size decreases, silty soils hold more water than coarse sands, through which water drains quickly.

Water capacity is not equivalent to water availability. Plant roots easily take up water that clings loosely to soil particles by surface tension, but water near the surface of sand and silt particles is bound tightly to the soil particles by stronger forces. Soil scientists measure the strength with which the cells of root hairs can absorb water from the soil in terms of equivalents of atmospheric pressure. Capillary attraction holds water in the soil with a force equivalent to a pressure of one-tenth to one-fifth atmosphere, about 1.5 to 3.0 pounds per square inch. (Sea level atmospheric pressure is 14.7 psi.) Water attracted to soil particles with less force than one-tenth atmosphere (water in the middle of large interstices between soil particles, hence at great distance from their surfaces) drains out of the soil, under the pull of gravity, into the ground water in the crevices of the bedrock below. The amount of water held against gravity by forces of attraction greater than one-tenth to one-fifth atmosphere is called the *field capacity* of the soil.

A force equivalent to one-tenth atmosphere is sufficient to raise a column of water about three feet above the water surface. We know that plant roots can exert a much greater pull on water in the soil, because water rises in the tallest trees to leaves more than 300 feet above the ground. In fact, plants can exert a pull of about fifteen atmospheres on soil water and, therefore, can take up water held in the soil by forces weaker than fifteen atmospheres. Water close to the surfaces of soil particles bound by forces of attraction greater than fifteen atmospheres is unavailable to plants, and the amount of water held by such forces is called the *wilting coefficient* or *wilting point* of the soil. (The wilting point is usually determined by using sunflower plants as the standard.) Once plants under drought stress have taken up all the water in the soil held by forces weaker than fifteen atmospheres, they can no longer obtain water, and they wilt, even though water remains in the soil.

As soil water is depleted, the remainder is held by increasingly stronger forces, on average, because a greater proportion of the water is situated close to the surfaces of soil particles. This relationship is shown in Figure 6-6 for a typical soil with a more or less even distribution of soil particle sizes from clay (up to 0.002 mm) through silt (0.002–0.05 mm) to sand (0.05–2.0 mm). Such soils are called *loams*. When saturated, this soil holds about 45 grams of water per 100 grams of dry soil (45 per cent water). The field capacity is 32 per cent, and the wilting coefficient 7 per cent. The available water is the field capacity minus the wilting coefficient, or 25 per cent, but water is more readily obtained by plants when the soil moisture is closer to the field capacity.

In soils with predominantly smaller particles, the surface area of the soil skeleton is relatively large, and both the wilting coefficient and field capacity are higher; a correspondingly larger proportion of soil water is held by

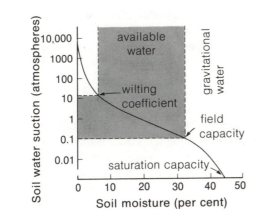

FIGURE 6-6 Relationship between water content of a loam soil and the average force of attraction of the soil water to soil particles (soil-water suction). The difference between the soil water content at field capacity (0.1 atmospheres) and the wilting coefficient (15 atm) is the water available to plants.

FIGURE 6-7 Relationship between soil-water content and soil-water suction for typical sand, loam, and clay soils. The available water (field capacity minus wilting coefficient) is greatest for the loam. Water availability is less for sandy soils because large pore spaces allow too free drainage, and for clay-based soils because the high particle surface area holds water too tightly.

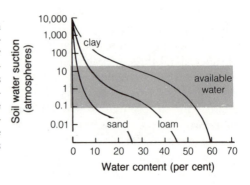

forces greater than fifteen atmospheres. Soils with predominantly larger skeletal particles have less surface area and larger interstices between particles. A larger proportion of the soil water is held loosely and is thus available to plants, but these soils have a lower field capacity. Availability of water to plants is maximum in soils with a variety of particle sizes between sand and clay (Figure 6-7).

Podsolization

Under mild, temperate conditions of temperature and rainfall, sand grains and clay particles are resistant to weathering and form stable components of the soil skeleton. Under conditions of high soil acidity, however, clay particles break down in the A horizon of the soil profile, and their soluble ions are transported downward and deposited in lower horizons. This process, known as *podsolization*, reduces the ion-exchange capacity, and, therefore, the fertility of the upper layers of the soil, by reducing the clay content.

Acid conditions are found primarily in cold regions where coniferous trees dominate the forests. The slow decomposition of acidic plant litter under conifer forests produces organic acids, which increase the acidity of the soil and promote breakdown of clay particles. In addition, rainfall usually exceeds evaporation during the year in regions of podsolization. Under these moist conditions, water continually moves downward through the soil profile, so there is little upward transport of new clay materials from the weathered bedrock below.

Podsolization advances farthest under spruce and fir forests in New England and the Great Lakes region and across a wide belt of southern and western Canada. A typical profile of a podsolized soil (Figure 6-8) has striking bands corresponding to regions of leaching (*eluviated* horizons) and redeposition (*illuviated* horizons). The topmost layers of the profile (O and A_1) are dark and rich in organic matter. These are underlain by a light-colored A_2 horizon, which has been leached of most of its clay content. As a result, the A_2 consists mainly of sandy skeletal material that holds neither water nor nutrients well. One usually finds a dark-colored band of deposition immediately under the eluviated A_2 horizon. This is the uppermost layer of the B horizon, where iron and aluminum oxides are redeposited, giving the layer a dark color. Other, more mobile minerals may accumulate to some extent in lower parts of the B horizon, which then grades almost imperceptibly into a C horizon and the parent material.

Laterization

Whereas acidic conditions foster the breakdown and eluviation of clay minerals (iron and aluminum oxides), a basic or alkaline soil reaction facilitates the removal of silica (SiO_2) from the soil. Leaching of silica, called *laterization*, occurs primarily in tropical regions, where it is localized in the soils under humid forests. Soil scientists do not fully understand why many tropical soils have low acidity (high alkalinity). It is known, however, that plant detritus decays rapidly in the tropics, owing to high temperatures and moisture. Humic acids, therefore, do not persist in tropical soils so long as they do in cooler regions. Furthermore, decomposition of plant litter is accomplished primarily by bacteria, which produce no acid. The plant litter of cool forests is decomposed in part by fungi, which themselves produce acids to aid the chemical breakdown of organic detritus.

FIGURE 6-8 Profile of a podsolized soil in Plymouth County, Massachusetts. The light-colored, eluviated A_2 horizon and the dark-colored, illuviated B_1 horizon immediately below it form distinct bands. Note the general absence of roots in the A_2 horizon compared with the lower B_1 horizon.

Laterization has an effect on the soil profile opposite to that of podsolization. Removal of silica from the top layers of the soil increases the proportion of iron and aluminum oxides, which give tropical soils their characteristic red color. If laterization proceeds far enough, all the silica disappears from the soil, including that in clay particles, leaving behind a material called *laterite*, which is more like concrete than soil.

Laterization normally does not alter soil completely in undisturbed tropical forests. Organic humus particles, which accumulate in the upper layers of the soil, maintain a soil structure and cation-exchange capacity suitable for root growth and plant nutrition. Strongly laterized layers may, however, form deeper in the soil profile.

Disturbance of tropical forests can have disastrous effects where soil is prone to laterization. Removal of trees for agriculture, lumber, and pulp exposes the soil to the drying effects of the sun. Evaporation of water from the ground surface frequently reverses the usual downward flow of water through the soil profile. Iron and aluminum oxides are then brought to the surface where they cement soil particles into a substance so hard that it is at best suitable for masonry. A completely laterized soil is nearly impervious to water and thus promotes surface runoff and erosion. Such disturbed soils are, of course, useless for agriculture, and their hardness and low water content slow the regeneration of natural vegetation.

Calcification and Salinization

Under arid conditions, where evaporation exceeds rainfall, water does not percolate completely through the soil. Calcium carbonate, dissolved in the topmost layers of the soil profile after a rainfall, is often redeposited in these same layers when water evaporates from the soil, or it may be transported downward to the lower limit of water penetration. The results of this process, called *calcification*, can be seen in the left-hand soil profile in Figure 6-2, in which a narrow, diffuse band of calcium carbonate (light-colored) has been redeposited about two feet below the soil surface. This horizon marks the lower limit of water percolation, immediately below which one finds relatively unweathered parent material. The depth of water penetration, hence the depth of the calcified layer, becomes less and less as rainfall diminishes.

Where ground water occurs close to the soil surface, dissolved minerals are drawn to the surface by evaporation and the upward pull of capillary movement. Evaporating water then leaves the minerals behind at the surface, sometimes forming thick crusts called *caliches* (Figure 6-9), which inhibit plant growth. A more serious reduction of soil fertility occurs when soluble neutral salts, such as sodium chloride (table salt, $NaCl$) and calcium sulfate ($CaSO_4$), accumulate in the soil and on its surface. This process, referred to as *salinization*, occurs where soil drainage is impeded and surface evaporation far exceeds percolation. Neutral salts reduce plant growth owing

FIGURE 6-9 An alkaline area devoid of plants in Chouteau County, Montana, where rising ground water has deposited a crust of calcium carbonate.

to the high salt concentration of soil water when it is available (saline soil conditions). When a salinized soil is relatively free of neutral salts but contains abundant sodium ion adsorbed to cation-exchange sites on soil micelles (sodic conditions), soil fertility is reduced twice over, first by increasing the alkalinity of soil and second by sodium toxicity.

In many desert basins, ground water is close enough to the surface to be drawn upward by evaporation. The resulting caliche and salt deposits form the "dry lakes" that are widespread in the Mohave Desert and Great Basin of the western United States. Standing water on the surface in such regions is usually so full of dissolved minerals that it is undrinkable. (For many early pioneers who crossed the deserts, the choice between dying of thirst or alkali poisoning must have been difficult.)

Irrigation can, indeed, make a desert bloom. Dry soils become highly fertile when they are irrigated because of the high concentration of adsorbed mineral ions in the upper layers of the profile. But the rich agricultural returns are often cut short by speeded salinization of the soil. Irrigation water is ordinarily obtained from rivers that, in dry regions, are usually loaded with silt and dissolved salts. Most water added to the soil by irrigation eventually evaporates, leaving behind, near the soil surface, the salts it carried. The ultimate result is similar to what happens when water naturally enters the soil profile from underlying ground water. Salts accumulate rapidly, and the soil soon becomes too alkaline for agriculture.

Vegetation and Soil Development

The initial weathering of bedrock and the secondary alteration of the soil profile by podsolization, laterization, and salinization primarily influence the inorganic composition of the soil. Yet many important characteristics of the soil, including its humus content and the availability of nitrogen and phosphorus, are determined largely by vegetation. Soil changes that can follow removal of vegetation in the tropics show dramatically the role of vegetation in maintaining a steady state in the soil. Denudation quickly alters the movement of water through the soil and rapidly changes patterns of leaching and deposition.

Vegetation exerts its most dramatic influence on the development of soils where the underlying parent material is freshly exposed. Primary soil development occurs where geologic agents remove layers of existing soil or add sediments over the top of existing soil horizons. Since the recession of the glaciers from the Great Lakes region 10,000 to 12,000 years ago, the surface level of the Great Lakes has periodically lowered, leaving behind a chronological series of sand dunes at the southern end of Lake Michigan. The value of these dunes to the study of ecological processes was first recognized by the pioneering plant ecologist Henry C. Cowles, who, in 1899, described changes in vegetation observed on progressively older dunes. A newly formed dune consists largely of sand (silica). Water percolates rapidly through the dune and, because clays are absent, quickly leaches out any mineral nutrients. The dune environment excludes all but the most hardy plants. Marram grass (genus *Ammophila*) colonizes the sand at an early stage by sending out rhizomes (horizontal roots) from plants growing in better soil at the edge of the dunes (Figure 6-10).

Once grasses become established, they stabilize the dune and begin to add organic detritus to the dune surface. By building up the humus content of the sand, dune grasses encourage true soil development. Grasses and shrubs dominate the first century of plant succession on dunes. These are followed by the establishment of pine and its rapid replacement by black oak at an age of 150 to 200 years.

Changes brought about by vegetation, and the ultimate attainment of a steady state in the soil, were described in a paper by Jerry Olson, published in 1958. Olson found that the humus added to the soil provided sites for cation exchange, just as the clay particles do in clay-based soils. Silt and clay particles eventually are added to the soil by wind deposition. The cation-exchange capacity of the soil increases rapidly for 500 to 1,000 years after dune stabilization, then levels off. Litter continues to accumulate on

FIGURE 6-10 Marram grass growing on dunes in Indiana Dunes State Park, Indiana. At left, plants are seen extending out over fresh sand. At right, sand has been removed to expose the underground rhizomes by which the grass spreads.

the forest floor, and humic acids eventually make the soil acidic. Hydrogen ions replace other cations (calcium, potassium, magnesium, etc.), until they occupy almost half the ion-exchange sites in the soil. As a result, soil fertility declines slowly, then levels off about 4,000 years after dune stabilization.

R. L. Crocker and J. Major, in 1955, described soil development on areas bared by receding glaciers at Glacier Bay, Alaska. The retreat of the edge of the glacier had been recorded for a century; thus Crocker and Major knew the exact age of each of their study sites. Unlike the Lake Michigan sand dunes, sediments left behind by receding glaciers contained abundant calcium and clay. Vegetation established itself rapidly, and decaying plant detritus changed the hydrogen-ion concentration of the soil from slightly alkaline to slightly acid in twenty years. Each species of plant influenced soil acidity differently, however. Alder thickets acidified the soil rapidly, but willow and cottonwood did so slowly. In any case, increasing soil acidity accelerated the removal of calcium, while accumulating detritus steadily added to the inorganic nitrogen content of the soil. These changes in turn influenced the suitability of the habitat for different species of plants and fostered further changes in vegetation, eventually leading to spruce forest. The relationship between vegetation and soil properties during the early development of soil on newly exposed sites dramatizes the more general interactive roles of the physical and biological world and the dynamic nature of the ecosystem itself.

We have seen how climate and geology affect soil properties. Geographical and temporal variations in all aspects of the physical world broadly influence the structure and functioning of the entire ecosystem, determining not only physical and chemical properties of the soils, but also levels of organic production, paths of energy flow and nutrient cycling, and the adaptations of plants and animals that give each habitat its characteristic appearance. In the next chapter, we shall examine the processes that determine climate variation over the surface of the earth.

7 | Variation in the Environment

EVERYDAY EXPERIENCE reveals variation in the natural world. Climate and the appearance of biological communities are closely linked. As one travels from east to west across the United States, the gradual change in climate is paralleled by change in vegetation. Tall forests of broad-leaved trees along the east coast are replaced by grasslands in the drier midwestern states and by desert shrublands in the arid Great Basin. Amidst the topographic diversity of the western mountains, one travels through altitudinal zones of vegetation from hot desert scrub to cool montane forests of aspen and spruce. Travelling from northern Canada to Panama, one encounters even more striking change.

Major patterns of climate on the earth's surface depend on the relation of the earth to the sun, placement of the continents and oceans, and the circulation of the wind and seas. But local variations in environment caused by geology and topography result in the diversification of communities within regions of uniform climate. Physical properties of the environment are further modified and diversified by vegetation and the activities of animals.

The Earth As a Heat Machine

A clue to diversity in the biological world comes from the study of variation in the physical world itself, which can be considered as a model of more complicated biological systems. The surface of the earth, its waters, and the atmosphere above it behave as a giant heat machine, obeying the same thermodynamic rules as do ecosystems and exhibiting a similar variety in time and space.

The simplest abstraction of a thermodynamic system might be a uniform hunk of rock in outer space. Lifeless and motionless, this alien world

71

intercepts light energy emanating from the sun, or some more distant star, and radiates energy into the black depths of space. The rock behaves as a simple ecosystem: energy is assimilated, energy is transformed, energy is lost from the system. When the rock absorbs sunlight, the molecules in the rock are caused to move more rapidly, and the rock heats up. The energy in light is transformed to heat energy. But the hotter an object, the more rapidly it loses heat to its surroundings, in this case empty space. A law of thermodynamics states that the distribution of energy in the universe tends to become more even with time. In other words, energy moves from points of high concentration to points of low concentration. As the rock heats up, it re-radiates the energy it received from the sun at a proportionately greater rate. When the rate of energy radiation has reached a level equal to the rate at which sunlight is received, the net energy balance of the rock is zero (energy loss equals energy gain), and the rock attains a constant steady-state temperature. And so it is with any thermodynamic model, including an ecosystem. Animals and plants liberate energy as heat by respiration at approximately the same rate that plants assimilate energy as light by pho- tosynthesis. Just as the temperature of the rock measures the energy stored in the system at any moment, the chemical energy in plants, animals, and organic detritus measures the energy stored in the ecosystem. Both will tend to achieve a steady state if left unperturbed.

The surface of the earth is much more complex than a homogeneous rock in space. As the earth's surface varies from bare rock to forested soil, open ocean, and frozen lake, its ability to absorb sunlight varies as well, thus creating differential heating and cooling. As with the rock in space, heat energy absorbed by the earth is eventually radiated back into space, but not before undergoing further transformations that perform the work of evaporating water and contributing to the circulation of the atmosphere and oceans. All these factors result in tremendous varieties of physical conditions over the surface of the earth, which, in turn, have fostered the diversification of ecosystems that we discover about us.

Global Climate Patterns

The earth's climate tends to be cold and dry toward the poles and hot and wet toward the equator. The sun exerts its greatest warming effect on the atmosphere, oceans, and land when it is directly overhead. The sun's warmth is diminished when it lies close to the horizon and its rays strike the surface at an oblique angle. Not only does a beam of sunlight spread over a greater area when the sun is low, it also travels a longer path through the atmosphere, where much of its light energy is either reflected or absorbed by the atmosphere and re-radiated into space as heat. The sun's highest position each day varies from directly overhead in the tropics to near the horizon in polar regions; hence the warming effect of the sun increases from the poles to the equator. This uneven distribution of the sun's energy over the surface of the earth creates major geographical patterns in temperature, rainfall, and wind.

Warming air expands, becomes less dense, and thus tends to rise. Its ability to hold water vapor increases, and evaporation is accelerated. The rate of evaporation from a wet surface nearly doubles with each ten degrees Celsius rise in temperature. The sun heats the atmosphere most intensely in the tropics. The warmed surface air picks up water vapor and begins to rise. As the moisture-laden air gains altitude and cools, its water vapor condenses into thick clouds that drench the tropical landscape with rain. Daily cycles of heating and cooling cause most tropical rain to fall during the afternoon and evening; in temperate areas, as well, summer thunder showers, resulting from strong vertical currents of warm, moist air, most often occur late in the day.

Because warm tropical air can hold much more water than temperate or arctic air, annual precipitation is greatest in tropical regions (Figure 7-1). The tropics are so wet not because more water occurs in tropical latitudes than elsewhere, but because water is cycled more rapidly through the tropical atmosphere. Cycles of evaporation and precipitation are driven by the sun, and it is the source of energy, not the quantity of water, that primarily determines latitudinal patterns in rainfall. The distribution of continental land masses exerts a secondary effect. Rainfall is more plentiful in the Southern Hemisphere because oceans and lakes cover a greater proportion of its surface (81 per cent, compared with 61 per cent of the Northern Hemisphere). Water evaporates more readily from exposed surfaces of water than from soil and vegetation.

Winds are driven by energy from the sun, just as the cycling of water in the atmosphere. Indeed, the two cannot be separated, and wind patterns exert a strong influence on precipitation. The mass of warm air that rises in the tropics eventually spreads to the north and south in the upper layers of the atmosphere. It is replaced from below by surface-level air from sub-

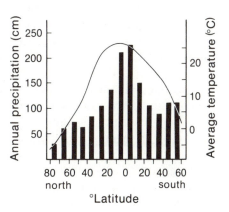

FIGURE 7-1 Average annual precipitation (vertical bars) and temperature (solid line) for 10° latitudinal belts within continental land masses. The figure represents averages for many localities, which obscures the great variation within each latitudinal belt.

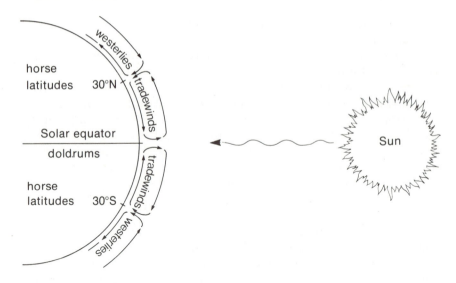

FIGURE 7-2   Simplified diagram of convection currents driven by the sun's energy in the atmosphere of tropical and subtropical latitudes.

tropical latitudes (Figure 7-2). The tropical air mass that rose under the warming sun cools as its heat is radiated back into space. By the time it has extended to 30° north and south of the solar equator,* the cooled air mass becomes so dense that it begins to sink back to the surface. It is relatively dry because condensation has removed much of its water, which fell as rain over the tropical regions where the air current originated. Its capacity to evaporate and hold water increases further as it sinks and warms. As the air strikes the earth's surface in subtropical latitudes and spreads to the north and south, it draws moisture from the land, creating zones of arid climate.

Convection currents in the atmosphere, driven by the sun's energy, redistribute heat and moisture about the surface of the earth. The region of rising air in the tropics—the doldrums—is one of high rainfall. Conversely, descending air robs the land of water, which is transported elsewhere by wind currents. The tradewinds, blowing steadily toward the equator from the dry horse latitudes, carry moisture picked up along the way into the tropics. Just as warm tropical air rises, cold air masses over the north and south polar regions descend and flow along the surface toward lower latitudes. When cold air meets a warmer air mass moving poleward across temperate latitudes, the warm moist air rises above the denser polar air and cools, bringing precipitation.

---

* The solar equator is the parallel of latitude that lies directly beneath the sun. The position of the solar equator varies seasonally from 23° north latitude on June 21 (the summer solstice) to 23° south latitude on December 21 (the winter solstice). The solar equator coincides with the earth's geographical equator at the equinoxes (March 21 and September 21).

Precipitation is distributed over the surface of the earth in such a way that most wet regions occur close to the equator, and the major deserts occupy a belt centered about 30° latitude north and south of the equator (Figure 7-3). Great names in deserts—the Arabian, Sahara, Kalahari, and Namib of Africa, the Atacama of South America, the Mohave and Sonoran of North America, and the Australian—all belong to regions within these belts.

Exceptions to this pattern are caused by major land masses. Mountains force air upward, causing it to cool and lose its moisture as precipitation on the windward side of the range. As the air descends the leeward slopes of the mountains and travels across the lowlands beyond, it picks up moisture and creates arid environments called *rain shadows* (Figure 7-4). The Great Basin deserts of the western United States and the Gobi Desert of Asia lie in the rain shadows of extensive mountain ranges.

The interior of a continent is usually drier than its coasts simply because the interior is farther removed from the major site of water evaporation, the surface of the ocean. Furthermore, coastal (maritime) climates are less variable than interior (continental) climates because the tremendous heat-storage capacity of water reduces temperature fluctuations. For example, the difference between the hottest and coldest mean monthly temperature near the Pacific coast of the United States at Portland, Oregon, is 16 C (28 F). Farther inland, this range increases to 18 C (33 F) at Spokane, Washington; 26 C (47 F) at Helena, Montana; and 33 C (60 F) at Bismarck, North Dakota.

Ocean currents also play a major role in transferring heat over the surface of the earth. In large ocean basins, cold water tends to move toward the tropics along the western coasts of the continents, and warm water tends to move toward temperate latitudes along the eastern coasts of con-

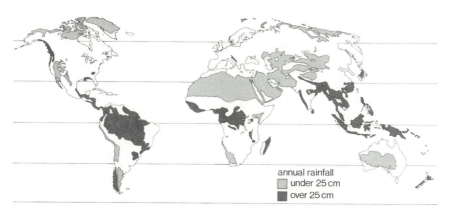

annual rainfall
☐ under 25 cm
■ over 25 cm

FIGURE 7-3   Distribution of the major deserts (regions with less than 25 centimeters [10 inches] annual precipitation) and wet areas (having more than 150 centimeters [80 inches] annual precipitation).

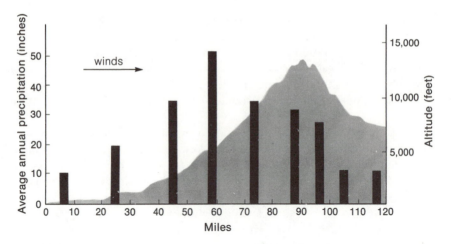

FIGURE 7-4 The influence of the Sierra Nevada mountain range on local precipitation and in causing a rain shadow to the east. Weather comes predominantly from the west (left) across the Central Valley of California. As moisture-laden air is deflected upwards by the mountains, it cools and its moisture condenses, resulting in heavy precipitation on the western slope of the mountains. As the air rushes down the eastern slope, it warms and begins to pick up moisture, creating arid conditions in the Great Basin.

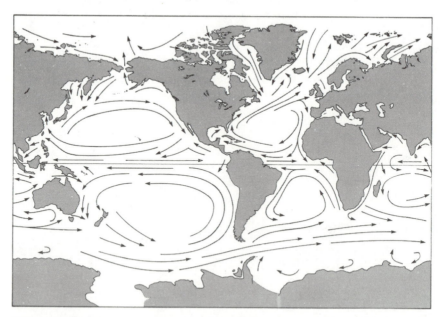

FIGURE 7-5 The major ocean currents. Water movement generally proceeds clockwise in the Northern Hemisphere and counterclockwise in the Southern Hemisphere.

tinents (Figure 7-5). The cold Humboldt Current moving north from the Antarctic Ocean along the coasts of Chile and Peru is partly responsible for the presence of cool deserts along the west coast of South America right to the equator, though these regions also lie within the rain shadow of the Andes Mountains. Conversely, the warm Gulf Stream, emanating from the Gulf of Mexico, carries a mild climate far to the north into western Europe.

The Changing Seasons

Although we may characterize a region's climate as hot or cold and wet or dry, regular cycles of change are as important aspects of climate as long term averages of temperature and precipitation. Periodic cycles in climate are based upon cyclical astronomical events: the rotation of the earth upon its axis causes daily periodicity in the environment; the revolution of the moon around the earth determines the periodicity of the tides; the revolution of the earth around the sun brings seasonal change.

The earth's equator is tilted slightly with respect to the path the earth follows in its orbit around the sun. As a result, the Northern Hemisphere receives more solar energy than the Southern Hemisphere during the northern summer, less during the northern winter. The seasonal change in temperature increases with distance from the equator (Figure 7-6). At high latitudes in the Northern Hemisphere, mean monthly temperatures vary by an average of 30 C (54 F), with extremes of more than 50 C (90 F) annually; the mean temperatures of the warmest and coldest months in the tropics differ by as little as two or three degrees.

Latitudinal patterns in rainfall seasonality are complicated by belts of wet and dry climate that move north and south with the changing seasons. Annual variation in rainfall is greatest in broad latitudinal belts lying about 20° north and south of the equator. As the seasons change, these regions are alternately crossed by the solar equator, bringing heavy rains, and by the subtropical high-pressure belts, bringing clear skies.

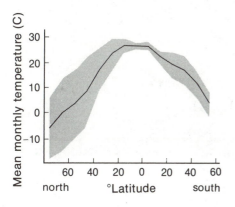

FIGURE 7-6 Annual range of mean monthly temperatures as a function of latitude. Note that seasonal variation is reduced in the Southern Hemisphere where there is a high ratio of water to land. The temperature range at a station may vary considerably from the mean for the latitudinal belt in which it is located.

Panama, at 9°N, lies within the wet tropics, but even there the seasonal movement of the solar equator profoundly influences the climate. The major tropical belt of high rainfall remains south of Panama during most of our winter, but it lies directly over Panama during the summer. Hence the winter is dry and windy, the summer humid and rainy. Panama's climate is wetter on the northern (Caribbean) side of the Isthmus, the direction of prevailing winds, than on the southern (Pacific) side. This rain-shadow effect is more pronounced in nearby western Costa Rica, where a high mountain range intercepts moisture coming from the Caribbean side of the Isthmus. The Pacific lowlands are so dry during the winter months that most trees lose their leaves. The tinder-dry forest and bare branches contrast sharply with the wet, lush, more typically tropical forest during the wet season (Figure 7-7).

Farther to the north, at 30°N in central Mexico, rainfall comes only during the summer when the solar equator reaches its most northward limit (Figure 7-8). During the rest of the year this region falls within the dry, subtropical high-pressure belt. The influence of the solar equator, bringing summer rainfall, extends into the Sonoran Desert of southern Arizona and

FIGURE 7-7   Kiawe forest on the island of Maui, Hawaii, during the peak of the dry season. Note the complete absence of leaves at this time. The grasses of the forest understory are tinder-dry.

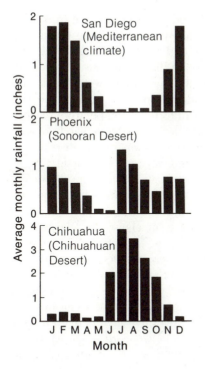

FIGURE 7-8 Seasonal occurrence of rainfall at three localities in western North America showing the summer rainy season of the Chihuahuan Desert, the winter rain—summer drought of the Pacific Coast (Mediterranean-climate type), and the combined climate pattern of the Sonoran Desert.

New Mexico. This area also receives moisture during the winter from the Pacific Ocean, carried by the southwesterly winds emanating from the subtropical high-pressure belt farther south. Southern California is beyond the summer rainfall belt and has a winter-rainfall–summer-drought climate, often referred to as a Mediterranean climate.

The seas are warmed by the sun just as the continents and the atmosphere are, but their great mass of water acts like a heat sink to dampen daily and seasonal fluctuations in temperature. Large seasonal changes in temperature are more often caused by seasonal movements of water masses of different temperature than by local heating and cooling. During the Panamanian dry season, roughly January to April, steady winds blowing in a southwesterly direction create strong *upwelling* currents in the Pacific Ocean along the southern and western coasts of Central America. During the upwelling period, warm surface water is blown away from the coast and cooler water moves upward from deeper regions to replace it. As a result, the annual range of water temperature on the Pacific coast of Panama is about three times that of the Caribbean coast.

Seasonal Cycles in Lakes

Small temperate-zone lakes are more sensitive than oceans to the changing seasons. Temperature cycles are an important force in the nutrient budgets of lakes because changes in temperature gradients from surface to bottom cause vertical mixing of the water twice each year, during the spring and the fall. In winter, the *temperature profile* of the lake is inverted, with the coldest water (0 C) at the surface, just beneath the ice. Because the density of water increases between freezing and 4 C, the warmer water sinks, and temperature increases (up to 4 C) with depth. In early spring, the sun warms the surface of the water gradually. Until surface temperature reaches 4 C, surface water tends to sink into the cooler layers immediately below. This minor vertical mixing creates a uniform temperature distribution throughout the water column. Without thermal layering to impede mixing, surface winds cause deep vertical movement of water in early spring (*spring turnover*), bringing nutrients from regions of decomposition in the bottom sediments to the surface.

As the sun rises higher each day and the air above the lake warms, surface water warms up faster than deeper water, creating a sharp zone of temperature change, called the *thermocline*, across which water does not mix. The warm surface water literally floats on the cooler water below. The depth of the thermocline varies with local wind patterns and with the depth and turbidity of the lake. Thermoclines occur anywhere between five and twenty meters below the surface; lakes less than five meters deep usually lack stratification. Thoreau's Walden Pond, in Concord, Massachusetts, develops a sharp thermocline between six and ten meters depth. In August, water temperatures decrease from 25 C at the top of the thermocline to 5 C at its bottom.

The thermocline demarcates an upper layer of warm water (the *epilimnion*) and a deep layer of cold water (the *hypolimnion*). Most of the primary production of the lake occurs in the epilimnion where sunlight is intense. Photosynthesis supplements mixing of oxygen at the lake surface to keep the epilimnion well aerated and thus suitable for animal life, but plants often deplete dissolved mineral nutrients, and thereby curtail their own productivity. The hypolimnion is cut off from the surface of the lake by sharp temperature *stratification*, and, being frequently below the *euphotic* zone of photosynthesis, animals and bacteria deplete the hypolimnion of oxygen, creating anaerobic conditions.

During the fall, surface layers of the lake cool more rapidly than deeper layers and, becoming heavier than the underlying water, begin to sink. This vertical mixing (*fall overturn*), persists into late fall, until the temperature at the lake surface drops below 4 C, and winter stratification ensues. Fall overturn produces greater vertical mixing of water than spring overturn because temperature differences in the lake are greater during summer stratification than during winter stratification. Fall overturn speeds the movement of oxygen to deep waters and rushes nutrients to the surface. Where the hypolimnion becomes fairly warm in midsummer, deep vertical mixing may take place in late summer when temperatures are still warm enough for plant growth. Infusion of nutrients into surface waters at this time often causes a later burst of phytoplankton population increase—the *fall bloom.* In deep cold lakes, vertical mixing does not penetrate to all depths until late fall or early winter when water temperatures are too cold to support plant growth.

Topographic and Geologic Influences

Variation in topography and geology can create variation in the environment within regions of uniform climate. In hilly areas, the slope of the land and its exposure to the sun influence the temperature and moisture content of the soil. Soils on steep slopes are well drained, often creating conditions of moisture stress for plants when the soils of nearby lowlands are saturated with water. In arid regions, stream bottomlands and seasonally dry river beds often support well-developed forests, which contrast sharply with the surrounding desert vegetation. Plant communities on shady and sunny sides of mountains and valleys frequently differ in accordance with the temperature and moisture regimes of each exposure. South-facing slopes are exposed to the direct-heating effect of the sun, which limits vegetation to shrubby, drought-resistant (*xeric*) forms. The corresponding north-facing slopes are relatively cool and wet and harbor moisture-requiring (*mesic*) vegetation (Figure 7-9).

FIGURE 7-9   The effect of exposure on the vegetation of a series of mountain ridges near Aspen, Colorado. The north-facing (left-facing) slopes are cool and moist, permitting the development of spruce forest. Shrubby, drought-resistant vegetation grows on the south-facing slopes.

FIGURE 7-10   Even in the tropics one may find communities dominated by cold temperatures. Biological communities stop abruptly at the snow line, at about 5,000 meters elevation in central Peru.

Air temperature decreases with altitude by about 6 C for each 1,000-meter increase in elevation. Even in the tropics, if one could climb high enough, one would eventually encounter freezing temperatures and perpetual snow. Where the temperature at sea level is 30 C, freezing temperatures would be reached at about 5,000 meters (16,000 feet). This, indeed, is the approximate altitude of the snow line in the Andes of central Peru (Figure 7-10).

A 6 C drop in temperature corresponds, in temperate latitudes, to an 800-kilometer (500-mile) increase in latitude. In many respects, the climate and vegetation of high altitudes resemble those of sea level localities at higher latitudes. Despite their similarities, however, alpine environments are usually less seasonal than their low-elevation counterparts at higher latitudes, even though average temperature and annual rainfall may be similar. Temperatures in tropical montane environments remain nearly constant over the year, and the occurrence of frost-free conditions at high altitudes allows many tropical plants and animals to live in the cool environments found there.

In the mountains of the southwestern United States, changes in plant communities with elevation create more or less distinct belts of vegetation, referred to as *life zones* by the early naturalist C. H. Merriam (1894). Merriam's scheme of classification included five broad zones that he named, from south to north (or low to high elevation): Lower Sonoran, Upper Sonoran, Transition, Canadian (or Hudsonian), and Arctic-Alpine.

At low elevations in the southwest, one encounters a cactus and desert-shrub association characteristic of the Sonoran Desert of northern Mexico and southern Arizona (Figure 7-11). In the woodlands along stream beds, plants and animals have a distinctly tropical flavor. Many hummingbirds and flycatchers, ring-tailed cats, jaguars, and peccaries make their only temperate-zone appearances in this area. At 2,500 meters (8,200 feet) higher, in the Alpine Zone, we find a landscape resembling the tundra of northern Canada and Alaska. By climbing 2,500 meters, we experience changes in climate and vegetation that would require a journey to the north of 2,000 kilometers (1,250 miles), or more, at sea level.

Local variation in the bedrock underlying a region promotes the differentiation of soil types and enhances biotic heterogeneity. In the northern Appalachian Mountains and in mountains near the Pacific coast of the United States, outcrops of serpentine (a kind of igneous rock) produce soils with so much magnesium that species of plants characteristic of surrounding soil types cannot grow. Serpentine *barrens*, as they are called, are usually dominated by a sparse covering of grasses and herbs, many of which are distinct *endemics* (species found nowhere else) that have evolved a high tolerance for magnesium. Depending on the composition of the bedrock and the rate of weathering, granite, shale, and sandstone also can produce a barren type of vegetation. The extensive pine barrens of southern New Jersey occur on a large outcrop of sand, which produces a dry, acid, infertile soil capable of supporting no more than knee-high pygmy forests of pines. Physical characteristics of the soil and of the underlying rock also influence drainage and the ability of the soil to hold moisture. The extensive pine forests found on the coastal plain of the southeastern United States grow on sandy soils that drain too well to support the growth of most broad-leaved trees. Climate, too, plays an important role in the weathering of rock and the formation of soils; in temperate and arctic regions of the Northern Hemisphere, glacial activity during the last 100,000 years has influenced soil characteristics over vast areas.

Integrated Descriptions of Climate

We find it easier to dissect climate into its component properties of temperature, humidity, precipitation, wind, and solar radiation, than to appreciate at once all the implications of these factors for the ecosystem. But we must also understand *interactions* among climate factors because these factors are clearly interdependent in their effect on life. For example, seasonal rainfall promotes plant growth more strongly during warm months than during cold months. Wind movement and solar radiation interact with temperature to determine thermal stress; temperature and humidity together influence water balance.

The *climograph* portrays seasonal changes in temperature and rainfall simultaneously. The climograph has a rainfall scale (horizontal axis) and a

Hudsonian Zone: Elev. 8,500'

Alpine Zone: Elev. 11,000'

Upper Sonoran Zone: Elev. 5,000'

Transition Zone: Elev. 6,500'

Lower Sonoran Zone: Elev. 3,000'

Upper Sonoran Zone: Elev. 4,000'

FIGURE 7-11 Vegetation types corresponding to different elevations in the mountains of southeastern Arizona. Lower Sonoran vegetation is mostly saguaro cactus, small desert trees such as paloverde and mesquite, numerous annual and perennial shrubs, and small succulent cacti. Agave, ocotillo, and grasses are conspicuous elements of the Upper Sonoran Zone, with oaks appearing toward its upper edge. Large trees are predominant at higher elevations: ponderosa pine in the Transition Zone; spruce and fir in the Hudsonian Zone. These gradually give way to bushes, willows, herbs, and lichens in the Alpine Zone above the treeline.

temperature scale (vertical axis); each month is plotted on the graph according to its average temperature and rainfall. Seasonal progression of climate is portrayed on the climograph by following the points for each month of the year in succession (Figure 7-12). The horizontal spread of months represents seasonal variation in rainfall; the vertical spread, variation in temperature.

The climograph permits a visual comparison of climates at different localities. We note immediately that the seasons in Panama bring marked variation in rainfall but little change in temperature, whereas the reverse characterizes New York City. During no months are the climates of the two localities similar, although July and August conditions in New York approach the April climate in Panama. Moving east to west across the United States from Cincinnati, Ohio, to Winnemucca, Nevada, climate becomes more arid but temperatures remain within the same range. The change in vegetation from deciduous, broad-leaved forest in Ohio, to short-grass prairie in Wyoming and desert shrubs in Nevada is thus dependent upon the water

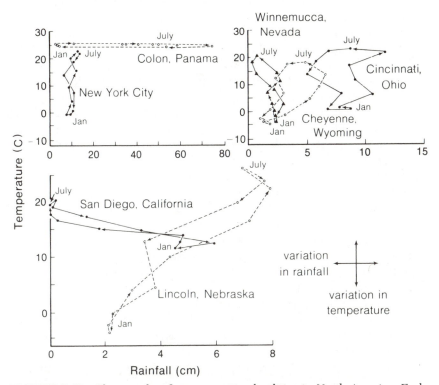

FIGURE 7-12 Climographs of representative localities in North America. Each point represents the mean temperature and rainfall for one month. Lines connecting the months at each locality indicate seasonal change in climate.

relations of different plant forms rather than temperature tolerance. Although San Diego's weather in January resembles that of Lincoln, Nebraska, in April, the overall climates differ as much as their vegetation. San Diego's Mediterranean-type climate, with hot, dry summers and cool, moist winters, favors slow-growing, drought-resistant shrubs (chaparral), while Lincoln is (or rather, was) surrounded by tall-grass prairie. Summer rainfall in the Great Plains states supports greater plant productivity than the winter rainfall of the West Coast because water is abundant on the prairies during the warm summer months, the best growing season. The winters, however, are too cold and dry to support shrubby, evergreen vegetation.

Although the climograph is useful for comparing localities, it fails to combine the effects of temperature and rainfall in any biologically meaningful way, and it does not show the cumulative effects of weather upon the environment. For example, during dry seasons both evaporation and *transpiration* (evaporation of water from leaves) remove water from the soil. If rainfall is insufficient to balance evaporation and transpiration losses, the water deficit in the soil steadily increases, perhaps for months at a time. In other words, soil water reflects last month's rainfall as well as more recent input to the soil.

In 1948, geographer C. W. Thornthwaite published a method for utilizing climate data to estimate the seasonal availability of water in the soil. He compared the rate at which water is drawn from the soil by plants and by direct evaporation with the rate at which it is restored by precipitation. The sum of evaporation and transpiration is the total *evapotranspiration* of the habitat. Evaporation and transpiration increase with temperature by a factor of nearly two for each 10 C rise in temperature, other things being equal, although the character of the soil and vegetation cover also influences water loss from the soil. In natural environments, evapotranspiration is at times limited by the availability of water in the soil. Potential evapotranspiration, which represents the amount of water that would be drawn from the soil if soil moisture were not limiting, can be calculated from temperature and precipitation. When precipitation input to the soil (rainfall minus surface runoff) exceeds potential evapotranspiration at all seasons, the soil will remain saturated with water throughout the year, as at Brevard, North Carolina (Figure 7-13). At Bar Harbor, Maine, precipitation falls below potential evapotranspiration during the warm summer months, and the soil is depleted of water during late summer and early fall. Canton, Mississippi, receives rainfall similar to that of Bar Harbor, but the hotter climate of Mississippi increases the potential evaporation of water from its soils, and serious water deficits are incurred during the summer months. Manhattan, Kansas, receives much less rain than Canton, Mississippi, but because the rainfall is concentrated during the summer period of maximum potential evapotranspiration, soil-water deficits are no more serious than in Mississippi. On the other hand, Grand Junction, Colorado, has a dry climate

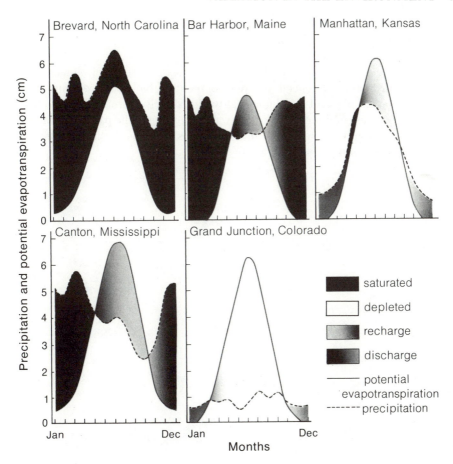

FIGURE 7-13   The relationship of precipitation and potential evapotranspiration to changes in the availability of soil moisture. When evapotranspiration exceeds precipitation, water is withdrawn from the soil until the deficit exceeds ten centimeters (four inches), the average amount of moisture that soils can hold.

where soils are depleted of water most of the year and rarely become saturated. Plant productivity is correspondingly low.

Because potential evapotranspiration increases with temperature, temperature and water stress go hand in hand. Thus, boreal regions receiving 25 to 50 centimeters of precipitation each year have a more favorable water budget for plant production than tropical regions with similar levels of precipitation.

Thornthwaite's analysis may lack the detail to predict local variation in soil moisture and plant production, but his graphs indicate the relative length and severity of seasonal drought. By keeping a running seasonal balance of gains and losses of water, one can appreciate the cumulative effects of climate on soil moisture.

In the next chapter, we shall see how major patterns of climate and topography have influenced the development of biological communities. Variety in the physical and biological world is derived from relatively few basic processes combined in different ways according to the particular nature of the environment, just as a few musicians can blend their individual contributions into an infinite variety of sounds.

8 | The Diversity of Biological Communities

BOTANISTS FREQUENTLY CONSTRUCT systems of classification for plant communities. Most of these schemes are based on vegetation structure—height of vegetation, leaf or needle structure, deciduousness, and dominant plant form. Such properties of plants are, in turn, adaptations to the physical environment where they live. We should not be surprised therefore to note the close correspondence between vegetation zones and climate. The natural vegetation of the United States, as classified by botanist A. W. Kuchler, conforms closely to variation in temperature and rainfall (Figure 8-1).

Major vegetation types are clearly discernible: tall forests, shrubland, and prairie are distinct; so are coniferous and broad-leaved forest types. The problem with classifications of vegetation, or with similar schemes for aquatic communities, is that many intermediates occur. In fact, most biological communities intergrade, sometimes almost imperceptibly, as the physical environment changes from one locality to the next. One is tempted to deal with these intermediates by making finer distinctions between communities. This practice can, however, lead to a bewildering variety of names for plant communities. A. W. Kuchler's somewhat conservative scheme lists 116 vegetation types in the United States alone.

Classifications of plant communities are most useful if they relate vegetation types to environment: temperature, rainfall, soil, and topography. In this chapter, we shall examine the influence of environment on vegetation structure and plant classification and survey the diversity of plant communities by way of a photographic essay.

Structural Schemes of Classification

The earliest traditions of vegetation classification were based on attempts to describe the major plant forms of each major association. These classifica-

89

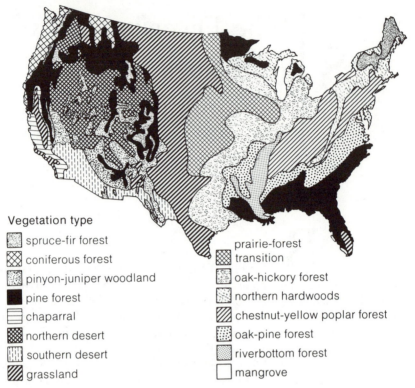

Vegetation type

spruce-fir forest

coniferous forest

pinyon-juniper woodland

pine forest

chaparral

northern desert

southern desert

grassland

prairie-forest transition

oak-hickory forest

northern hardwoods

chestnut-yellow poplar forest

oak-pine forest

riverbottom forest

mangrove

FIGURE 8-1   Vegetation map of the United States.

tions embraced complete analysis of the *species* composition of communities (*floristic* analysis) on the one hand, and description of plant *forms*, regardless of the particular species, on the other hand. Floristic analysis proved useful in restricted areas where botanists knew all the species and where minor differences between communities involved the replacement of species by similar species with slightly different ecological requirements. Floristic analysis is completely unworkable on a global scale because biogeographical barriers restrict the distributions of individual species. Functionally similar forests in Europe and the United States have very few species in common. Floristic differences between vegetational counterparts in California and Australia would be even greater. The tropics pose still greater problems for floristic analysis because of their great species diversity. Few botanists can recognize a majority of the hundreds of kinds of trees in a tropical forest and, to make matters more difficult, many species can be distinguished only when they are in flower or fruit.

Difficulties of floristic analysis for world-wide vegetation classification are partly overcome by analysis of form and function of plants rather than their scientific names. Numerous sets of symbols were devised to describe such characteristics as plant size, life form, leaf shape, size, and texture, and per cent of ground coverage. Kuchler worked out a system of letter and number symbols which could be combined into a formula describing the characteristics of plant formations. Thus M6iCXE5cD3i6H2pL1c represents an oak-yew woodland and E4hcD2rGH2rL1c(b) a madrone-holly scrub. A similar symbolic method of description, devised by Pierre Dansereau, portrays vegetation formations by use of lollipop and ice-cream cone shaped figures with internal shading and symbols varying according to the nature of the plant (Figure 8-2).

Kuchler's and Dansereau's methods are primarily descriptive. They are too complex to be used as a hierarchical scheme of classification although they have found application where classification of only the predominant features of a plant formation is desired.

In 1903, the Danish botanist Christen Raunkiaer proposed to classify plants according to the position of their buds (regenerating parts), and found that the occurrence of his major categories corresponded closely to climatic conditions. Raunkiaer distinguished five principal life forms (see also Figure 8-3):

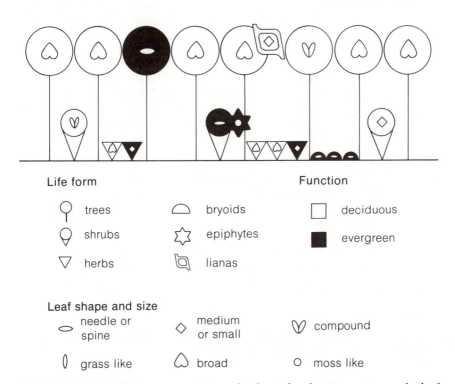

FIGURE 8-2 Symbolic representation of a forest by the Dansereau method of vegetation classification.

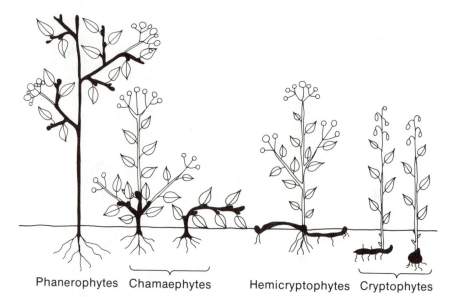

Phanerophytes Chamaephytes   Hemicryptophytes Cryptophytes

FIGURE 8-3 Diagrammatic representation of Raunkiaer's life forms. Unshaded parts of the plant die back during unfavorable seasons, while the solid black portions persist and give rise to the following year's growth. Proceeding from left to right, the buds are progressively better protected.

phanerophytes (from the Greek *phaneros*, visible) carry their buds on the tips of branches, exposed to extremes of climate. Most trees and large shrubs are phanerophytes. As one might expect, this plant form dominates in moist, warm environments where buds require little protection.

chamaephytes (from the Greek *chamai*, on the ground, dwarf) comprise small shrubs and herbs which grow close to the ground (prostrate life form). Proximity to the soil protects the bud. In regions of heavy snowfall, the buds are protected beneath the snow from extreme air temperatures. Chamaephytes are most frequent in cool, dry climates.

hemicryptophytes (from the Greek *kryptos*, hidden) persist through the extreme environmental conditions of the winter months by dying back to ground level where the regenerating bud is protected by soil and withered leaves. This growth form is characteristic of cold, moist zones.

cryptophytes are further protected from freezing and desiccation by having their buds completely buried beneath the soil. The bulbs of irises and daffodils are representative of cryptophyte plants, and are also found in cold, moist climates.

therophytes (from the Greek *theros*, summer) die during the unfavorable season of the year and do not have persistent buds. Therophytes are regenerated solely by seeds, which easily resist extreme cold and drought. The therophyte form includes most annual plants and occurs most abundantly in deserts and grasslands.

The proportional occurrence of Raunkiaer's life forms in various climatic regions is summarized in Figure 8-4. Life form and climate go closely together. Phanerophytes dominate vegetation forms in warm, moist environments, being replaced by chamaephytes, hemicryptophytes, and cryptophytes in temperate and arctic regions. Deserts have a large proportion of therophytes.

The Holdridge classification

The botanist L. R. Holdridge has proposed a classification of the world's plant formations based solely on climate (Figure 8-5). Holdridge considers temperature and rainfall to prevail over other environmental factors in determining vegetation form, although soils and exposure may exert strong influences on plants within each climate zone.

Holdridge's scheme incorporates the biological effects of climate on vegetation. As in Thornthwaite's analysis of climate (page 86), temperature and rainfall are seen as interacting to define humidity provinces. The dividing lines between humidity provinces are determined by critical ratios of potential evapotranspiration to precipitation. Potential evapotranspiration is in turn a function of temperature. Therefore the humidity provinces relate temperature and rainfall to the water relations of plants in a way that is meaningful. Holdridge's formula indicates, for example, that the availability of moisture to plants in wet tundra, with an annual rainfall of 25 centimeters (cm) and average temperature near freezing, is similar to that in a wet tropical forest, with 400 cm precipitation and an average temperature of 27 C.

Holdridge relates differences between plant formations to percentage

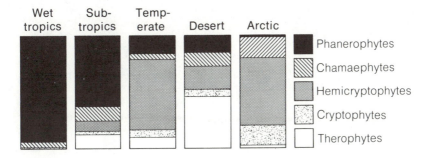

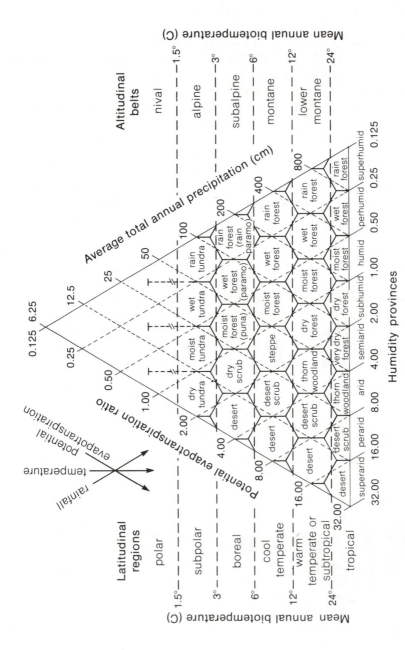

FIGURE 8-5 The Holdridge scheme for the classification of plant formations. *Mean annual biotemperature* is calculated from monthly mean temperatures after converting means below freezing to 0 C. The *potential evapotranspiration ratio* is the potential evapotranspiration divided by the precipitation; the ratio increases from humid to arid regions.

differences between their climates. The temperature or rainfall of each zone is either twice or one-half that of the adjoining zone. Thus 25 cm annual precipitation can make as big a difference in the vegetation of arid regions as 250 cm does between plant formations in the humid tropics. It seems intuitively reasonable that a little rainfall should stimulate desert annuals much more than it would rain forest trees. Experiments under controlled conditions largely support this notion.

Holdridge also constructs his temperature scale with biological considerations in mind. He assumed that biological activity ceases below 0 C. The temperature of any month whose mean was below freezing was set at 0 C for calculating mean annual temperature. Furthermore, because small increases in temperature affect biological systems more at low temperatures than at high temperatures, Holdridge set the temperature boundaries of his life zones at 1.5, 3, 6, 12, and 24 C, each temperature being twice the previous one. A factorial scale of temperature is consistent with increases in rate of evaporation and rate of biological activity in relation to increasing temperature.

Simple climate classifications, such as the Holdridge scheme, are far from ideal. In regions with similar mean rainfall and temperature, differing seasonal patterns of precipitation and temperature can create differences in vegetation structure. Topography, soil, and fire can also influence the development of vegetation types. Still, it is fair to say that climatic schemes of life zone classification do emphasize the pervasive influence of temperature and moisture on plant formations.

A Survey of Biological Communities

If a random sample of terrestrial localities is placed on a graph according to the mean annual temperature and rainfall of each locality, the points fall within a triangular area whose three corners represent warm moist, warm dry, and cool dry environments (Figure 8-6). Cold regions with high rainfall are conspicuously absent; water does not evaporate rapidly at low temperature and the atmosphere in cold regions has little water vapor. But because of the depressing effect of low temperature on evaporation, a little water goes a long way. In the tropics, 20 inches of rainfall can support little more than a desert scrub-type vegetation, but the same 20 inches permits the development of an impressive coniferous forest in Canada.

Plant ecologist R. H. Whittaker has combined several structural classifications of plant communities into one scheme, which he has transposed onto a graph of temperature and rainfall (Figure 8-6). Within the tropical and subtropical realms, with mean temperatures between 20 and 30 C, vegetation types grade from true rain forest, which is wet throughout the year, to desert. Intermediate climates support seasonal forests, in which some or all trees lose their leaves during the dry season (see Figure 7-7), and low dry forests or scrublands with many thorntrees. As aridity increases, shrubs appear farther apart, exposing large patches of bare ground. This vegetation characteristic is most highly developed in true deserts.

The range of plant communities in temperate areas follows the same pattern as tropical communities, with the same basic vegetation types distinguishable in both. In colder climates, however, the range of precipitation from one locality to another is so narrow that vegetation types are poorly differentiated on the basis of climate. Where mean annual temperatures are below −5 C, Whittaker lumps all plant associations into one type—tundra. The whole scale of moisture gradients represented in the tropics is compressed into a narrow band in the arctic. Which is tundra: rain forest or desert? Water abounds on moist tundra, but because it is frozen most of the year, and permanently frozen a few feet below the soil surface, plants cannot obtain it.

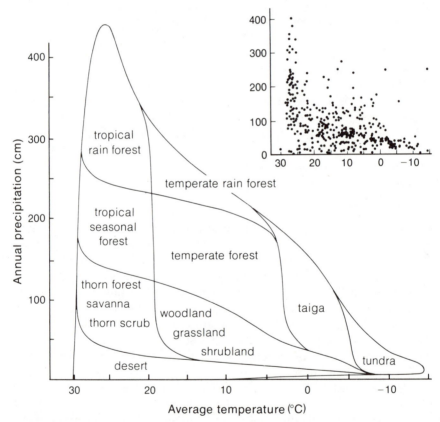

FIGURE 8-6 Whittaker's classification of vegetation types superimposed on the range of terrestrial climates. In climates intermediate between those of forested and desert regions, fire, soil, and climate seasonality determine whether woodland, grassland, or shrubland develops. The inset at the upper right shows the average annual temperature and rainfall for a large sample of localities more or less evenly distributed over the land area of the earth, excluding high mountains.

At the drier end of the rainfall spectrum within each temperature range, fire plays a distinct role in shaping the form of plant communities. In the African savannas and midwestern American prairies, frequent fires kill the seedlings of trees and prevent the establishment of tall forest, for which favorable conditions otherwise exist. Burning favors the growth of perennial grasses with extensive root systems that can survive fires. After an area has burned over, the grass roots send up fresh shoots and quickly revegetate the surface.

Like all classifications, exceptions to the scheme outlined in Figure 8-6 appear frequently. Boundaries between vegetation types are at best fuzzy. All plant forms do not respond to climate in the same way. For example, some species of Australian eucalyptus trees can form forests under climate conditions which would support only shrubland or grassland on other continents. Plant communities are affected by factors other than temperature and rainfall. We have already examined the influence of topography, soils, fire, and seasonal variation, but the point deserves emphasis.

A Survey of Biological Communities

The photographs on the pages that follow illustrate most of the important plant formations in temperate North America and tropical Central and South America. They are arranged to show the influence of temperature and moisture on vegetation structure. In addition, there are photographs of many important fresh-water and marine communities.

Environments Without Life

Life can gain a foothold in regions with almost any combination of temperature and moisture found on earth, providing the moisture is available and other nutrients are present. But life is excluded from a few environments. The extreme cold on the slopes of Mount McKinley, Alaska (below), freezes life to a standstill. Water occurs only as ice and is therefore unavailable to

plants. Water is a problem as well on the shifting sand dunes of Death Valley, California (below). The little rain that falls either evaporates or percolates through the coarse sand. Temperatures at White Sands, New Mexico (above), are favorable for life, and the region's rainfall supports desert shrubs in the surrounding valley, but the pure gypsum sand (calcium sulfate) does not contain the nutrients needed to support life.

The Humid Tropics

Year-round warm temperatures and plentiful moisture in the humid tropics create conditions for the most luxuriant and diversified communities in the world. Vegetation forms include vines that drape the trees in a lowland forest in Panama (right), and air plants that clothe trees in a mist-enshrouded cloud forest in Guatemala (above). Because soils are impoverished of nutrients except near the surface, root systems of tropical trees tend to be shallow and the trunks of many trees are buttressed for support (left).

Tropical Mountains

Temperature decreases about 6 C for each 1,000-meter increase in elevation. Plant productivity parallels the lower temperatures of montane habitats, creating cold and almost barren deserts in the tropics. The mean
annual temperature and rainfall would support a forest or woodland in
seasonal temperate climates with warm summers, but the year-round cold
of tropical mountains does not permit such luxurious growth. On the paramo
of the high Andes in Colombia at about 3,700 meters (top left), the temperature hovers around 5 C throughout the year. One is struck by the
paucity of life forms and by the silence, broken only by the relentless wind.
At the same elevation in Costa Rica on the Cerro de la Muerte (Mountain
of Death, bottom left), the ever-present fog slips among dwarfed plants,
whose small thick leaves are clustered tightly around the plant stem for
protection from the cold wind (above). The bare patch of rock shows the
thinness of the soil layer in the tropical montane habitat.

Subtropical Deserts

Belts of seasonally hot, dry climate girdle the earth at about 30° north and south of the equator. These are harsh environments where only a few drought-adapted species of plants and animals thrive. Whereas light and nutrients are critical in the humid tropics, the bare ground exposed in deserts testifies that these resources go wanting where rainfall limits plant growth. Cacti have greatly reduced leaves to decrease water loss. Their thick, succulent stems have taken over the function of photosynthesis. Numerous thorns hinder desert animals from getting at their stored water. Desert shrub habitats of the Sonoran Desert of Arizona and northern Mexico (top left) are among the most diverse vegetation types of arid regions. Giant saguaro cacti and paloverde trees dominate the landscape. The joshua tree, a treelike yucca, occurs primarily in the Mohave Desert of southern California (bottom left). An extremely dry habitat near the Gulf of California in northern Sonora, Mexico (below), supports only two kinds of large plants. Note the wide spacing between individuals. Desert plants do not tolerate close proximity because their extensive root systems compete for water.

Temperate Woodland and Shrubland

In temperate habitats with better water relations and lower summer temperatures than deserts, succulent cacti are replaced by bushes, shrubs, and small trees. The wide spacing and low growth form of plants in the Great Basin region exemplified in Zion National Park (above) indicate that water is still a critical factor. At higher elevations, in Coconino National Forest of Arizona, an open woodland dominates the landscape (top right). Juniper woodland develops at about 2,000-meters (6,000 to 7,000 feet) elevation in this area, where snow covers the ground for much of the winter and summers are cool. The milder Mediterranean climate of the southern California coast, characterized by warm, dry summers and cool, moist winters, supports a characteristic dense shrubland called chaparral (bottom right). In moist canyons and valleys, oak woodland tends to replace chaparral species, but frequent fires often prevent this natural succession and maintain the fire-adapted chaparral vegetation.

Temperate Forests

Tall forests of broad-leaved, deciduous trees occur throughout the temperate zone where rainfall is plentiful and winters are cold. Oak, beech, maple, hickory, and other hardwoods dominate temperate forests. Seasonal patterns of summer activity and winter dormancy are characteristic. The stand of Indiana hardwoods dominated by white oak has a well-developed understory of sugar maple and smaller shrubs (above). In the Appalachian Mountains

of West Virginia, red spruce and hemlock occur with broad-leaved trees to form mixed forests (bottom left). In the southeastern United States, sandy soils are too poor for broad-leaved trees. Pines are widely distributed in vast forests that are managed and harvested for paper pulp. In Florida, the palmetto frequently forms a dense understory (above). In the northern United States and Canada, and in mountainous regions of the west, birch and aspen, frequently mixed with spruce and fir, represent the farthest incursion of broad-leaved forests into cold regions (below).

Temperate Grasslands

Grasslands occur under a variety of temperate climates with cold winters and summer drought. True prairie, remnants of which can be found in Kansas (above), Texas (below), and other midwestern states, is characterized by grasses with extensive root systems. Tall grass prairies grow on fertile

soil and are probably maintained by periodic fires that keep trees from becoming established. Farther to the west, lower rainfall supports sparser vegetation, the shortgrass prairies to the east of the Rocky Mountains (above) and in western interior valleys (below). These grasslands are delicate and they are sensitive to plowing and overgrazing.

Fresh-Water Habitats

Fresh water covers a small fraction of the earth's surface, yet fresh-water habitats display remarkable diversity. Variation in water movement, mineral and oxygen content of the water, and size and shape of the stream or lake basin all contribute to this variety. Communities in deep lakes and fast-moving streams consist mostly of phytoplankton and thin layers of diatoms on the surfaces of rocks. Vegetation shows above the surface only where water is shallow and still, as in an artificially flooded marsh in Maine (left), or a cattail marsh in New York (above). Floating water hyacinths choke a deeper channel in Louisiana, buoyed up by gas trapped in their stems (right).

Temperate Montane Environments

Montane habitats are much colder and are often drier than the surrounding lowlands. Trees reach their upper limit of elevation at about 3,000 meters (10,000 feet) in the Cascade Mountains of Oregon (top left). Above timberline, snow persists well into summer in habitats that can support only the low grassy vegetation characteristic of the alpine tundra, as in the Rocky Mountains of Colorado at 3,700 meters (12,000 feet) elevation (bottom left). Lichens are the first plants to colonize bare rock surfaces in these habitats (right) and start the slow process of soil formation. Wind-driven ice strips bark and branches from trees near the timberline in Colorado (below).

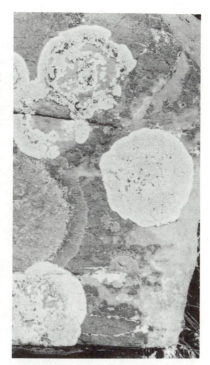

Temperate Conifer Forests

Forests of pine, spruce, fir, hemlock, redwood, and others grow under a variety of temperature, soil, moisture, and fire conditions that favor drought-resistant needle-leaf species over less tolerant broad-leaved trees. Poor soils and frequent fires favor pines throughout much of the southeastern United States. Dry summers and cold winters characterize the environments of coniferous forests at high elevations in western mountains. Pinyon pine-juniper-cedar woodland is found near Flagstaff, Arizona, where the climate is too dry to support a closed forest (bottom left). A moister site in Inyo National Forest, California, is dominated by tall Jeffrey pines (top left). Undergrowth is sparse in the dry, acid soil. Abundant winter rainfall and cool, foggy summers create ideal conditions for redwoods in the temperate rain forests of northern California (above). They bear little resemblance to the humid forests of the tropics because they lack diversity in species and plant forms and are relatively unproductive.

117

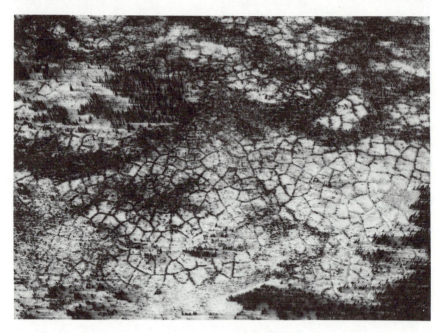

The Arctic Tundra

Permanently frozen soils underlie the arctic tundra habitat. Warm summer temperatures thaw the ground to a depth of a few inches or feet, briefly creating a shallow, often waterlogged layer of soil on which arctic vegetation develops. Repeated freezing and thawing creates characteristic polygonal patterns in the ground surface of some areas (above). At Cold Bay, Alaska,

lichens, mosses, and grasses are found on the hummocky, frost-heaved soil (bottom left). Kettle lakes formed by the melting of large blocks of ice, left by retreating glaciers, are a prominent feature of the Kuskokwim River Delta, Alaska (above). Montane tundra in the arctic is better drained than lowland habitats and spruce trees occasionally get a foothold in protected valleys, such as these near Mount McKinley, Alaska (below).

The Land Meets the Sea

The topography of coastal areas often determines the character of plant communities at the edge between the land and the sea. Shallow, sloping, sandy shores, like those of Cape Cod, Massachusetts (above), create shifting dune habitats colonized by a few species of plants that stabilize the dune and allow other species to establish a foothold. Salt marshes develop in

more protected bays and in river estuaries, as at Barnstable Harbor on Cape Cod (bottom left). In the tropics, such protected habitats are usually invaded by mangrove trees, creating forests in standing water at the edge of the ocean, as in Biscayne Bay, Florida (above). At the rockbound coast of Maine, an abrupt meeting of land and sea tolerates little intermingling of the two environments (below).

The Marine Environment

The little explored mantle of water covering most of the surface of the earth contains a wide variety of habitats and life forms. Open waters create a vast realm for the tiny phytoplankton and zooplankton, the fish that exploit them (above left), and the sea birds that eat the fish (below). Other fish are more like terrestrial grazers and predators, feeding on algae and small animals near the bottom and among corals (above right). Subtidal tropical habitats on the Caribbean coast of Panama are dominated by corals, including the elk horn coral in shallow, rough water (top right) and the important reef-building coral at a depth of 15 meters (50 feet; bottom right). Reef-building corals are restricted to sunlit depths because they rely on symbiotic green algae in their tissues for much of their nutrition.

9 | Primary Production

ORGANISMS NEED ENERGY to move, grow, and maintain the functions of their bodies. Energy to support these activities enters the ecosystem as light, which plants convert to chemical energy during photosynthesis. The rate at which plants assimilate the energy of sunlight is called *primary productivity*. It is important to realize that primary production underlies the entire trophic structure of the community. The energy made available by photosynthesis drives the machinery of the ecosystem. The flux of energy through populations of herbivores, carnivores, and detritus feeders, and the biological cycling of nutrients through the ecosystem are ultimately tied to the primary productivity of plants. In this chapter, we shall consider how light, temperature, water, and nutrients influence the rate of photosynthesis, and thereby determine the productivity of natural communities.

Plants Assimilate Energy by Photosynthesis

During photosynthesis, plants capture light energy and transform it into chemical energy in the form of organic compounds. These compounds may be stored conveniently, and their energy later released to meet the demands of biological processes. Photosynthesis chemically unites two common inorganic compounds, carbon dioxide and water, to form glucose, a simple sugar. The overall chemical equation for photosynthesis is

$$6CO_2 \quad + \quad 6H_2O \quad \longrightarrow \quad C_6H_{12}O_6 \quad + \quad 6O_2$$

| six molecules of carbon dioxide | plus | six molecules of water | yield, with an input of light energy | one molecule of glucose sugar | plus | six molecules of oxygen |

Photosynthesis requires a net energy input in the form of light equivalent

124

to 9.3 kilocalories* per gram of carbon assimilated. Because the light energy is converted to chemical energy, the metabolic breakdown of sugar into its inorganic components makes available 9.3 kilocalories per gram of carbon converted to carbon dioxide; this energy can then be utilized to perform such functions as muscle contraction and biosynthesis.

Photosynthesis is an uphill process; it requires an input of energy to drive the chemical reaction. The mere presence of the ingredients for photosynthesis is not sufficient to make the reaction occur. Carbon dioxide, water vapor, and sunlight occur together in the earth's atmosphere, yet we are not deluged by a constant rain of glucose. The probability of the appropriate chemical reaction occurring there is very low. In plant cells, pigments and enzymes bring molecules and energy together so as to make the chemical reactions of photosynthesis highly probable.

All green plants have identical photosynthetic reactions. Hydrogen atoms, usually obtained from water, combine with carbon and oxygen to form a sugar molecule. Some bacteria have evolved alternative biochemical pathways for producing glucose. For example, photosynthetic sulfur bacteria obtain the hydrogen needed to form sugars from hydrogen sulfide (H_2S) rather than from water. The resulting chemical equation is

$$12H_2S + 6CO_2 \rightarrow C_6H_{12}O_6 + 6H_2O + 12S$$

Production of sugars by sulfur bacteria is a true photosynthetic process, however, because the source of energy is light.

Production

Plants are made up of more than just glucose. Various biochemical processes use glucose both as a source of energy and as a building block to construct other, more complex organic compounds. Rearranged and joined together, sugars become fats, oils, and cellulose, the basic structural material of the plant cell wall. Combined with nitrogen, phosphorus, sulfur, and magnesium, sugars are used to produce an array of proteins, nucleic acids, and pigments. Plants cannot grow unless they have all these basic building materials. Chlorophyll contains an atom of magnesium in addition to nitrogen and carbon atoms, just as hemoglobin contains an atom of iron. Even though all other necessary materials might be present in abundance, a plant lacking magnesium could not produce chlorophyll, and thus could not grow. Plants clearly cannot function in the absence of water, either. The amount

---

* The kilocalorie (kcal) is a measure of heat energy. It is defined as the energy required to raise the temperature of one kilogram (2.2 pounds or 1.06 quarts) of water one degree Celsius. The kilocalorie, sometimes referred to as the large calorie (Cal), well-known to diet watchers, equals 1,000 small calories (cal). Some familiar benchmark figures may prove useful. Average people utilize 2,000 to 3,000 kcal of energy daily. One kilowatt of power is equivalent to the expenditure of 860 kcal per hour; one horsepower is equivalent to 642 kcal per hour.

of water required by photosynthesis is a drop compared to the bucket needed to balance transpiration. Water makes up the bulk of plant tissues and is the essential medium that makes nutrients available.

Thus the basic equation for production must include mineral inputs as well as the raw materials for photosynthesis. Many biological reactions are involved in production, but we can summarize production by a word equation:

carbon dioxide + water + minerals (in the presence
of light and within the proper temperature range) →
plant production + oxygen + transpired water

The sugars produced by photosynthesis are not wholly incorporated into plant biomass; that is, all do not serve to increase the size and number of plants. Some must be oxidized to release energy for biosynthesis and maintenance. Plants are no different from animals in this respect. They need energy to keep going.

We must distinguish two measures of production. *Gross production* refers to the total energy assimilated by photosynthesis. *Net production* refers to the accumulation of energy in plant biomass, that is to say, plant growth and reproduction. The difference between the two measures is accounted for by *respiration* by the plant. Gross production (photosynthesis) is allocated between respiration and net production. In most studies of plant productivity, particularly in terrestrial habitats, ecologists measure net rather than gross production because the techniques are less difficult and because net production indicates the quantity of resources available to heterotrophic consumers in the ecosystem.

Measurement of Primary Production

Net production can be expressed conveniently as grams of carbon assimilated, dry weight of plant tissues, or the energy equivalent of dry weight. Ecologists use these indices interchangeably. The energy content of an organic compound depends primarily on its carbon and nitrogen content. The proportion of carbon by weight in most plant tissues is close to the proportion of carbon in glucose, 40 per cent.* When plants convert sugars to fats and oils, oxygen is biochemically stripped from the molecules, thereby increasing the proportion of carbon. The fat tripalmitin ($C_{51}H_{98}O_6$), for example, contains 76 per cent carbon by weight. Fats and oils contain more than twice as much energy per gram as sugars and are, therefore, widely used by plants and animals for energy storage.

The energy content of a substance is estimated by burning a sample in

* The relative weights of atoms of hydrogen, carbon, and oxygen are 1, 12, and 16, respectively. The proportion by weight of carbon in glucose ($C_6H_{12}O_6$) is therefore $72/180 = 0.40$.

a device called a bomb calorimeter. The guts of a calorimeter are a small chamber where the sample is burned. Oxygen is forced under high pressure into the chamber to ensure complete combustion. The chamber is surrounded by a water jacket that absorbs the heat produced. The increase in temperature of a known amount of water in the jacket provides a direct estimate of the heat energy released by combustion.

The photosynthetic combination of carbon dioxide and water requires an energy input of 9.3 kcal for each gram of carbon assimilated. The complete oxidation of a carbon compound to carbon dioxide and water should, therefore, release exactly 9.3 kcal per gram of carbon oxidized. In practice, the biochemical rearrangements involved in making most complex organic compounds alter energy values slightly. As a result, ecologists rely on established values for energy content obtained directly from calorimeters. Generally accepted amounts of energy released in oxidation are 4.2 kcal per gram of carbohydrate (sugars, starch, cellulose), 5.7 kcal per gram of protein, and 9.5 kcal per gram of fat.

The equation for production suggests several possible methods for measuring the primary productivity of natural habitats. Uptake of carbon dioxide and mineral nutrients, production of plant biomass, and release of oxygen are all proportional to production. Water flux would not provide a useful measure of photosynthesis because water is too abundant in the plant and the environment, and, depending upon soil moisture, temperature, and humidity, its uptake and transpiration vary independently of the rate of photosynthesis. Uptake of carbon dioxide and production of oxygen and organic matter can be measured more reliably.

Primary production in terrestrial ecosystems is usually estimated by the annual increase in plant biomass (net production). Yearly growth of annual plants is measured by cutting, drying, and weighing the plants at the end of the growing season. The harvest method is commonly used for crop and field plants in temperate regions, where most plants die back to the ground each year. Because root growth is usually ignored—roots are difficult to remove from most soils—harvesting measures the *net annual aboveground productivity* (NAAP), which is perhaps the most commonly used basis for comparing the productivity of terrestrial communities.

The harvest method has several inherent problems. Herbivores harvest some of the net production. Root growth, as we have just noted, is difficult to measure, though the root systems of annual plants can sometimes be separated from the soil by painstaking washing. But the roots of perennials continue to grow each year, so that their biomass represents the accumulation of many years' growth. The difficulty in measurement created by root production in field habitats is compounded by branch and trunk growth in forests. Harvesting leaf fall and clippings of new twigs allows only a partial estimate of production. The annual growth of woody parts is often calculated by relating tree girth to total biomass, and then measuring annual increments in the girth of living trees. To arrive at a measurement of total biomass, a series of trees of increasing size is cut down and divided into trunk, branch, and sometimes, root components, which are then dried in large ovens and weighed. The annual increase in girth of living trees can then be converted to an increase in total weight. Leaves, flowers, and fruits, which are renewed each year, are collected at the end of the growing season and dried. Their weight is then added to the growth of woody parts to complete the estimate of production.

In aquatic habitats, plant production can be measured by gas exchange. The concentration of dissolved oxygen in water is so low that the input of oxygen by photosynthesis adds substantially to the oxygen already present. Under natural conditions, most of the oxygen produced by photosynthesis is either consumed by animals or bacteria, or escapes into the atmosphere. Ecologists evade problems by measuring production within sealed bottles. At desired depths beneath the surface of a natural body of water, samples of water containing phytoplankton are suspended in *light bottles,* which are clear and allow sunlight to enter, and *dark bottles,* which are opaque and exclude light. In the light bottles, photosynthesis and respiration occur together, and part of the oxygen released into the water in the bottles is consumed. Photosynthesis does not occur in the dark bottles, but respiration does consume oxygen. The change in oxygen concentration in the light bottle provides a measure of net production; by adding to that measure the oxygen removed from the dark bottle, we obtain a measure for gross production. The calculations are summarized: photosynthesis *minus* respiration (light bottle) *plus* respiration (dark bottle) *equals* gross production. The estimate for net production obtained from the light bottle includes the respiration of plants, animals, and bacteria. Only the estimate for gross production is really valid as a measure of plant productivity.

The light-and-dark-bottle technique is restricted to short-term measurements in small parts of an aquatic ecosystem. The technique cannot be applied easily to benthic algae or to whole systems. Ecologist Howard T. Odum partly solved this problem of measuring the production of entire stream communities. Rather than employ light and dark bottles, he compared the change in the oxygen content of the stream water during the day and the night, correcting for the exchange of oxygen between the stream and the atmosphere. By combining Odum's method with both light-and-dark-bottle techniques and conventional harvest methods for large seaweeds, ecologists have obtained reasonably accurate measurements of aquatic production.

For measuring photosynthesis in terrestrial ecosystems, carbon dioxide exchange is more useful than oxygen exchange because carbon dioxide is the rarer gas in the atmosphere. Small changes in carbon dioxide concentration are relatively easy to measure (the atmosphere contains only 0.03 per cent CO_2), and leaks in sampling chambers do not produce large errors. Measurement of production by carbon dioxide exchange resembles the light-and-dark-bottle technique. A portion of a habitat, or even of an individual plant, is enclosed in an air-tight chamber, and the decrease in carbon dioxide during the day is compared to the increase in carbon dioxide, owing to respiration alone, during the night. Gross production can be measured accurately in this way, although attempts to measure production in whole forest canopies have been fraught with technical difficulties, including leakage and problems of air conditioning large plastic enclosures, which have forced ecologists to fall back on more conventional harvest techniques.

The use of radioactive atoms of carbon, particularly the isotope ^{14}C, provides a useful variation on the gas exchange method of measuring productivity. When a known amount of radioactive carbon is added, in the form of carbon dioxide, to an air-tight enclosure, plants assimilate the radioactive carbon atoms in the same proportion in which they occur in the air in the chamber. The rate of carbon fixation is calculated by dividing the amount of radioactive carbon in the plant by the proportion of radioactive carbon dioxide in the chamber at the beginning of the experiment. Thus, if a plant assimilates 10 milligrams of ^{14}C in an hour, and the proportion of radioactive carbon dioxide in the plant chamber is 0.05 (5 per cent), we calculate that the plant has assimilated carbon at the rate of 200 milligrams per hour ($10 \div 0.05$). Plant respiration eventually releases some of the assimilated carbon as carbon dioxide, which the plant can reassimilate. Measured over a one- to three-hour period, uptake of radioactive carbon allows a reliable estimate of gross productivity. After one to two days, uptake and release of radioactive carbon approach a steady state, and estimates represent net production more nearly than gross production.

Plants use nutrients other than carbon dioxide and water to synthesize organic compounds. The disappearance of dissolved nitrates and phosphates from aquatic environments can sometimes be used as a relative measure of net production, but only under restricted conditions: Growth must occur rapidly, and plants must convert inorganic nutrients into biomass much more quickly than they are made available by decomposition of dead plants or by mixing with deep water. When production and decomposition balance each other in a steady state, decomposition releases inorganic nutrients at the same rate that they are assimilated by photosynthesis, and the concentration of dissolved nutrients does not change. Nor are nutrients necessarily accumulated by plants in fixed proportion relative to rates of production. Algae are known to take up more phosphorus when dissolved phosphates are plentiful than when they are scarce. Conversely, plants sometimes leak dissolved minerals into the environment. Many physical and chemical processes, particularly erosion, upwelling, and sedimentation, also influence nutrient concentrations in aquatic systems. Conditions permitting reliable estimation of productivity from the disappearance of inorganic nutrients usually occur only during algal blooms, which follow the quiescent winter period in temperate and arctic lakes and oceans.

A final method for estimating plant production is based on the idea that chlorophyll determines the rate of photosynthesis. Marine algae assimilate a maximum of 3.7 grams of carbon per gram of chlorophyll per hour. The productivity of a marine habitat may be estimated if the concentration of chlorophyll at different depths and the decrease in light intensity with depth are known. Although the chlorophyll method lacks the precision of gas-exchange methods, it does provide a simple and rapid index to the productivity of oceans and lakes.

Several methods for measuring the productivity of aquatic ecosystems were compared by Canadian ecologist Ian McLaren at Ogac Lake, a land-locked fiord on Baffin Island. Primary production was measured throughout the growing season by the uptake of radioactive carbon in light bottles suspended beneath the fiord surface, but concentrations of chlorophyll, nitrates, phosphates, and dissolved oxygen were also monitored (Figure 9-1). The daily productivity of the fiord increased rapidly in early summer as ice disappeared from the surface and light began to penetrate the fiord's depths. Chlorophyll concentration paralleled the increase in productivity. Nitrate and phosphate concentrations declined throughout the summer. Dissolved oxygen increased in spring with the burst of plant production, but as the season progressed, increased zooplankton respiration obscured any direct relationship between oxygen and production. A surge in production in late summer was, curiously, unrelated to any of the factors that McLaren monitored.

Light and Photosynthesis

By subjecting the photosynthetic machinery of plants to varied levels of light intensity, plant physiologists have determined the influence of light on productivity. At relatively low intensities, usually less than one-fourth the intensity of bright sunlight, the rate of photosynthesis is directly proportional to light intensity. Brighter light saturates the photosynthetic pigments, however, and the rate of photosynthesis increases more slowly or levels off. In many algae, very bright light reduces photosynthesis because it deactivates the photosynthetic apparatus.

The response of photosynthesis to light intensity can be characterized by two reference points. The first, the *compensation point*, is the level of light intensity at which photosynthetic assimilation of energy just balances respiration. Above the compensation point, the plant has a positive energy balance; below it, the plant suffers a net energy loss. The second reference point is the *saturation point*, above which rate of photosynthesis no longer responds to increasing light intensity.

Species differ in their response to light intensity according to their nature and the habitat in which they live. Among terrestrial plants, the compensation points of species that normally grow in full sunlight occur between 0.02 and 0.03 kcal m^{-2} min^{-1}; the compensation points of shade species are usually below 0.01 kcal m^{-2} min^{-1}. The saturation point, at which photosynthesis reaches a maximum, occurs in several groups of ma-

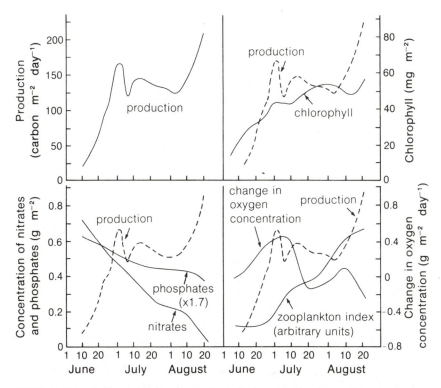

FIGURE 9-1   Relationship between phytoplankton production and concentrations of chlorophyll, nitrates, phosphates, oxygen, and zooplankton in Ogac Lake, Baffin Island.

rine phytoplankton when light intensity falls between 0.5 and 2 kcal m^{-2} min^{-1}. Above this point, photosynthesis declines rapidly. Although oak and dogwood leaves become light saturated at intensities similar to those that saturate algae, supersaturation does not depress photosynthetic activity in these species, and loblolly pine is fully light saturated only on the brightest days. In general, the saturation points of sun species occur between 0.4 and 0.6 kcal m^{-2} min^{-1}, and those of shade species occur between 0.09 and 0.11 kcal m^{-2} min^{-1}.

The sunlight that strikes the surface of a leaf is made up of a spectrum of light of different wavelengths, which we perceive as different colors. Not all colors of light are utilized in photosynthesis. Green leaves contain several pigments, particularly *chlorophylls* (green) and *carotenoids* (yellow), that absorb light and harness its energy. Carotenoids, which give carrots their orange color, absorb light primarily in the blue and green regions of the spectrum (see Figure 4-1) and reflect light with yellow and orange wavelengths. Chlorophyll absorbs light in the red and violet portions of the spectrum and reflects green, the color we perceive in leaves. The absorption spectra of whole leaves resemble the combined absorption spectra of photosynthetic pigments, but organic compounds not involved in photosynthesis evidently absorb considerable orange light. As one might expect, light under the canopy of a forest is relatively rich in the green and infrared, but poor in the red-orange and blue portions of the spectrum that are most effective in photosynthesis. Leaves of different species have different absorption spectra. Fig leaves, being thick and heavily pigmented, absorb 85 per cent of green light (550 nm), the wavelength absorbed *least* efficiently. Tobacco leaves absorb only 50 per cent of green light.

Temperature and Photosynthesis

Temperature and light intensity normally have a close relationship to natural systems, and so their effects on photosynthesis are difficult to separate. By controlling these factors in the laboratory, one can, however, assess their separate influences. Photosynthesis is relatively insensitive to temperature at low light intensities, where light constitutes a limiting factor, but at moderate light intensity, photosynthetic rate increases two to five times for each 10 C rise in temperature.

Like most other physiological functions, photosynthesis is greatest within a narrow range of temperature, above which its rate declines rapidly. Because leaves absorb light, their temperatures can become great enough during the middle of the day that photosynthesis is effectively prohibited; rate of photosynthesis then reaches a peak in mid-morning and a second peak in mid-afternoon (Figure 9-2). As one would expect, the optimum temperature for photosynthesis varies with the environment, from about 16 C in many temperate species to as high as 38 C in tropical species. The optimum temperature also varies with light intensity in some species, such as the alpine heath *Loiseleuria* in Austria. Net production depends on rate of respiration as well as rate of photosynthesis. In general, respiration increases steadily with increasing leaf temperature.

Photosynthetic efficiency is a useful index of rates of primary production of plant formations under natural conditions. The photosynthetic efficiency is the per cent of incident visible radiation that is converted to net primary production during seasons of active photosynthesis. Where water and nutrients do not limit plant production, photosynthetic efficiency varies between 1 and 2 per cent of available light energy.

Water and Transpiration Efficiency

Because photosynthesis requires gas exchange across the surface of the leaf, productivity also parallels the rate of transpiration of water from the leaf surface. As the moisture content of soil decreases, plants have greater difficulty removing water from the soil and leaves must close their stomata to reduce water loss. When soil moisture is reduced to the wilting point,

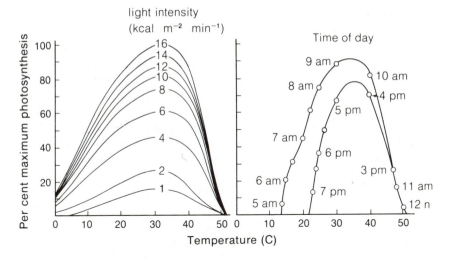

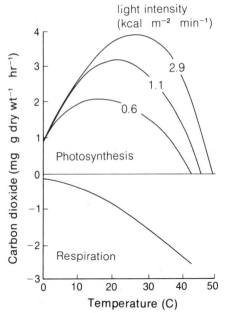

FIGURE 9-2 Net photosynthetic rate as a function of leaf temperature and light intensity in a typical desert bush (upper left) and the heath *Loiseleuria* (lower left). The daily course of photosynthetic rate in the desert bush (upper right) shows a dip in midday because the leaves become excessively hot.

leaves are effectively shut off from the surrounding air and photosynthesis slows to a standstill. Rate of photosynthesis is, therefore, closely tied to the plant's ability to tolerate water loss, to the availability of moisture in the soil, and to the influence of air temperature and solar radiation on rate of evaporation. Humid environments favor high rates of photosynthesis by reducing transpiration from leaves.

Agronomists have devised *transpiration efficiency* as an index of the drought resistance of plants. Transpiration efficiency is the ratio between net production and transpiration, expressed as grams of production per 1,000 grams of water transpired. Most plants have transpiration efficiencies of less than 2 grams of production per 1,000 grams of water transpired; some drought-resistant crops have transpiration efficiencies of 4. High efficiencies result from morphological and physiological adaptations to reduce evaporation and leaf temperature. Regardless of their adaptations to conserve water, plants cannot escape the physical reality that if they must reduce transpiration during drought, they must also reduce the gas exchange necessary for photosynthesis, and consequently reduce their productivity.

Solutions to the opposing problems of water retention and gas exchange have led to a variety of biochemical modifications of the assimilation of carbon dioxide by plants. In most temperate zone plants, carbon dioxide enters photosynthetic cells by diffusion. But because the level of carbon dioxide in the atmosphere is low (0.03 per cent) and the affinity of carbon dioxide for the enzymes that assimilate it in photosynthesis is weak, carbon dioxide uptake is extremely inefficient. Many plants in hot, dry climates have greatly increased affinity for carbon dioxide owing to a highly efficient carbon dioxide uptake step. As a result, these plants—called C_4 plants because carbon dioxide is first assimilated into a four-carbon carbohydrate— take up carbon dioxide from the atmosphere more rapidly and, therefore, with relatively less water loss. This is illustrated in Figure 9-3, in which rate of carbon dioxide assimilation is plotted as a function of carbon dioxide concentration in the atmosphere surrounding the plant. The honey-sweet plant *Tridestromia oblongifolia* (Amaranthus family), with C_4 metabolism, reaches maximum assimilation at less than 0.01 per cent carbon dioxide whereas the C_3 saltbush plant *Atriplex glabriuscula* does not attain its potential rate of carbon dioxide assimilation until the concentration of carbon dioxide is greater than its normal concentration in the atmosphere (0.03 per cent).

A disadvantage of C_4 metabolism is that the assimilation of carbon dioxide is low at low temperatures (Figure 9-3). The temperature optimum of C_4 plants is usually close to the maximum tolerated—about 45 C. That of C_3 plants usually lies between 20 C and 30 C. As a result, the proportions of C_4 and C_3 plants in a community vary inversely in relation to the average temperature during the growing season. C_4 plants predominate in hot climates, and C_3 species in cool climates.

Nutrients and Production

Most habitats respond to artificial applications of fertilizers by increased primary production. No matter what the natural fertility of the habitat, nutrient availability interacts with water, temperature, and light to determine levels of production. Nutrient limitation is probably most strongly felt in aquatic habitats, particularly the open ocean, where the scarcity of dissolved minerals reduces production far below terrestrial levels. The abun-

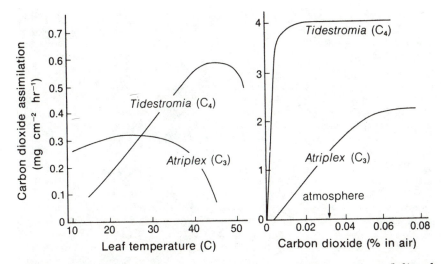

FIGURE 9-3   Rate of photosynthesis as a function of leaf temperature (left) and concentration of carbon dioxide (right) in a C₃ plant, *Atriplex glabriuscula,* and a C₄ plant, *Tidestromia oblongifolia.*

dance of nutrients in lakes and oceans depends on upwelling currents, water depth, proximity to coastlines and rivers, and drainage patterns of nearby land masses. Recent intense fertilization of inland and coastal water by sewage and runoff from fertilized agricultural land has greatly increased aquatic production in some inland and coastal waters. Such relationships will be discussed in greater detail in Chapter 11.

Even oxygen can limit terrestrial plant production under some circumstances. Roots require oxygen in the soil for metabolism and growth. If the soil is nonporous, or if it is completely saturated with stagnant water, available oxygen can be reduced below the point required to support plant growth. In the waterlogged soils of swamps, many plants have structures for obtaining oxygen for their roots directly from the atmosphere. For example, the knees of cypress trees are projections of the roots above the surface of the water (Figure 9-4). Cypress knees allow free exchange of gases between the root system and the atmosphere. The air roots, or pneumatophores, of the white mangrove *Avicennia,* whose root system grows in anaerobic mud, also provide direct gas exchange between the roots and the atmosphere.

Production in Terrestrial Ecosystems

The favorable combination of intense sunlight, warm temperature, and abundant rainfall makes the humid tropics the most productive terrestrial ecosystem on earth, square mile for square mile. Low winter temperatures

FIGURE 9-4 Cypress knees in a drained swamp in South Carolina. Under normal water levels in the swamp the knees would project above the surface, providing an avenue of gas exchange between the roots and the air.

and long nights curtail production in temperate and arctic ecosystems. Lack of water limits plant production in arid regions where light and temperature are otherwise favorable for plant growth.

The production of litter (falling leaves, fruits, and branches) provides an index to plant production in forests at various latitudes. Annual litter production varies between 900 and 1,500 grams per square meter per year ($g\ m^{-2}\ yr^{-1}$) in equatorial forests (0° to 10° latitude), 400 to 800 $g\ m^{-2}\ yr^{-1}$ in warm temperate forests (30° to 40°), 200 to 600 $g\ m^{-2}\ yr^{-1}$ in cool temperate localities (40° to 60°) and 0 to 200 $g\ m^{-2}\ yr^{-1}$ in arctic or alpine communities (60° to 70°). These data demonstrate clearly that the production of litter decreases with distance from the equator. Litter production is known to represent about 40 per cent of total net annual aboveground productivity (NAAP) in temperate forests, so aboveground primary production is about 2.5 times the litter fall.

Within a given latitude belt, where light and temperature do not vary appreciably from one locality to the next, net production is directly related to annual precipitation. Studies of net aboveground productivity in grasslands around the world show that in dry areas (Idaho, South Dakota, southwestern Africa) production varies between 90 and 200 $g\ m^{-2}\ yr^{-1}$, in moister areas (including South Carolina, Louisiana, Japan, and parts of India), where grassland is maintained by agricultural practices and is not the predominant plant association, primary production varies between 250 and 500 g m^{-2} yr^{-1}. Although plant production initially increases rapidly as annual precipitation increases, productivity tends to level off in temperate habitats with more than 100 cm (39 inches) of rain annually, presumably because the availability of light and nutrients becomes limiting (Figure 9-5).

Ecologists Robert Whittaker and Gene Likens have recently estimated net primary production for representative terrestrial and aquatic ecosystems (Table 9-1). Their estimates are based on many studies using a wide variety of techniques and are probably reasonably close to true values. Production in terrestrial habitats decreases dramatically from the wet tropics to temperate regions, even more so where the climate is too dry or too cold to support forests. Swamp and marsh ecosystems occupy the interface between terrestrial and aquatic habitats, where plants are as productive as in tropical forests. Maximum rates of production in marshes have been reported to be as high as 4,000 g m^{-2} yr^{-1} in temperate regions and 7,000 g m^{-2} yr^{-1} in the tropics. Marsh plants are highly productive because their roots are frequently under water and their leaves extend into the sunlight and air, obtaining benefits of both aquatic and terrestrial living. In addition, rapid decomposition by bacteria of detritus that washes into marshes releases abundant nutrients.

The productivity of agricultural land usually falls somewhat below the productivity of natural vegetation in the same area because croplands are plowed each year and laid bare early and late in the growing season, when undisturbed habitats continue to be productive. Furthermore, most agricultural plantings consist of a single species, which cannot exploit the resources of the land as efficiently as a mixture of species with differing ecological requirements.

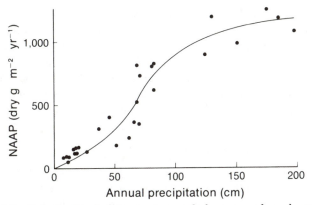

FIGURE 9-5 Relationship between net annual aboveground production (NAAP) and rainfall; data are primarily from temperate areas.

TABLE 9-1 Average net primary production and related dimensions of the earth's major habitats.

Habitat	Net primary production (g m^{-2} yr^{-1})	Biomass (kg m^{-2})	Chlorophyll (g m^{-2})	Leaf surface area (m^2 m^{-2})
Terrestrial				
Tropical forest	1,800	42	2.8	7
Temperate forest	1,250	32	2.6	8
Boreal forest	800	20	3.0	12
Shrubland	600	6	1.6	4
Savanna	700	4	1.5	4
Temperate grassland	500	1.5	1.3	4
Tundra and alpine	140	0.6	0.5	2
Desert				
Cultivated land	650	1	1.5	4
Swamp and marsh	2,500	15	3.0	7
Aquatic				
Open ocean	125	0.003	0.03	—
Continental shelf	360	0.01	0.2	—
Algal beds and reefs	2,000	2	2.0	—
Estuaries	1,800	1	1.0	—
Lakes and streams	500	0.02	0.2	—

Irrigation and the application of fertilizers can increase agricultural yields two to threefold over world averages. Sugar cane, a common tropical crop, has a world average production of about 1,700 g m^{-2} yr^{-1}. Intensively cultivated sugar cane in the Hawaiian Islands has double the average world yield and a maximum productivity of about 7,000 g m^{-2} yr^{-1}. Poor crop management, conversely, can lead to soil deterioration and reduced production.

Net primary production of temperate zone cereal crops (wheat, corn, oats, and rice), hay, and potatoes varies between 250 and 500 g m^{-2} yr^{-1}; sugar beets commonly attain twice that productivity. These values are compared to the estimates of Whittaker and Likens for temperate forests (600 to 2,500 g m^{-2} yr^{-1}) and temperate grasslands (150 to 1,500 g m^{-2} yr^{-1}). The productivity of all agricultural land varies between 100 and 4,000 g m^{-2} yr^{-1}, depending on the crop, with an average of 650 g m^{-2} yr^{-1}.

Production in Aquatic Ecosystems

The open ocean is a virtual desert, where scarcity of mineral nutrients—not water—limits productivity to one-tenth or less that of temperate forests (Table 9-1). Upwelling zones (where nutrients are brought up from the depths by vertical currents) and continental-shelf areas (where exchange between shallow bottom sediments and surface waters is well developed) support greater production, averaging 500 and 360 g m^{-2} yr^{-1}, respectively. In shallow estuaries, coral reefs, and coastal algae beds, production approaches that of adjacent terrestrial habitats, with averages approaching 2,000 g m^{-2} yr^{-1}. Primary production in fresh-water habitats is similar to that of comparable marine habitats.

Availability of nutrients largely determines variation in the production of aquatic ecosystems. Light apparently does not limit production within the euphotic zone. As much as 95 per cent of the incident radiation penetrates the surface of the water and is available to plants. Variation in the depth to which light penetrates, a function of the clarity of the water, influences the depth to which photosynthesis occurs but does not affect annual production per square meter of ocean surface. In clear water, algae are spread thinly throughout the deep euphotic zone; in turbid water, they are concentrated closer to the surface.

Temperature evidently does not influence the production of marine habitats. Although the photosynthetic rates of individual plants may be depressed by cold temperatures, marine algae attain great density in cold water, enough so that arctic oceans are as productive as warm tropical seas. In cold temperate waters, large seaweeds produce as much biomass per square meter of marine habitat as in the Indian Ocean and Caribbean Sea. In Nova Scotia, the species *Laminaria* alone attains a productivity of about 1,500 g m^{-2} yr^{-1}, a respectable value for a temperate forest!

The Annual Net Primary Production of the Earth

The most productive habitats attain photosynthetic efficiencies of 1 to 2 per cent, but so much of the earth's surface lacks optimum conditions for plant growth that only one-tenth of 1 per cent of the light energy striking the earth's surface is assimilated by plants. The energy value of sunlight reaching the outer atmosphere of the earth directly under the sun (the *solar constant*) is about 10^7 (10 million) kcal m^{-2} yr^{-1}. The angle of incident radiation varies, however, with time of day and season, and over a year's time, all points on the earth are shrouded by night for an equivalent of six months. If the total energy reaching the outer atmosphere were spread evenly over the surface of the earth, each square meter would receive one-quarter of the solar constant, or 2.5×10^6 kcal m^{-2} yr^{-1}. In fact, the earth's surface does not actually receive that much solar energy during the year, and its distribution is not uniform. Perhaps 40 per cent of the total light income of the earth is absorbed by the atmosphere and re-radiated back into space as heat. Part is also reflected and scattered by particles of dust in the atmosphere and reflected by surfaces of water, rock, and vegetation.

The annual energy income of a particular locality varies with latitude and cloud cover. Temperate localities usually receive between 2×10^5 and 2×10^6 kcal m^{-2} yr^{-1}, which represents 2 to 20 per cent of the solar constant. Only half the light energy can be assimilated by plants; the other half lies outside the absorption spectrum of plant pigments.

Whittaker and Likens estimated the total annual primary productivity of the earth as 162×10^{15} (million billion) grams (about 730×10^{15} kcal), of which terrestrial habitats are responsible for two-thirds. The average productivity of terrestrial areas (720 g m^{-2} yr^{-1} or $3,200$ kcal m^{-2} yr^{-1} represents assimilation of about 0.3 per cent of the light reaching the surface. The overall photosynthetic efficiency of aquatic habitats is less than one-quarter that of terrestrial communities.

The distribution of production among the major vegetation zones of the earth is a function of the local productivity of these zones, determined by light, temperature, rainfall, and nutrients, and their total surface area (Figure 9-6). Tropical forests cover only 5 per cent of the earth's surface, but they account for almost 28 per cent of the total production. Temperate forests and the open ocean, representing 2.4 and 63 per cent of the surface area, are responsible for 9.2 and 25 per cent of the earth's productivity. Although inshore waters (estuaries, algal beds, and reefs) occupy only 0.4

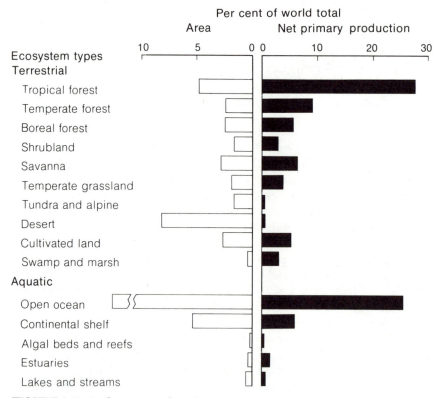

FIGURE 9-6 Surface area and total annual productivity of the major ecosystems. Values are expressed as a percentage of the total for the earth.

per cent of the earth's surface, they account for 2.3 per cent of the earth's productivity, and 6.7 per cent of all aquatic production. Swamps and marshes have similarly high productivity for their small area.

In this chapter we have examined basic patterns of primary production over the surface of the earth. Productivity is greatest where light, warmth, water, and mineral nutrients are all abundant. Decreasing light and temperature reduce production of habitats distant from the tropical zones. Lack of moisture further restricts production in arid regions. In some aquatic habitats, scarcity of nutrients imposes the greatest limitation to production, particularly in the open ocean, but inshore areas with abundant nutrients can match or exceed nearby terrestrial communities for organic production.

Net productivity measures the overall energy input to the ecosystem. This production is eventually consumed and dissipated by herbivores, by the carnivores that eat them, and by the myriad detritus feeders that scavenge dead debris. The precise pathways that energy follows on its route through the ecosystem and the time required for its journey are the subject of the next chapter.

10 | Energy Flow in the Community

PHOTOSYNTHESIS and net primary production make energy available to the community. Herbivorous animals eat plants, carnivores eat herbivores, carnivores are in turn eaten by other carnivores, and so on through a series of steps that together form a *food chain*. Each step in the food chain represents a *trophic level*. The feeding relationships of organisms impart a trophic structure to the community through which energy flows.

Energy flux is the only sound currency in the economics of ecosystem function; biomass and numbers are static descriptions of the community frozen in an instant of time. The dynamics of the community are measured in terms of change—rates of energy and nutrient transferral from organism to organism through the structure of the food web. The unique status of energy as an ecological currency has greatly stimulated the study of community energetics. In this chapter, we shall study the pathways of energy through the community and how the physical environment of the community influences energy flow.

The food chain illustrated in Figure 10-1 oversimplifies nature in several ways. First, few carnivores feed on a single trophic level; second, many organisms (or their parts in the case of plants) die of causes other than predation, and their remains are consumed by detritus-feeding organisms; third, most energy assimilated by a trophic level is dissipated as heat because biological processes are energetically inefficient and organisms utilize energy to maintain themselves as well as to grow and reproduce. Energy incorporated into growth and reproduction, the food of the next higher trophic level, is a small percentage of the total food eaten.

Some Definitions

Ecologists have applied many different terms to the feeding relationships of organisms. Such terms are usually introduced to clear up confusion and give precise definitions to the activity of organisms. But they often have the opposite effect. Plants and animals have been functioning perfectly well without having names applied to their activities. We should take this cue and concentrate on understanding the feeding relationships within a community, rather than try to categorize them.

We can arrange most of the troublesome terms in pairs with opposite meanings. *Producer* and *consumer* refer to different activities performed by the same organisms, although *primary producer* is usually reserved for

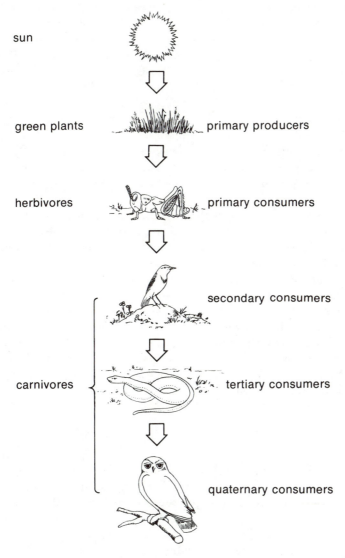

FIGURE 10-1   A simplified terrestrial food chain showing the sequence of trophic levels.

green plants and *primary consumer* for herbivores. Production refers to the assimilation of materials into the body of the organism, that is, growth and reproduction. Whether the source of energy for production is sunlight or chemical energy in food distinguishes primary producers from secondary, tertiary, quaternary, etc., producers. Just as all organisms produce, they also consume. Plants and animals both must metabolize assimilated substances to keep themselves going. Ecologists apparently will not consider the assimilation of sunlight by plants as an act of consumption. Herbivores, therefore, are designated primary consumers by default.

Autotroph and *heterotroph* distinguish organisms that convert inorganic forms of energy to organic forms from those whose sole source of energy is organic matter. Thus autotrophs are primary producers. Green plants are *photo*autotrophs because they assimilate the energy of sunlight. The blue-green alga *Beggiatoa* is partly a *chemo*autotroph because it can derive energy from the oxidation of hydrogen sulfide to sulfur.

Herbivore and *carnivore* refer to eaters of plants and animals, respectively, although carnivore is sometimes reserved for meat eaters. Other terms such as *insectivore* and *piscivore* (fish eater) are used. Herbivores also come in a variety of forms—nectivores, frugivores, grazers, browsers, and so on—depending on what they eat and how they eat it. *Predator* distinguishes organisms that consume whole prey from *parasites*, organisms that live on living prey, or *hosts*. Hence seed eaters are properly called predators because they destroy the tiny plant embryo in the seed, and many grazers and browsers are properly called parasites, because they consume leaves or buds without killing the tree. Mosquitos and vampire bats would fit equally well into the category of browsers.

Biophage and *saprophage* distinguish organisms that eat living prey from those that go the easy route of eating dead prey. Saprophages also may be called *detritivores* as they are in this book. Detritivores, or saprophages, are often referred to as decomposers. While detritivores *do* decompose organic compounds into simpler inorganic compounds, this function is not unique to detritivores. All organisms metabolize organic foods to obtain usable energy and release carbon dioxide and water as end products. All organisms therefore are decomposers.

Food Chains and Food Webs

Trophic levels provide a simple framework for understanding energy flow through the ecosystem. We should be able to determine the trophic level of the energy in a particular chemical bond by the number of times it has changed hands since it was first put together in a green plant. But energy follows complicated paths through the trophic structure of the ecosystem, and we cannot easily follow a particular chemical bond.

Many carnivores eat fruit and other plant materials when their normal prey is not readily available. Some carnivores eat other carnivores as well as herbivores. Whether an owl eats herbivorous field mice or insectivorous shrews is of little consequence nutritionally to the owl; both prey have about the same food value and require similar effort to capture. Large owls sometimes eat snakes or small weasels, which themselves feed on mice and shrews. Owls are not known to feed on plant material, but foxes, which normally prey on small mammals, often eat fruit. It is therefore not unusual for predators to feed on three or four trophic levels. Plants are not always the *victims* of animals; some partly carnivorous species are known—Venus's fly traps, pitcher plants, and sundews. Most herbivores, however, particularly those that eat leafy vegetation, are so specialized in their food habits that they are incapable of carnivory. A cow's teeth can grind tough plant material to shreds but could not get a grip on a rabbit's flesh, much less tear it apart. (As if a cow could even catch a rabbit.)

The great variety of feeding relationships within the community links species into a complex *food web*. Each species feeds on many kinds of prey (Figure 10-2). Within the context of this complexity, trophic levels become abstract concepts, useful for understanding general patterns of structure and energy flow through the ecosystem, but usually quite useless as categories for assignment of individual species.

Pyramids of Productivity, Biomass, and Numbers

The amount of energy metabolized by each trophic level decreases as energy is transferred from level to level along the food chain. Green plants, the primary producers, constitute the most productive trophic level. Herbivores are less productive, and carnivores still less. The productivity of each trophic level is limited by the productivity of the trophic level immediately below it. Because plants and animals expend energy for maintenance, less and less energy is made available, through growth and reproduction, to each higher trophic level. Of the light energy assimilated by a plant, 30 to 70 per cent is used by the plant itself for maintenance functions and the energetic costs of biosynthesis. Herbivores and carnivores are more active than plants and expend even more of their assimilated energy. As a result, the productivity of each trophic level is usually no more than 5 to 20 per cent of that of the level below it. The percentage transfer of energy from one trophic level to the next is called the *ecological efficiency*, or *food chain efficiency*, of the community.

The pioneer ecologist Charles Elton, in 1927, suggested that if each trophic level in the community were represented by a block whose size corresponds to the productivity of the trophic level, and if the blocks were then stacked on top of each other with the primary producers at the bottom, we would obtain a characteristic pyramid-shaped structure (Figure 10-3). The structure of the pyramid varies from community to community, depending on the ecological efficiencies of the trophic levels. In the particular case pictured in Figure 10-3, these efficiencies are 20, 15 and 10 per cent.

FIGURE 10-2 Some of the feeding relationships in a simple food web.

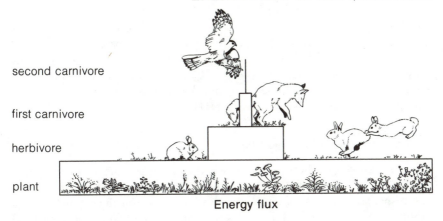

second carnivore

first carnivore

herbivore

plant

Energy flux

FIGURE 10-3   An ecological pyramid representing the net productivity of each trophic level in the ecosystem. This particular structure represents ecological efficiencies of 20, 15, and 10 per cent between trophic levels, but these values vary widely between communities.

Herbivore production is therefore 20 per cent of plant production; first-level carnivore production is 15 per cent of herbivore production and only 3 per cent of plant production (15 per cent of 20 per cent equals 3 per cent); second-level carnivore production is only 0.3 per cent of plant production. These values are probably unrealistically high compared to most natural communities, but they illustrate the universal decrease in the availability of energy at progressively higher trophic levels.

As energy availability decreases, the biomass and numbers of individuals on each trophic level usually decrease as well, although no law of energetics prevents a reverse trend. The biomass structure of the community resembles the pyramid of productivity in most terrestrial communities. If one were to collect all the organisms in a grassland, the plants would far outweigh the grasshoppers and ungulates that eat the plants. The herbivores in turn would outweigh the birds and large cats at the first carnivore level, and these too would outweigh their predators, if there were such. An individual lion may be heavy, but lions are spread out so thinly that they do not count for much on a gram-per-square-meter basis.

The pyramid of biomass is sometimes turned upside down in aquatic plankton communities. Algae must be more productive than the tiny animals that eat them; the laws of energetics cannot be violated. But the phytoplankton are sometimes consumed so rapidly that their numbers are kept small by herbivorous zooplankton. Intensive grazing reduces phytoplankton biomass, but the algae are so productive that they can often support a larger biomass of herbivores under optimum conditions for growth.

The pyramid of numbers is even more shaky than the pyramid of biomass. Disease organisms (parasitic bacteria and protozoa, for example), mosquitos, ants, and others are certainly more numerous than the organisms they feed on, even though all of them together do not weigh as much as their prey or hosts. A single tree may be host to thousands of aphids, caterpillars, and other herbivorous insects.

Detritus Pathways of Energy Flow

Many ecologists attribute a distinctive role to organisms which consume dead plant and animal matter. It is frequently said that these detritus eaters are responsible for breaking down dead organic remains, which would otherwise accumulate, and for releasing their nutrients so they can be used again by plants. Detritus-consuming organisms *do* have this function in the ecosystem, but this view of a special role in the community is misleading for two reasons. First, as we have seen, all organisms "decompose" organic matter. Terrestrial mammals consume 20 to 200 grams of food for every gram of body weight produced. The remainder is metabolized and used as a source of energy. Required minerals and other nutrients are retained and incorporated into body tissue, but most ingested food is returned to the environment in an inorganic form: carbon dioxide and water are exhaled; water and various mineral salts are excreted in sweat and urine.

Second, detritivores have no special purpose different from any other organism in the overall function of the ecosystem. The individual detritivore obtains energy and nutrients in its food just like herbivores and carnivores. And just like all other organisms, detritivores leave behind undigestible remains, breakdown products of metabolism, and excess minerals that they cannot use. Detritivores eat the garbage of the ecosystem for the same reason herbivores and carnivores eat fresh food: to make a living. It's all a matter of taste.

Detritivores include such diverse species as carrion eaters—crabs, vultures, and the like—whose freshly dead food differs little from the live prey eaten by carnivores—and bacteria and fungi, which are biochemically specialized to consume certain organic materials and waste products that are particularly difficult for most organisms to digest.

From the standpoint of energy use detritus feeders are not readily distinguishable from other kinds of consumers. Detritivores have better pickings in some communities than others. As we shall see below, terrestrial plants produce large quantities of indigestible supportive tissue, most of which is consumed after death by organisms of decay in the soil. More than 90 per cent of the net primary production of a forest is consumed by detritivores, and less than 10 per cent by herbivores. Aquatic plants are more digestible by herbivores, and the detritus pathways are correspondingly less prominent.

The Individual Link in the Food Chain

Once food is eaten, its energy follows a variety of paths through the organism (Figure 10-4). Not all food can be fully digested and assimilated. Hair, feathers, insect exoskeletons, cartilage and bone in animal foods, and cellulose and lignin in plant foods cannot be digested by most animals. These materials are either egested by defecation or regurgitated in pellets of undigested remains. Some egested wastes are relatively unaltered chemically during their passage through an organism, but nearly all are mechanically broken up into fragments by chewing and by contractions of the stomach and intestines and are thereby made more readily usable by detritus feeders.

Organisms use most of the food energy that they assimilate into their bodies to fulfill their metabolic requirements: performance of work, growth, and reproduction. Because biological energy transformations are inefficient, a substantial proportion of metabolized food energy is lost, unused, as heat. Organisms are no different from man-made machines in this respect. Most of the energy in gasoline is lost as heat in a car's engine rather than being transformed into the energy of motion. In natural communities, energy used to perform work or dissipated as heat cannot be consumed by other organisms and is forever lost to the ecosystem.

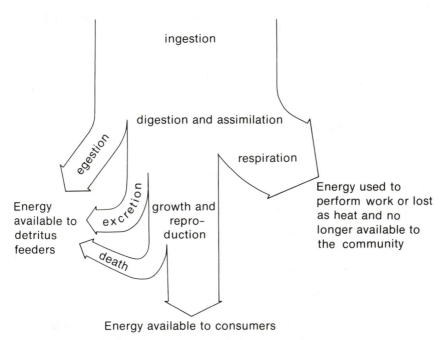

FIGURE 10-4   Partitioning of energy within a link of a food chain.

Proteins create special metabolic problems because they contain nitrogen. Nitrogen in excess of requirements for growth and body maintenance is usually excreted in an organic form—ammonia in most aquatic organisms, urea and uric acid in most terrestrial organisms. Excreted nitrogenous waste products therefore represent a loss of potential chemical energy. But, because these waste products can be metabolized by specialized microorganisms, they enter the detritus pathways of the community. We shall examine the path of nitrogen compounds through the ecosystem in greater detail in the next chapter.

Assimilated energy that is not lost through respiration or excretion is available for the synthesis of new biomass through growth and reproduction. Populations lose some biomass by death, disease, or annual leaf drop, which then enters the detritus pathways of the food chain. The remaining biomass is eventually consumed by herbivores or predators, and its energy thereby enters the next higher trophic level in the community.

Energetic Efficiencies

The movement of energy through the community depends on the efficiency with which organisms consume their food resources and convert them into biomass. This efficiency is referred to as the food chain, or *ecological efficiency*. Ecological efficiencies are determined by both internal, physiological characteristics of organisms and their external, ecological relation-

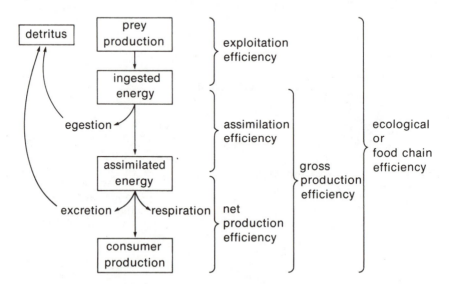

FIGURE 10-5   Diagram of the partitioning of energy in a link of the food chain and the energetic efficiencies associated with each metabolic step through the organism. Detritus produced by egestion and excretion of unusable food remains on the trophic level of the prey and is available for consumption by other organisms.

TABLE 10-1 Definitions of energetic efficiencies.

1. Exploitation efficiency $= \dfrac{\text{Ingestion of food}}{\text{Prey production}}$

2. Assimilation efficiency $= \dfrac{\text{Assimilation}}{\text{Ingestion}}$

3. Net production efficiency $= \dfrac{\text{Production (growth and reproduction)}}{\text{Assimilation}}$

4. Gross production efficiency $= (2) \times (3) = \dfrac{\text{Production}}{\text{Ingestion}}$

5. Ecological efficiency $= (1) \times (2) \times (3) = \dfrac{\text{Consumer production}}{\text{Prey production}}$

ships to the environment. To understand the biological basis of ecological efficiency, one must dissect the individual link of the food chain into its component parts (Figure 10-5). Ecological efficiency depends on the efficiencies of three major steps in energy flow: exploitation, assimilation, and net production (Table 10-1). The product of the *assimilation* and *net production efficiencies* is the *gross production efficiency*—the percentage of ingested food converted to consumer biomass. The product of the exploitation and gross production efficiencies is the food chain, or ecological efficiency—the percentage of available food energy in prey converted into consumer biomass.

Pigeonholing nature into a stilted and often unnatural framework by categorization often produces more problems than solutions. Ecological efficiencies are no exception. Egested and excreted energy are the square pegs one tries to fit into round holes in this case. On what trophic level do we place egested and excreted detritus? If put with the prey trophic level, it must be classed as unexploited energy and figured into the exploitation efficiency. If put with the consumer trophic level, egested and excreted matter properly should be figured into gross production efficiency, thereby increasing overall ecological efficiency. Most ecologists either place detritus in a special food category all its own, belonging to no trophic level, or ignore the problem altogether.

Above all, energetic efficiencies, as ecologists define them, correspond to quantities that are readily measured. Practicality is often the architect of concept. The study of energy transformation by plants and animals does, however, provide a useful insight into the basis of ecological efficiency and the trophic structure of the community.

Plants are terrible predators. Perhaps we should say that light is an elusive prey. Light can pass right through a leaf, it can be absorbed by the wrong molecules, or it can be quickly converted to heat and slip away before the plant can harness its energy. As a form of energy, light differs so much from organic molecules that the conversion of energy from one form to the other is inefficient. Once light energy *has* been harnessed, plants can utilize it efficiently for production.

Animals have different problems. With their prey in hand—or in mouth—its energy can be assimilated efficiently by rearranging chemical bonds rather than by converting one form of energy to another. Yet animals expend so much energy in maintaining their delicate—compared to plants—bodies and in pursuing prey that relatively little assimilated energy ends up as production.

Energetic Efficiencies in Plants

The absorption of light by green plants is a form of ingestion and may be used as a basis for calculating the energetic efficiency of the plant trophic level. Probably 20 to 30 per cent of all light striking the earth's surface in productive habitats is absorbed by green plants. Most of the remainder is either reflected or absorbed by bare earth and water. Because plants incorporate little absorbed energy into biomass, gross production efficiency is usually less than 2 per cent.

Most of the light energy absorbed by plants is converted directly to heat and lost by re-radiation, convection, or transpiration. In a classic study of the energy relations of a corn field, E. N. Transeau determined that only 1.6 per cent of the light energy incident on the field was assimilated by photosynthesis. The remainder was either converted to heat (54 per cent) or absorbed by water, either in the soil or in leaves, in evaporating and transpiring the 15 inches of rainfall received during the growing season (44.4 per cent). Although plants convert 1 to 2 per cent of absorbed light into chemical energy under suitable natural conditions, photosynthesis can achieve a maximum efficiency of 34 per cent under the most favorable laboratory conditions.

Plants use relatively less assimilated energy for maintenance than animals because they do not move or maintain body temperatures, and they "feed" continuously during daylight periods. Estimates of net production efficiency vary between 30 and 85 per cent depending on the habitat (Table 10-2). Rapidly growing vegetation in temperate zones, whether natural fields, crops, or aquatic plants, exhibits uniformly high net production efficiency (75 to 85 per cent). Lower values characterize tropical vegetation. The cause of low net production efficiencies in tropical aquatic communities (40 to 50 per cent) is unclear. Bright light tends to saturate the photosynthetic mechanism and perhaps reduce the rate of photosynthesis without reducing respiration. High temperature also accelerates plant respiration more rapidly than it accelerates photosynthesis. The combination of these factors in tropical waters would increase the rate of respiration relative to photosynthesis and thereby lower net production efficiency.

Wet tropical forests also exhibit low net production efficiencies, probably

TABLE 10-2   Net production efficiency (net primary production/gross primary production) of several plants and plant communities.

Plant community	Locality	Net production efficiency (%)
Terrestrial		
Perennial grass and herb	Michigan	85
Corn	Ohio	77
Alfalfa		62
Oak and pine forest	New York	45
Tropical grasslands		55
Humid tropical forest		30
Aquatic		
Duckweed	Minnesota	85
Algae	Minnesota	79
Bottom plants	Wisconsin	76
Phytoplankton	Wisconsin	75
Sargasso Sea	Tropical Atlantic Ocean	47
Silver Springs	Florida	42

around 30 per cent. (Measurements of forest productivity is exceedingly difficult and most published figures should be viewed as tentative.) Variation in energetic efficiency of terrestrial habitats roughly parallels the ratio of photosynthetic to supportive tissue in different plants. Leaves comprise 1 to 10 per cent of the aboveground biomass of forest trees compared to 20 to 60 per cent for shrubs and over 80 per cent for most herbs. Nonphotosynthetic living parts of plants respire, though the outer bark and dead cores of trunks and branches do not. Roots certainly contribute substantially to plant respiration.

If the ratio between photosynthetic and supportive tissue determines net production efficiency, young plants that have not yet developed extensive supportive or root tissue should exhibit higher efficiencies than large, mature plants. Comparative data for different vegetation types support this hypothesis. Further evidence comes from studies of seasonal change in the net production efficiency of a field of lespedeza (a member of the pea family). The energetic efficiency fell from 75 per cent in April to 15 per cent in August, paralleling a similar decline in growth rate and increase in the proportion of root and stem biomass.

Energetic Efficiencies in Animals

Assimilation efficiency varies with quality of the diet of animals. In particular, animal food is digested more easily than plant food. Assimilation efficiencies of predatory species vary between 60 and 90 per cent of the food consumed, with insectivores occupying the lower end and meat and fish eaters the upper end of the range. The tough chitinous exoskeleton of insects, which constitutes a large portion of the weight of many species, resists digestion and therefore reduces the assimilation efficiencies of insect-eating species.

The proportion of cellulose, lignin, and other indigestible materials influences the nutritional value of plant foods. Trunks and branches of trees consist mostly of cellulose and lignin, which lack nitrogen and many essential minerals. Species of wood-boring beetles that live off the woody parts of plants either grow very slowly or restrict their feeding to the layer of nonwoody, living cells immediately under the bark.

Leaves contain between 2 and 4 per cent protein and thus are more suitable than wood as a food for herbivores, though plants have devised many defensive mechanisms to protect leaves, including a variety of toxins. The leaves of oaks and other trees have tannins that prevent herbivores from digesting their proteins. Seeds are the most desirable plant food because they are provisioned with the nutrients to get a plant started in life—nutrients that are equally well-suited to sustaining herbivores. Pine nuts, for example, contain about 50 per cent oil, 30 per cent protein, and 5 per cent sugars.

Assimilation efficiencies of herbivores parallel the nutritional quality of their food: up to 80 per cent for seed diets, up to 60 per cent for young foliage, 30 to 40 per cent for most mature foliage, and 10 to 20 per cent or less for wood, depending on its state of decay.

Net production efficiency of animals is inversely related to activity. Maintenance activities, and heat production by warm-blooded animals, require energy that otherwise could be utilized for growth and reproduction. Active, terrestrial, warm-blooded animals exhibit low net production efficiencies: birds less than 1 per cent because they maintain uniformly high activity; small mammals with high reproductive rates (rabbits and mice, for example) up to 6 per cent. Man maintains the net production efficiency of beef cattle at as much as 11 per cent by slaughtering them soon after, or even before, growth is completed. More sedentary, cold-blooded animals, particularly aquatic species, channel as much as 75 per cent of their assimilated energy into growth and reproduction, which approaches the maximum biochemical efficiency of growth.

The efficiency of biomass production within a trophic level (gross production efficiency) is the product of assimilation efficiency and the net growth efficiency (Figure 10-6). Gross production efficiencies of few warm-blooded, terrestrial animals exceed 5 per cent, and those of some birds and large mammals fall below 1 per cent. The gross production efficiencies of insects lie within the range of 5 to 15 per cent, and some aquatic animals exhibit efficiencies in excess of 30 per cent. Net production efficiency tends to be inversely related to assimilation efficiency, especially in aquatic animals, but ecologists do not fully understand the basis for this relationship.

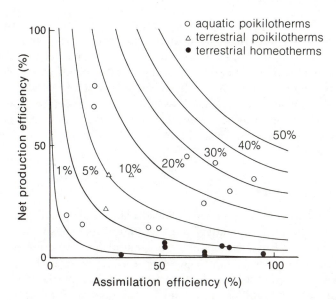

FIGURE 10-6 Relationships between assimilation efficiency and net production efficiency for a variety of animals. Gross production efficiencies are indicated by the curved lines on the graph.

Exploitation Efficiency

The efficiency of energy conversion within a trophic level (gross growth efficiency) does not fully describe the flow of energy between trophic levels. One must also include the efficiency with which consumers exploit their food resources. Unless organic material is steadily accumulating in an ecosystem, exploitation efficiencies on each trophic level, including detritivores, account for all the net production of the next lower trophic level. Peat bogs are an exception to this rule; a large fraction of plant production sinks to the bottom of the bog where acid, anaerobic conditions prevent its decay. Nonetheless, most ecosystems exhibit steady-state conditions in which all production is eventually consumed or transported out of the system by wind or water currents.

Energy flow between trophic levels represents the sum of the feeding activities of many species. Individual populations usually consume only a small fraction of available food resources. Carnivores, seed eaters, and aquatic herbivores are most efficient; the commonest species consume 10 to 100 per cent of the food available to them. Terrestrial herbivores usually consume only 1 to 10 per cent of the leafy vegetation.

Different prey are caught with different success. A study in Alberta, Canada, revealed that red-tailed hawks captured 20 to 60 per cent of local ground squirrel populations during the summer months, but only 1 per cent of snowshoe hares and 1 to 3 per cent of ruffed grouse. Other predators—foxes, weasels, owls, snakes—also hunt the same prey and ensure the transfer of their biomass to the next trophic level. Plants and prey that escape herbivores and predators eventually die and their chemical energy enters the next trophic level by way of the detritus pathway.

One predator's failure is another's success, but the relative feeding efficiency of consumers influences two important aspects of community energetics: the proportion of energy that travels through detritus pathways in the ecosystem, and the time energy remains in the system before it is dissipated as heat.

Detritus Pathways in the Ecosystem

Detritus feeders consume remains of dead plants and animals, undigested or partially digested fecal matter, and excreted nitrogenous waste products of protein metabolism—any nonliving organic material that can be metabolized to provide energy. Detritus feeding is inversely related to the digestibility of fresh food materials. Detritivores are not so prominent in planktonic aquatic communities as in terrestrial communities, where they consume as much as 90 to 95 per cent of net primary production.

Large carrion eaters and scavengers—vultures and crows on land, and crabs in the sea—draw one's attention to detritivores as members of natural communities, but most dead organic matter is consumed by the unnoticed worms, mites, bacteria, and fungi that teem under the litter of the forest floor and in the mucky sediments at the bottom of streams, lakes, and the sea.

Of all the detritus-based communities, the organisms that consume the litter of leaves and branches on the forest floor are probably best known. Herbivores consume less than 10 per cent of the production of a broad-leaved forest—except during outbreaks of defoliating insects, such as gypsy moths. The remainder of the production drops to the forest floor each year as old leaves and branches, or accumulates in roots, trunks, and branches where it escapes consumers until the tree finally dies.

The breakdown of leaf litter occurs in three ways: (1) leaching of soluble minerals and small organic compounds from leaves by water, (2) consumption of leaf material by large detritus feeding organisms (millipedes, earthworms, woodlice, and other invertebrates), and (3) eventual breakdown of organic compounds to inorganic nutrients by specialized bacteria and fungi.

Between 10 and 30 per cent of the substances in newly fallen leaves dissolve in cold water; leaching rapidly removes most of these from the litter. As soil microorganisms decompose the litter further, they produce many small organic and inorganic molecules, which are also exposed to leaching if they are not first assimilated by detritivores.

The role of large organisms in the breakdown of leaf litter has been demonstrated by enclosing samples of litter in mesh bags with openings large enough to let in microorganisms and small arthropods such as mites and springtails, but small enough to keep out large arthropods and earthworms (Figure 10-7). Large detritus feeders assimilate no more than 30 to 45 per cent of the energy available in leaf litter, and even less from wood. They nonetheless speed the decay of litter by microorganisms because they macerate the leaves in their digestive tracts, breaking the litter into fine particles and exposing new surfaces for microbial feeding.

Leaves from different species of trees decompose at different rates, depending on their composition. In eastern Tennessee, weight loss of leaves during the first year after leaf fall varies from 64 per cent for mulberry, to 39 per cent for oak, 32 per cent for sugar maple, and 21 per cent for beech. The needles of pines and other conifers also decompose slowly. Differences between species depend to a large extent upon the lignin content of the leaves. Lignin is the substance that gives wood many of its structural qualities, and it is even more difficult to digest than cellulose. Conifer needles typically contain 20 to 30 per cent lignin, broad leaves 15 to 20 per cent.

The toughness of some types of plant litter, particularly wood, points up the unique role of the fungi as detritivores. The familiar mushrooms and shelf fungi are merely fruiting structures produced by the mass of the fungal organism deep within the litter or wood (Figure 10-8). Most fungi consist of a vast network, or mycelium, of hyphae, thread-like elements which can penetrate the woody cells of plant litter that bacteria cannot reach. Fungi are also distinguished by secreting enzymes and acids into the substrate itself, digesting organic matter even at a distance. The fungal hypha is like a biochemical blowtorch, cutting its way deep into wood and opening the way for bacteria and other microorganisms. Fungi are prominent in woody litter that is not attacked readily by larger detritus feeders. Bacteria occur more abundantly where detritus has been mechanically broken up by earthworms and large arthropods.

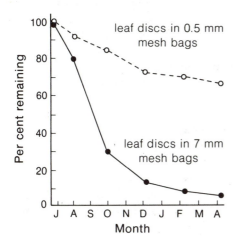

FIGURE 10-7 Percentage consumption of leaf area by detritus feeders. Leaves were enclosed in mesh bags with either large (7 mm) or small (0.5 mm) openings. The small openings admitted bacteria, fungi, and small arthropods, but excluded most large detritus feeders such as earthworms and millipedes.

FIGURE 10-8   Shelf fungi speed the decomposition of a fallen log. The brackets are fruiting structures produced by the fungal hyphae, the whole mass of which is called the mycelium. The hyphae grow throughout the interior of the log, slowly digesting its structure.

The Time Scale of Energy Flow

Characteristics of plants that inhibit digestion by animals slow the passage of assimilated energy through the ecosystem. For a given level of productivity, the *transit time* of energy in the community and the storage of chemical energy in living biomass and detritus are directly related: the longer the transit time, the greater the accumulation of living and dead biomass. Energy storage in the ecosystem has important implications for the stability of the biological community and its ability to withstand perturbation.

The average transit time of energy in living organisms in the community is equal to the energy stored in the system as biomass divided by the rate of energy flow through the system, or

$$\text{transit time (yrs)} = \frac{\text{biomass (g m}^{-2})}{\text{net productivity (g m}^{-2} \text{ yr}^{-1})}$$

(The transit time defined by this equation is sometimes referred to as the *biomass accumulation ratio*.) According to Whittaker and Likens, wet tropical forests produce an average of 2,000 grams of dry matter per square meter per year and have an average living biomass of 45,000 g m^{-2}. Inserting these values into the equation for average transit time we obtain 22.5 years (45,000/2,000). Average transit times presented for representative ecosystems in Table 10-3 vary from more than 20 years in forested terrestrial environments to less than 20 days in aquatic planktonic communities. These figures underestimate the total transit time of energy in the ecosystem, however, because they do not include the accumulation of dead organic matter in the litter.

An estimate of the transit or residence time of energy in accumulated litter can be obtained by an equation analogous to the biomass accumulation ratio

$$\text{transit time (yrs)} = \frac{\text{litter accumulation (g m}^{-2})}{\text{rate of litter fall (g m}^{-2}\text{ yr}^{-1})}$$

The average transit time for energy varies from a minimum of 3 months in wet tropical forests to 1 to 2 years in dry and montane tropical forests, 4 to 16 years in pine forests of the southeastern United States, and more than 100 years in montane coniferous forests. Warm temperature and abundance of moisture in lowland tropical regions create optimum conditions for rapid decomposition. Because most of the energy assimilated by forest communities is dissipated by detritus feeders (most of a tree being unavailable to herbivores for the several reasons we have seen), the average transit time of energy in the litter must be added to the transit time in living vegetation to obtain a complete estimate of the persistence of assimilated energy in the ecosystem.

More direct estimates of the rate of energy flow can be obtained by using radioactive tracers. Energy itself cannot be followed directly, but organic compounds containing energy can be labeled with a radioactive element, and their movement followed. In radioactive tracer studies, plants

TABLE 10-3 Average transit time of energy in living plant biomass (biomass/net primary production) for representative ecosystems.

System	Net primary production (g m^{-2} y^{-1})	Biomass (g m^{-2})	Transit time (yrs)
Tropical rain forest	2,000	45,000	22.5
Temperate deciduous forest	1,200	30,000	25.0
Boreal forest	800	20,000	25.0
Temperate grassland	500	1,500	3.0
Desert shrub	70	700	10.0
Swamp and marsh	2,500	15,000	6.0
Lake and stream	500	20	0.04 (15 days)
Algal beds and reefs	2,000	2,000	1.0
Open ocean	125	3	0.024 (9 days)

(or water if an aquatic system is being studied) are labeled with a radioactive isotope, usually of phosphorus (^{32}P), applied in a phosphate solution. Consumer species are collected at intervals after the initial labeling and are examined with a radiation counter.

Eugene Odum and his coworkers at the University of Georgia have followed the movement of radioactive phosphorus through components of an old-field community. They labeled the dominant plant, telegraph weed (*Heterotheca*), with drops of radioactive phosphate solution placed directly on the leaves, and then collected insects, snails, and spiders at intervals of several days for about five weeks (Figure 10-9). Certain herbivores, notably crickets and ants, began to accumulate radioactive phosphorus within a few days of its initial application, and attained peak amounts within two weeks. Other herbivorous insects and snails accumulated peak amounts of the tracer at two to three weeks. Ground-living, detritus-feeding insects (carabid beetles, tenebrionid beetles, and gryllid crickets) and predatory spiders did not accumulate peak amounts of tracer until three weeks after the start of the experiment. Thus the phosphorus label appeared in herbivores first and in detritivores and predators later, as the investigators undoubtedly had hoped.

The movement of radioactive phosphorus through components of the old-field community gives some indication of the time required by labelled substances to reach various trophic levels. Most of the energy assimilated into a trophic level is dissipated by respiration before it reaches the next level. Respired energy therefore has a shorter residence time in the ecosystem than the food energy that eventually reaches higher trophic levels. Energy appears to move between trophic levels via the herbivorous insect pathway in an average of a few weeks. Most plant production in terrestrial communities is not, however, consumed by herbivores, rather it is stored and consumed by detritus feeders over a prolonged period. Based on the biomass accumulation ratio, a minimum estimate of average transit in temperate grasslands is on the order of three years (Table 10-3).

A radioactive tracer experiment performed on a small trout stream in Michigan gave results comparable to those of Odum's old-field study. Radioactive phosphate was added to the stream at one point and its accumulation in plants and animals downstream from the release site was monitored for two months. The median time for each population to accumulate its maximum concentration of radioactive phosphorus varied from a few days for aquatic plants to one to two weeks for filter feeders and other herbivores, three to four weeks for omnivores, four to five weeks for detritus feeders, and four weeks to more than two months for most predators. The results suggest that most of the energy assimilated in aquatic ecosystems is dissipated within a few weeks, although a small portion may linger for months in the predator food chain, and perhaps for years in organic sediments on the bottoms of streams and lakes.

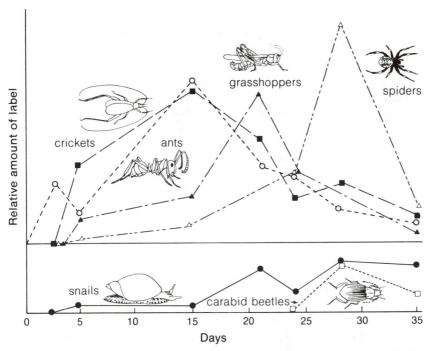

FIGURE 10-9 Accumulation of radioactive phosphorus by animals on different trophic levels after initial labeling of plants. The experiment was performed in an old-field ecosystem.

Community Energetics

The measurement of energy flow through an entire community is a complex and virtually impossible task. Yet the total flux of energy and the efficiency of its movement determine the basic trophic structure of the community: number of trophic levels, relative importance of detritus and predatory feeding, steady-state values for biomass and accumulated detritus, and turnover rates of organic matter in the community. Each of these properties, in turn, influences the inherent stability of the community. Detritus pathways stabilize communities because detritivores do not affect the abundance of their "prey" directly. Accumulation of biomass and detritus, coupled with low turnover rates, stabilizes communities by ironing out short-term fluctuations in the physical environment that affect energy flux. Ecologists have not yet agreed on the influence of predation and competition at higher trophic levels on community stabilty. Some argue that the increased complexity of a diverse and trophically stratified community adds stability to the system. Others contend that biological interactions build lag times into the response of the system to the physical environment, thereby promoting instability. Ecosystem stability will be considered more fully toward the end of this book. In any event, the trophic structure of the community constitutes an important component of the regulation of ecosystem function.

The flow of energy through a community is not unlike the flow of money through a checking account. Income must equal or exceed expenditure over a long period to keep the checking account open and the community viable. The balance undergoes normal periodic fluctuations corresponding to pay day and bills due, just as the community undergoes regular daily and seasonable fluctuations in energy income and expenditure. An unexpected disaster shakes the bank account just as a hurricane or cold snap shakes the community. Under such circumstances, a large balance and hockable assets are as helpful as stored biomass and detritus in meeting disaster.

The red and black ink of community energy balance can be summarized by the following sources of income and expenditure:

Income	Expenditure
Assimilated light energy	Respiration
Transport in	Transport out

All the energy assimilated or transported into the community is either dissipated by respiration or is carried out of the system by wind, water currents, or gravity.

It is difficult to avoid drawing boundaries when we discuss community structure and function, particularly when we wish to distinguish aquatic and terrestrial communities. The lake community clearly extends to the edge of the lake, and no further. Smaller distinctions of communities within the aquatic or terrestrial realms are drawn more arbitrarily. We may find it difficult to point to the exact place where the desert shrub community stops and the grassland community takes up. We can be sure, however, of the boundaries around a study area, and ecologists often define the community pragmatically as a place with a characteristic vegetation type.

Regardless of the kind of boundary we imagine around a community, energy frequently finds its way in and out of communities across these boundaries. Movement, or transport, of energy between communities accounts for a varying proportion of the community's income and expenditures. Terrestrial communities receive most of their energy income from light, although they lose energy when plant and animal detritus falls, blows, or washes into rivers and lakes. Transport of material between different terrestrial communities is probably negligible.

Some isolated communities rely primarily on detritus produced elsewhere and transported into the community. For example, in Root Spring, near Concord, Massachusetts, herbivores consume 2,300 kcal m^{-2} yr^{-1}, but only 655 kcal are produced by aquatic plants; the balance is transported into the spring by leaf fall from nearby trees. Life in caves and abyssal depths of the oceans, to which no light penetrates, subsists entirely on energy transported into the system by wind, water currents, and sedimentation.

The relative importance of predatory- and detritus-based food chains varies greatly among communities. Predators are most important in plankton communities; detritus feeders consume the bulk of production in terrestrial communities. As we have seen earlier, the proportion of net production that enters the herbivore-predator segment of the food web depends on the relative allocation of plant tissue between structural and supportive functions on one hand and growth and photosynthetic functions on the other. Herbivores consume 1.5 to 2.5 per cent of the net production of temperate deciduous forests, 12 per cent in old-field habitats, and 60 to 99 per cent in plankton communities (Table 10-4). The low level of herbivory in old fields may not truly represent natural habitats because most large North American herbivores became extinct during the last 15,000 years. Large grazing mammals today consume between 28 and 60 per cent of the net production of African grasslands; beef cattle consume 30 to 45 per cent of the net production of managed range lands in the United States.

In 1942, Raymond Lindeman made one of the earliest attempts to describe the energy flow in an entire community. He chose Cedar Bog Lake, Minnesota, a relatively small, self-contained system, for his study.

TABLE 10-4 Exploitation of net primary production as living vegetation by primary consumers.

Community	Characteristics of primary producers	Exploitation by herbivores (%)*
Mature deciduous forest	Trees; large amount of non-photosynthetic structure; low turnover rate	1.5 to 2.5
Thirty-year-old Michigan field	Perennial forbs and grasses; medium turnover rate	1.1
Desert shrub	Annual and perennial herbs and shrubs; low turnover rate	5.5**
Georgia salt marsh	Herbaceous perennial plants; medium turnover rate	8
Seven-year-old South Carolina fields	Herbaceous annual plants; medium turnover rate	12
African grasslands	Perennial grasses; high turnover rate	28 to 60**
Managed rangeland	Perennial grasses, high turnover rate	30 to 45**
Ocean waters	Phytoplankton; very high turnover rate	60 to 99

* Aboveground production only for terrestrial systems.
** Grazing by mammals only.

Lindeman estimated energy flow from the harvestable net production of each trophic level and from laboratory determinations of respiration and assimilation. The animals and plants collected at the end of the growing season constituted the net production of the trophic levels to which each species was assigned:

Trophic level	Harvestable production (kcal m^{-2} yr^{-1})
Primary producers (green plants)	704
Primary consumers (herbivores)	70
Secondary consumers (carnivores)	13

Lindeman estimated the energy dissipated by respiration from ratios of respiratory metabolism to production measured in the laboratory: 0.33 for aquatic plants, 0.63 for herbivores, and 1.4 for the more active carnivores in the lake. The gross production of carnivores was calculated as the sum of their harvestable production (13 kcal m^{-2} yr^{-1}) and respiration (13 × 1.4 = 18 (kcal m^{-2} yr^{-1}), or 31 kcal m^{-2} yr^{-1} (Table 10-5). Lindeman determined that predation on secondary consumers was negligible. The gross production of primary consumers was similarly calculated as the sum of their harvestable production (70 kcal m^{-2} yr^{-1}), respiration (70 × 0.63 = 44 kcal m^{-2} yr^{-1}), and the consumption of primary consumers by secondary consumers (34 kcal m^{-2} yr^{-1}). (Lindeman assumed that because secondary consumers have assimilation efficiencies of 90 per cent they must consume 3 kcal m^{-2} yr^{-1} over and above their gross production of 31 kcal m^{-2} yr^{-1}.) Therefore, the gross production of primary consumers was 148 kcal m^{-2} yr^{-1} which corresponded in turn to the removal of net primary production by herbivores. Assuming the assimilation of herbivores feeding on plant material to be 84 per cent, Lindeman calculated that herbivores consumed, but did not as-

TABLE 10-5   An energy flow model for Cedar Lake Bog, Minnesota.

Energy (kcal m^{-2} yr^{-1})	TROPHIC LEVEL		
	Primary producers	Primary consumers	Secondary consumers
Harvestable production*	704	70	13
Respiration	234	44	18
Removal by consumers			
assimilated	148	31	0
unassimilated	28	3	0
Gross production (totals)	1,114	148	31

* Does not include net production removed by consumers. Actual net production, including removal by consumers, was 879 kcal m^{-2} yr^{-1} primary producers, 104 kcal m^{-2} yr^{-1} for primary consumers, and 13 kcal m^{-2} yr^{-1} for secondary consumers.

TABLE 10-6   A comparison of energy flow models for Cedar Lake Bog, Minnesota and Silver Springs, Florida.

	Cedar Lake Bog	Silver Springs
Incoming solar radiation (kcal m^{-2} yr^{-1})	1,188,720	1,700,000
Gross primary production (kcal m^{-2} yr^{-1})	1,113	20,810
Photosynthetic efficiency (%)	0.10	1.20
Net production efficiency (%)		
Producers	79.0	42.4
Primary consumers	70.3	43.9
Secondary consumers	41.9	18.6
Exploitation efficiency (%)*		
Primary consumers	16.8	38.1
Secondary consumers	29.8	27.3
Ecological efficiency (%)		
Primary consumers	11.8	16.7
Secondary consumers	12.5	4.9

* Based on assimilated energy rather than ingested energy. Assimilation efficiencies were probably above 80 per cent, so values are not much below actual exploitation efficiencies.

similate $(0.16/0.84) \times 148 = 28$ kcal m^{-2} yr^{-1}, making the gross primary production 1,114 kcal m^{-2} yr^{-1}.

Studies of community energetics have come a long way since Lindeman's pioneering venture. A more recent study by Howard T. Odum on another small aquatic ecosystem at Silver Springs, Florida, employed more refined techniques. Gross production of aquatic plants was estimated by gas exchange rather than by the harvest method. Odum also accounted for the inflow of energy in the form of detritus from tributary streams and the surrounding land. The community energetics of Cedar Bog Lake and Silver Springs are compared in Table 10-6. The more southern location and warmer temperatures of Silver Springs probably accounts for its greater primary production and the lower net production efficiencies of its inhabitants. Herbivores consumed little of the net primary production of Cedar Bog Lake; most was deposited as organic detritus in lake sediments. Consequently, the exploitation efficiency of primary consumers was lower in Cedar Bog Lake than in Silver Springs. In spite of high respiratory energy losses in Silver Springs and quantities of production transported out of the system at Cedar Bog Lake, exploitation efficiencies varied in both locations between 15 and 40 per cent and the overall ecological efficiency of energy transfer between trophic levels varied between 5 and 17 per cent.

Ecological efficiencies are usually lower in terrestrial habitats and a useful rule of thumb states that the top carnivores in terrestrial communities can feed no higher than the third trophic level on the average, whereas aquatic carnivores may feed as high as the fourth or fifth level. This is not to say that there can be no more than three links in a terrestrial food chain;

some energy may travel through a dozen links before it is dissipated by respiration. These high trophic levels probably do not, however, contain enough energy to fully support a predator population.

We can crudely estimate the average length of food chains in a community from the net primary production, average ecological efficiency, and average energy flux of predator populations. Because the energy reaching a given trophic level is the product of the net primary production and the intervening ecological efficiencies, the appropriate equation for calculating the average trophic level that a community can support is

$$\text{trophic level} = 1 + \frac{\log (\text{predator ingestion} \div \text{net primary production})*}{\log (\text{average ecological efficiency})}$$

Using this equation and some rough estimates for the values needed on the right-hand side of the equation, we can calculate average number of trophic levels as about seven for marine plankton communities, five for inshore aquatic communities, four for grasslands, and three for wet tropical forests (Table 10-7). These estimates should be taken with a grain of salt, to be sure, but they do indicate how measurements of energetics for individual species and trophic levels can be used to determine the overall trophic structure of the community.

We have seen how the quality of food and allocation of energy to various functions by organisms create patterns of energy flow through communities. These patterns differ most between aquatic and terrestrial environments because of basic differences in the adaptations of organisms to each of these realms. In aquatic ecosystems, energy flows rapidly and is transferred efficiently between trophic levels, thereby permitting long food chains. In terrestrial ecosystems, some energy is dissipated rapidly, making energy

---

* We note that the energy $E(n)$ available to a predator on the nth trophic level may be calculated by the equation

$$E(n) = NPP \cdot Eff^{n-1}$$

where NPP = net primary production. Eff is the geometric mean ecological efficiency of trophic levels 1 to n, and $n - 1$ is the number of links in the food chain between trophic levels 1 and n. We now rearrange the equation above to the following form

$$Eff^{n-1} = \frac{E(n)}{NPP}$$

take the logarithm of both sides of the equation

$$(n - 1) \log (Eff) = \log (E[n]/NPP)$$

and rearrange to obtain

$$n = 1 + \frac{\log (E[n]/NPP)}{\log (Eff)}$$

TABLE 10-7 Community energetics and the average number of trophic levels in various communities. Values for production, predator energy flux, and ecological efficiencies are rough estimates based on many studies.

Community	Net primary production (kcal m^{-2} yr^{-1})	Predator ingestion (kcal m^{-2} yr^{-1})	Ecological efficiency (%)	Number of trophic levels
Open ocean	500	0.1	25	7.1
Coastal marine	8,000	10.0	20	5.1
Temperate grassland	2,000	1.0	10	4.3
Tropical forest	8,000	10.0	5	3.2

transfer between trophic levels relatively inefficient, and the remainder is stored for long periods as supportive tissue in plants and as organic detritus in the soil.

Each parcel of energy assimilated by plants travels through the ecosystem only once. Energy is dissipated as heat, a form which plants cannot harness for primary production. As we shall see, however, in the next chapter, mineral nutrients are continually recycled through the ecosystem. These cycles differ greatly among the elements, but they share two fundamental properties: first, the cycles are tied to, and are driven by, energy flux through the ecosystem; second, nutrients alternate between inorganic and organic forms through the complimentary processes of assimilation and decomposition.

11 | Nutrient Cycling

NUTRIENTS, UNLIKE ENERGY, are retained within the ecosystem where they are continually recycled between living organisms and the physical environment. Because plants and animals can use only those nutrients that occur at or near the surface of the earth, persistence of life requires that the materials assimilated by organisms eventually become available to other organisms. Each chemical element follows a unique route in its cycle through the ecosystem, as we shall see below, but all cycles are driven by energy, and their elements alternate between organic and inorganic forms as they are assimilated and excreted by plants and animals.

Nutrient cycles of communities sometimes become unbalanced, and nutrients accumulate in or are removed from the system. For example, during periods of coal and peat formation, dead organic materials accumulate in the sediments of lakes, marshes, and shallow seas where anaerobic conditions prevent their decomposition by microorganisms. In other instances, under intensive cultivation or after removal of natural vegetation, erosion can wash away nutrient-laden layers of soil that took years to develop. Most ecosystems exist in a steady state in which outflow of nutrients from the system is balanced by inflow from other systems, from the atmosphere, and from the rock beneath the system. Furthermore, gains and losses are usually small compared to the rate at which nutrients are cycled within the system.

Energy Flow and Nutrient Cycling

Exchanges of nutrients between living organisms and inorganic pools are about evenly balanced in most communities. Carbon and oxygen are recycled by the complementary processes of photosynthesis and respiration. Nitrogen, phosphorus, and sulfur follow more complex paths through the ecosystem, aided along the way by microorganisms with specialized metabolic capabilities.

We may describe the ecosystem as being divided into *compartments*

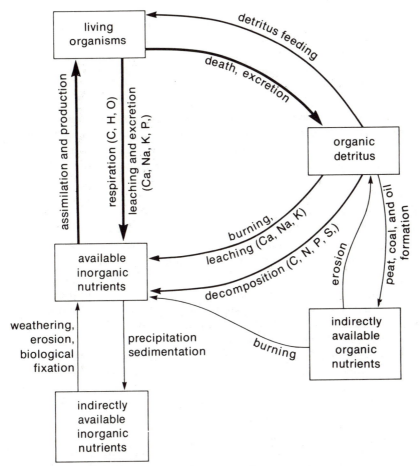

FIGURE 11-1   A compartment model of the ecosystem with some of the most important routes of mineral exchange indicated.

through which material passes and within which material may remain for varying periods (Figure 11-1). Most of the mineral cycling in the ecosystem involves three active compartments: living organisms (or biomass), dead organic detritus, and available inorganic minerals. Two additional compartments—indirectly available inorganic minerals and organic sediments—are peripherally involved in nutrient cycles, but exchange between these compartments and the rest of the ecosystem occurs slowly compared to exchange among active compartments.

The processes responsible for the movement of mineral nutrients within the ecosystem are indicated in Figure 11-1. Assimilation and production cause elements to move from the inorganic to the organic compartment. Primary production by plants is the most important component of this step in the cycling of carbon, oxygen, nitrogen, phosphorus, and sulfur, but animals assimilate many such essential minerals as sodium, potassium, and calcium directly from the water they drink.

Respiration returns some carbon and oxygen directly to the pool of available inorganic nutrients, perhaps after being cycled within the living biomass compartment several times in predator food chains. Calcium, sodium, and other mineral ions excreted or leached out of leaves by rainfall or the water surrounding aquatic organisms are recycled rapidly. Most of the carbon and nitrogen assimilated into living biomass is transferred by death and excretion into the detritus compartment. Some nutrients in detritus may be returned to the biomass compartment by detritus feeders, but all eventually are returned to the pool of available inorganic minerals by leaching and decomposition. Exchange between the actively cycled pools of minerals and the vast reservoirs of indirectly available nutrients locked up in the atmosphere, limestone, coal, and the rocks forming the crust of the earth, occurs slowly, primarily by geological processes.

Elements usually occur in different forms in the air, soil, water, and in living organisms (often referred to as the atmosphere, lithosphere, hydrosphere, and biosphere). For example, oxygen occurs as oxygen molecules (O_2) and as carbon dioxide (CO_2) both in gaseous form in the atmosphere and in dissolved form in water, but it also combines with hydrogen to form water (H_2O). Oxygen appears in the form of oxides (iron oxide, Fe_2O_3) and salts (calcium carbonate, $CaCO_3$) in the lithosphere. The rate at which an element is transferred between its inorganic forms and thus its availability in inorganic forms to living organisms vary greatly. The largest pool of oxygen—more than 90 per cent of the oxygen near the surface of the earth—occurs as calcium carbonate in sedimentary rocks, particularly limestone. Except for minute quantities released by volcanic activity, the oxygen in limestone and other sedimentary rock is virtually unavailable to the biosphere. In contrast to oxygen, nitrogen is most abundant in its gaseous form (N_2) in the atmosphere, but plants assimilate nitrogen primarily from nitrates (NO_3^-) in the soil or in water. Despite its abundance, atmospheric nitrogen plays a minor role in short-term nutrient cycling.

The assimilation and decomposition processes that cycle nutrients through the biosphere are closely linked to the acquisition and release of energy by organisms. The paths of nutrients, therefore, parallel the flow of energy through the community. (As we saw in the last chapter, radioactive elements incorporated into organic compounds can be used to follow the path of energy through the community.) The carbon cycle is most closely linked to the transformation of energy in the community because organic carbon compounds contain most of the energy assimilated by photosynthesis. Most energy-releasing processes, of which respiration is the most important, release carbon as carbon dioxide. When organisms metabolize organic compounds containing nitrogen, phosphorus, and sulfur, these elements are often retained in the body for the synthesis of structural proteins, enzymes, and other organic molecules that make up structural and functional components of living tissue. Consequently, nitrogen, phosphorus, and sulfur pass through each trophic level somewhat more slowly than the average transit time of energy.

Movement of oxygen and hydrogen in the ecosystem is overwhelmingly influenced by the water cycle. Organisms lose water rapidly by evaporation and excretion; body water may be replaced hundreds or even thousands of times during an organism's lifetime. When discussing the oxygen cycle, ecologists usually distinguish pathways involving chemical assimilation of oxygen into organic compounds and those involving movement of water.

The water, or hydrological, cycle does, however, demonstrate the basic features of all nutrient cycles—that they are approximately balanced on a global scale, and that they are driven by energy—and so we shall let it provide a model.

The Water Cycle

Although water is chemically involved in photosynthesis, most of the water flux through the ecosystem occurs through evaporation, transpiration, and precipitation. Evaporation and transpiration correspond to photosynthesis in that light energy is absorbed and utilized to perform the work of evaporating water and lifting it into the atmosphere. The condensation of water vapor in the air, which eventually causes rainfall, releases the potential energy in water vapor as heat, much as respiration by plants and animals releases energy. The water cycle is outlined in Figure 11-2.

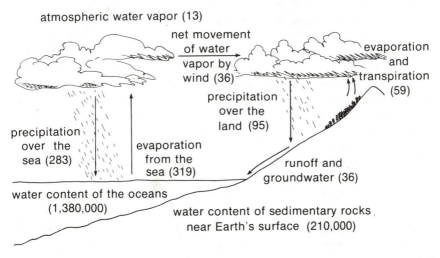

FIGURE 11-2 The water cycle, with its major components expressed on a global scale. All pools and transfer values (shown in parentheses) are expressed as billion billion (10^{18}) grams, and billion billion grams per year.

Because more than 90 per cent of the earth's water is locked up in rocks in the core of the earth and in sedimentary deposits near the earth's surface, this water enters the hydrological cycle in the ecosystem very slowly through volcanic outpourings of steam. Hence the great reservoirs of water in the earth's interior contribute little to the movement of water near the earth's surface.

Precipitation over the land surface of the earth exceeds evaporation and transpiration from terrestrial habitats. Whereas 23 per cent of the earth's precipitation occurs over land surfaces, only 16 per cent of the water vapor in the earth's atmosphere comes from the continents. The oceans exhibit a corresponding deficit of rainfall compared to evaporation. Much of the water vapor that winds carry from the oceans to the land condenses over mountainous regions where rapid heating and cooling of the land create vertical air currents. The net flow of atmospheric water vapor from ocean to land areas is balanced by runoff from the land into ocean basins.

We can estimate the role of plant transpiration in the hydrological cycle from the organic productivity of terrestrial habitats and the transpiration efficiency of plant production. The primary production of terrestrial habitats is about 1.1×10^{17} grams (g) of dry material per year, and approximately 500 g of water are transpired for each gram of production. Terrestrial vegetation, therefore, transpires 55×10^{18} g of water annually, nearly the total evapotranspiration from the land. Although this figure may overestimate transpiration somewhat, plants clearly play the major role. The influence of vegetation on water movement is best shown by removing it. At Hubbard Brook, New Hampshire, experimental cutting of all trees from small watersheds increased the flow of streams draining the clear-cut areas by more than 200 per cent. This excess would normally have been transpired, as water vapor from leaves, directly into the atmosphere.

We may calculate the amount of energy required to drive the hydrological cycle by multiplying together the energy required to evaporate 1 g of water (0.536 kcal) and the total annual evaporation of water from the earth's surface (378×10^{18} g). The product, about 2×10^{20} kcal, represents about one-fifth of the total energy income in light striking the surface of the earth. The remaining energy is absorbed and reradiated as heat. Of the total incident radiation in temperate forests, about 40 per cent is absorbed by plants, and two-thirds of the absorbed energy (one-fourth of the incident energy) is dissipated through evapotranspiration.

Evaporation, not precipitation, determines the flux of water through the hydrological cycle. Ignoring the relatively minor inputs of energy to create wind currents and to heat water vapor in the atmosphere, the absorption of light energy by liquid water is the major point at which an energy source is geared to the water cycle. Furthermore, the ability of the atmosphere to hold water vapor is limited. An increase in the rate of evaporation of water into the atmosphere eventually results in an equal increase in precipitation.

The water vapor in the air at any one time corresponds to an average of

2.5 centimeters (1 inch) of water spread evenly over the surface of the earth. An average of 65 cm (26 in) of rain falls each year, which is twenty-five times the average amount of water vapor in the atmosphere. The steady-state content of water vapor in the atmosphere, referred to as the atmospheric pool, is therefore recycled twenty-five times each year. Conversely, water has an average transit time of about two weeks. The water content of soils, rivers, lakes, and oceans is a hundred thousand times greater than that of the atmosphere. Rates of flux through both pools are the same, however, because evaporation equals precipitation. The average transit time of water in its liquid form at the earth's surface (about 3,650 years) is, therefore, 100,000 times longer than in the atmosphere.

The Oxygen Cycle

Next to nitrogen, oxygen is the most abundant element in the atmosphere, accounting for 21 per cent of its volume. But because oxygen is abundant and ubiquitous in the terrestrial environment, ecologists do not pay as much attention to the oxygen cycle as they do to the cycles of scarcer nutrients— carbon, nitrogen, phosphorus, and so on. The oxygen cycle is relatively simple, but it exhibits the basic characteristics of nutrient cycling in the ecosystem.

The atmosphere contains about 1.1×10^{21} g of oxygen. Much more is bound up in water molecules, mineral oxides, and salts in the earth's rocky crust, but this tremendous pool of oxygen is not directly available to the ecosystem. Terrestrial plants probably assimilate close to 10^{17} g of carbon in gross production. Because photosynthesis releases two atoms of oxygen for each atom of carbon fixed, and because oxygen weighs 16/12 as much as carbon, atom for atom, green plants release about 2.7×10^{17} g of oxygen each year. This amount corresponds to about 1/2,500 of the oxygen in the atmosphere.

The oxygen cycle is more complicated than the complementary equations for photosynthesis and respiration suggest. The molecule of oxygen (O_2) released by photosynthesis derives both its atoms from water. But, of the oxygen required by respiration, half is released to the ecosystem as water, and half is released as carbon dioxide. Therefore, the oxygen cycle involves pools of oxygen in the atmosphere and water bodies of the earth, and the complete cycle takes a very long time.

The Carbon Cycle

The biological cycling of carbon in the ecosystem is more direct than the cycling of oxygen. The carbon cycle involves only organic compounds and carbon dioxide (Figure 11-3). Photosynthesis and respiration fully complement each other. Photosynthesis assimilates carbon entirely into carbohydrate; respiration converts all the carbon in organic compounds to carbon

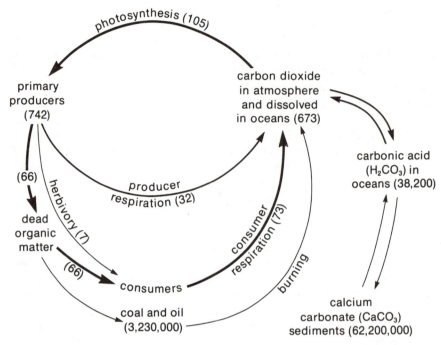

FIGURE 11-3 The global carbon cycle, with some estimated pools and annual transfer rates. Values are in million billion (10^{15}) grams and million billion grams per year.

dioxide. Large inorganic pools of carbon—atmospheric carbon dioxide, dissolved carbon dioxide, carbonic acid, and carbonate sediments—enter the carbon cycle to different degrees (Figure 11-3). The carbon in igneous rocks, calcium carbonate (limestone) sediments, coal, and oil is exchanged with other more active pools so slowly that these sources have little influence on the short-term functioning of the ecosystem.

Plants assimilate about 105×10^{15} g of carbon each year, of which about 32×10^{15} g are returned to the carbon dioxide pool by plant respiration. The remainder, 73×10^{15} g, supports the respiration and production of animals, bacteria, and fungi in herbivore- and detritus-based food chains. Anaerobic respiration (without oxygen) produces a small quantity of methane (CH_4) that is converted to carbon dioxide by a photochemical reaction in the atmosphere. Plants and animals annually cycle between 0.25 and 0.30 per cent of the carbon present in carbon dioxide and carbonic acid in the atmosphere and oceans, hence the total active inorganic pool is recycled every 300 to 400 years ($1 \div 0.003$ to $1 \div 0.0025$). Because the atmosphere and oceans exchange carbon dioxide slowly, they may be considered as separate pools over short periods. Terrestrial ecosystems annually cycle about 12 per cent of the carbon dioxide in the atmosphere; the transit time of atmospheric carbon is, therefore, about eight years ($1 \div 0.12$).

The combustion of coal and oil adds carbon dioxide to the atmosphere. Although man's present use of fossil fuels amounts to less than 2 per cent of the carbon cycled through the ecosystem each year, combustion of fuels adds carbon dioxide to the atmosphere over and above that utilized by photosynthesis. The carbon dioxide content of the atmosphere has in fact risen during this century, and it can be expected to rise even more rapidly in the future. Scientists cannot agree on the implications of increased atmospheric carbon dioxide for air temperature or for plant production, yet man is unquestionably shifting the steady-state balance of the ecosystem.

The Nitrogen Cycle

The path of nitrogen through the ecosystem differs from that of carbon in several important respects. First, the immense pool of nitrogen (N_2) in the atmosphere (3.85×10^{21} g) cannot be assimilated by most organisms. Second, nitrogen is not directly involved in the release of chemical energy by respiration; its role is linked to protein molecules and nucleic acids, which provide structure and regulate biological function. Third, the biological breakdown of nitrogenous organic compounds to inorganic forms requires many steps, some of which can be performed only by specialized bacteria (Figure 11-4). Fourth, most of the biochemical transformations involved in the decomposition of nitrogenous compounds occur in the soil, where the

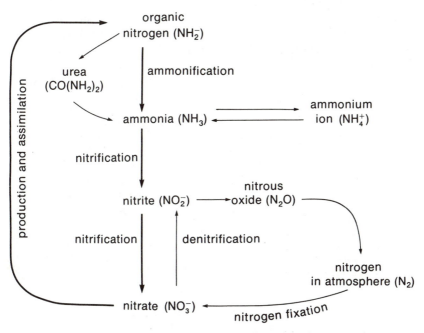

FIGURE 11-4 Basic biochemical steps in the nitrogen cycle.

solubility of inorganic nitrogen compounds influences the availability of nitrogen to plants.

Living tissues contain slightly more than 3 per cent of the nitrogen in active pools in the ecosystem. The rest is distributed between detritus and nitrates (NO_3^-) in the soils and ocean, with smaller amounts in the intermediate stages of protein decomposition—ammonia, NH_3, and nitrites, NO_2^- (Table 11-1). Plants assimilate 86×10^{14} g of nitrogen annually, less than 1 per cent of the active pool; the overall cycling time of nitrogen, therefore, exceeds one hundred years.

The nitrogen cycle involves the stepwise breakdown of organic nitrogen compounds by many kinds of organisms until nitrogen is finally converted to nitrate. Of the forms of nitrogen in the soil that are available to plants, ammonia (NH_3) or the ammonium ion (NH_4^+) would seem to be the most desirable because their conversion to organic compounds requires the least chemical work. Ammonia is, however, unsuitable as a source of nitrogen in the soil because it is toxic to plant tissues in high concentrations and is not persistent in the soil. Ammonia dissolves easily in water and is quickly leached out of soil. Under acid soil conditions, ammonia is converted to the ammonium ion. Although the positively charged ion can adhere to the surface of the clay-humus micelle, it is easily displaced by hydrogen ions in acid soils, and thus it is readily leached out by water. (Some types of clay minerals actually adsorb ammonium ions into their crystal framework so tightly that they are inaccessible both to leaching and to plants.) The soil ammonia that escapes leaching is readily attacked by specialized bacteria which obtain energy when they oxidize the nitrogen in ammonia to nitrites and nitrates. Negatively charged nitrite and nitrate ions do not bind to clay particles at all, hence they are susceptible to leaching. Once nitrates are produced in the soil they are quickly assimilated by plant roots. Storage of nitrogen in terrestrial ecosystems occurs primarily in organic detritus. Most

TABLE 11-1   Percentage distribution of nitrogen among active pools, and annual transfer rates. The pools taken together contain about 10^{18} grams of nitrogen.

Pool	Nitrogen (10^{14}g)	Transfer rate (per cent per year)
Organic forms		
Plants	342 ⎫	25
Animals	11 ⎭	
Detritus	6,100	1.4
Inorganic forms in soils and oceans		
Ammonia (NH_3)	286	30
Nitrites (NO_2^-)	138	63
Nitrates (NO_3^-)	4,180	2.1

of the nitrogen in aquatic ecosystems occurs as dissolved nitrates and in the detritus in sediments.

The biochemical reactions of nitrogen compounds are highly varied because nitrogen can combine with other elements in several different ways. The most important processes in the nitrogen cycle are the breakdown of organic nitrogen compounds by *ammonification* and *nitrification,* the reduction of nitrate and nitrite to nitrogen (N_2) and its release into the atmosphere by *denitrification,* and the biological assimilation of atmospheric nitrogen by *nitrogen fixation.*

Denitrification removes nitrogen from active pools in the soil and surface waters and releases it into the atmosphere; nitrogen fixation brings atmospheric nitrogen back into active circulation through the ecosystem. Although these processes are minor compared to the overall cycling of nitrogen in the ecosystem, nitrogen fixation can assume local importance where soils lack adequate nitrogen for normal plant growth, as we shall see below.

In its organic form, nitrogen occurs in amine groups (NH_2), or some variation, combined with other organic molecules. Animals rid their bodies of excess nitrogen by detaching the amines from organic compounds and excreting them relatively unchanged, primarily as ammonia, NH_3, or urea, $CO(NH_2)_2$. Soil microorganisms readily convert urea to ammonia by hydrolysis

$$CO(NH_2)_2 + H_2O \rightarrow 2NH_3 + CO_2$$

This reaction does not, however, release energy to perform biological work.

Some specialized but ubiquitous bacteria can release the chemical energy contained in the amine group by a series of nitrifying steps that require oxygen. *Nitrosomonas* transforms the ammonia ion to nitrite; *Nitrobacter* completes the nitrification process by oxidizing nitrite to nitrate.

Nitrification represents a critical step in the nitrogen cycle, ultimately determining the rate at which nitrates become available to green plants and thereby influencing the productivity of the habitat. Any soil condition that inhibits bacterial activity—high acidity, poor soil aeration, low temperature, and lack of moisture—also inhibits nitrification. Slow nutrient release in the soil of cold and dry regions may depress plant productivity beyond the direct effect of cold and dryness on photosynthesis. Furthermore, if the organic detritus in the soil has a low nitrogen content compared to its content of carbon, bacteria assimilate all the nitrogen into their cell structure rather than using some as a substrate for metabolism. Nitrogen thus becomes tied up in bacteria biomass rather than being made available to plants. The ratio of carbon to nitrogen in detritus influences the rate of bacterial decomposition. At one extreme, mulberry leaves (C/N ratio = 25) support an abundant microflora of bacteria and fungi and decompose rapidly; at the other extreme, loblolly pine (C/N ratio = 43) inhibits the activity of microorganisms and decomposes slowly (Table 11-2).

TABLE 11-2   Average weight loss, carbon/nitrogen (C/N) ratio, and microflora in four species of decaying leaves in forest litter at Oak Ridge, Tennessee; November 1960 to November 1961.

Species	Weight loss (%)	C/N ratio	Bacteria colonies (millions/g dry weight)	Fungi colonies (thousands/g dry weight)	Ratio of bacteria to fungi
Red mulberry	90	25	698	2,650	264
Redbud	70	26	286	1,870	148
White oak	55	34	32	1,880	17
Loblolly pine	40	43	15	360	42

Denitrification transforms nitrate to nitrogen in a series of steps

$$NO_3^- \rightarrow NO_2^- \rightarrow N_2O \rightarrow N_2$$

each of which releases oxygen. Nitrous oxide (N_2O) and nitrogen molecules (N_2) escape into the atmosphere and leave the active nitrogen pools. Denitrification also occurs purely by chemical means, independently of microorganisms. For example, the reaction of nitric acid (HNO_3) and urea to release nitrogen occurs in acid soils

$$2\ HNO_3 + CO(NH_2)_2 \rightarrow CO_2 + 3\ H_2O + 2\ N_2$$

Micoorganisms utilize nitrate or nitrite in denitrification for the same purpose that most animals and plants utilize oxygen for respiration. Nitrate is a more powerful oxidizer than even oxygen, and it is used to release the chemical energy of carbohydrates under anaerobic conditions. The overall reaction can take the form

$$C_6H_{12}O_6 + 6\ KNO_3 \rightarrow 6\ CO_2 + 3\ H_2O + 6\ KOH + 3\ N_2O$$

in which the carbon in glucose is oxidized to carbon dioxide and the nitrogen in nitrate is reduced to that in nitrous oxide. Like nitrogen, sulfur also has many oxidation states and, in anaerobic habitats, it plays an important role in bacterial photosynthesis, by the oxidation of hydrogen sulfide (H_2S) to sulfur (S), and in respiration by the reverse reaction.

Nitrogen fixation is energetically expensive and requires considerable chemical work, although it requires no more energy than does conversion of an equivalent amount of nitrate, NO_3^-, to ammonia, NH_3. For the price of the chemical energy in a glucose molecule ($C_6H_{12}O_6$), some blue-green algae and the bacterium *Azotobacter* can assimilate eight nitrogen atoms (ignoring the inefficiency of biochemical transformations). Nitrogen fixation requires specialized biochemical machinery that is apparently unavailable to higher plants. Nonetheless, many legumes such as alfalfa and peas, plus scattered species in other plant groups, have entered into symbiotic relationships with nitrogen-fixing bacteria. Even some marine algae, lichens, and shipworms have symbiotic, nitrogen-fixing bacteria or blue-green algae. Peas develop clusters of nodules throughout their root system that become infected by *Azotobacter* (Figure 11-5). The relationship benefits both par-

FIGURE 11-5 The root system of an Austrian winter pea plant, showing the clusters of nodules that harbor symbiotic nitrogen-fixing bacteria.

ties: the plant furnishes glucose to the bacteria, and the bacteria assimilate nitrogen from the soil atmosphere for plant uptake.

In nitrogen-deficient habitats, nitrogen fixation is a critical factor in plant production. The nitrogen-fixing capabilities of certain plants have become widely exploited in agriculture to restore soil fertility after farmland has been planted with soil-depleting crops like corn. Nitrogen-accumulating plants (usually peas or alfalfa) are planted in rotation with corn, and then plowed under the soil, increasing its nitrogen and humus content and water retention. It has been estimated that about 175×10^{12} g of atmospheric nitrogen are fixed annually, about half of it in agricultural land. The total amounts to about 2 per cent of the nitrogen assimilated by plants annually.

Assimilation of nitrogen from the atmosphere probably constitutes a more important agent promoting soil fertility than its annual rate indicates. Nitrogen-fixing microbes are widespread in natural habitats, even on the leaves of trees. If we stop to consider for a moment that parental rocks underlying most soils are completely devoid of nitrogen, we realize that most of the nitrogen in active pools in the ecosystem must have originated through nitrogen fixation.

Phosphorus Cycle

Oxygen, carbon, and nitrogen cycles demonstrate the basic features of mineral cycles in the ecosystem, but other elements—particularly phosphorus, potassium, calcium, sodium, sulfur, magnesium, and iron—play important roles in ecosystem function. Other elements—such as cobalt, aluminum, and manganese—may influence ecosystem dynamics in ways still undiscovered.

Ecologists have studied the role of phosphorus in the ecosystem most intensively because organisms require phosphorus at a high level (about one-tenth that of nitrogen) as a major constituent of nucleic acids, cell membranes, energy-transfer systems, bones, and teeth. Phosphorus is important for a number of other reasons. It is thought to limit plant productivity in many aquatic habitats, and the influx of phosphorus to rivers and lakes in the form of sewage (particularly from phosphate detergents) and runoff from fertilized agricultural lands stimulates the production of aquatic habitats to undesirably high levels. Also, ecologists can measure concentrations of phosphorus easily and use one of its isotopes as a radioactive tracer in the ecosystem.

The phosphorus cycle has fewer steps than the nitrogen cycle: plants assimilate phosphorus as phosphate ion ($PO_4^=$) directly from the soil or water; animals eliminate excess organic phosphorus in their diets by excreting phosphorus salts in urine; phosphatizing bacteria convert the organic phosphorus in detritus to phosphate in the same way. Phosphorus does not enter the atmosphere in any form other than dust. The phosphorus cycle therefore involves only the soil and water of the ecosystem.

In spite of the relative simplicity of the phosphorus cycle, many environmental factors influence the availability of phosphorus to plants. In the presence of abundant dissolved oxygen, phosphorus readily forms insoluble compounds that precipitate and remove phosphorus from the pool of available nutrients. If such conditions persist, deposits of phosphate accumulate and eventually form phosphate rock, which returns to active pools in the ecosystem very slowly by erosion—or by artificial fertilization of crops and disposal of phosphate detergents in sewage (fertilizers and detergents themselves deriving their phosphorus content from phosphate rock).

Acidity also affects the availability of phosphorus to plants. Phosphate compounds of sodium and calcium are relatively insoluble in water. Under alkaline conditions phosphate ions ($PO_4^{\equiv}$) readily combine with sodium or calcium ions to form insoluble compounds. Under acid conditions, phosphate is converted to highly soluble phosphoric acid. At intermediate levels of acidity, phosphate ions form compounds with intermediate solubility, as shown below:

	Increasing acidity $\longrightarrow$			
Ionic form	$PO_4^{\equiv}$ $\rightarrow$	$HPO_4^{=}$ $\rightarrow$	$H_2PO_4^{-}$ $\rightarrow$	H_3PO_4
	$\downarrow$	$\downarrow$	$\downarrow$	
Salt	Na_3PO_4	$CaHPO_4$	NaH_2PO_4	None
Solubility	(slightly soluble)	(insoluble)	(soluble)	(very soluble)

Although acidity increases the solubility of phosphate in the laboratory, high acidity reduces the availability of phosphorus in the ecosystem because of its reaction with other minerals. In acid environments, aluminum, iron, and manganese become soluble and reactive, forming chemical complexes that bind phosphorus and thereby remove it from the active pool of nutrients. The acid conditions in bogs and in soils of cold, wet regions remove phosphorus in this way and reduce the fertility of these habitats. Phosphorus is most readily available in a narrow range of acidity, just on the acid side of neutrality.

Phosphorus and Eutrophication

Ecologists classify natural bodies of water between two extremes on the basis of their nutrient content and organic productivity. On one hand, *oligotrophic* (from the Greek, meaning little-nourished) habitats have low nutrient content and harbor relatively little plant and animal life. The water in oligotrophic lakes and rivers is clear and unproductive. On the other hand, *eutrophic* (from the Greek, meaning well-nourished) lakes and rivers are rich in nutrients and support an abundant flora and fauna. Eutrophic habitats are excellent fisheries because of their high productivity. In fact, lakes and ponds are often artificially fertilized to increase fish production:

Harvests vary between one and seven pounds of fish per acre per year (0.2 to 1.6 kcal m^{-2} yr^{-1}) from the oligotrophic Great Lakes; small eutrophic lakes in the United States yield up to 160 pounds per acre per year (36 kcal m^{-2} yr^{-1}); in Germany and the Philippines, where ponds are artificially fertilized, yields of 1,000 pounds per acre per year (200 kcal m^{-2} yr^{-1}) are not uncommon. Primary production of the phytoplankton increases with eutrophy in a similar progression: 7 to 25 g carbon m^{-2} yr^{-1} in oligotrophic lakes, 75 to 250 g carbon m^{-2} yr^{-1} in naturally eutrophic lakes, and 350 to 700 g carbon m^{-2} yr^{-1} in lakes polluted by sewage and agricultural runoff.

Eutrophication does not by itself constitute a major problem for the aquatic ecosystem. High rates of production and rapid nutrient cycling are to be expected where the mineral resources for production are abundant. In naturally eutrophic lake and stream ecosystems, most of the energy and minerals come from within the system—*autochthonous* inputs—in the form of primary production. In culturally eutrophic lakes and streams, external sources of mineral nutrients and organic matter—*allochthonous* inputs—contribute to the productivity of the system. (In the course of attaching special names to almost everything, ecologists have frequently turned to Greek or Latin roots for terms. *Chthonos* is Greek for "of the earth," *auto* means "the same," and *allo*, "other" or "different," referring to internal and external inputs.)

Whereas naturally eutrophic systems are usually well balanced, the addition of artificial nutrients can upset the natural workings of the community and create devastating imbalances in the ecosystem. Algal blooms are among the most noticeable of these effects. The combination of high nutrient loads and favorable conditions of light, temperature, and carbon dioxide stimulate rapid algal growth. Algal blooms are a natural response of algae to their environment. But when the environment changes and no longer can support dense algal populations, the algae that accumulate during the bloom die and begin to decay. The ensuing rapid decomposition of organic detritus by bacteria robs the water of its oxygen, sometimes so thoroughly depleting the water of oxygen that fish and other aquatic animals suffocate. The nutrient imbalance in culturally eutrophic systems stems from the addition of nutrients at seasons when nutrients are less available in naturally eutrophic waters, primarily during the summer peak of plant production. During less productive seasons, phosphorus is readily absorbed by benthic bacteria and sediments at the bottoms of lakes, and its concentration in lake water is thus quickly reduced.

Early studies of eutrophication suggested that algal blooms occurred only when the concentration of phosphorus was greater than 0.01 milligrams per liter of water. These findings strongly implicated phosphorus as the prime factor limiting the productivity of many aquatic communities. Although ecologists have since argued this point, recent experiments on small Canadian lakes demonstrate that adding carbon (as sucrose) and nitrogen

FIGURE 11-6 Experimental lake demonstrating the crucial role of phosphorus in eutrophication. The near basin, fertilized with carbon (in sucrose) and nitrogen (in nitrates), exhibited no change in organic production. The far basin, separated from the first by a plastic curtain, received phosphate in addition to carbon and nitrogen and was covered by a heavy bloom of blue-green algae within two months.

(as nitrate) does not stimulate algal blooms without simultaneously adding phosphate (Figure 11-6).

In spite of the disturbing effects of outside enrichment, culturally eutrophied lakes can recover their original condition if inputs are shut off. Apparently the sediments at the bottom of most lakes have a high affinity for phosphates and quickly remove them from active pools in the surface waters of the lake. Diversion of sewage from Lake Washington, in Seattle, quickly reversed the eutrophication process. Similar recovery could be expected from other bodies of water if inputs from sewage and agricultural runoff were reduced.

In their normal development, lakes usually proceed from oligotrophic to eutrophic stages. New lake basins are deep and devoid of nutrients. As sediments wash into a lake, nutrients are added to the water. As the lake begins to fill in and become shallow, exchange between the bottom and surface water accelerates, returning the nutrients to the active pool near the surface and enriching the water. In extreme old age, lakes fill in completely and are succeeded by the local terrestrial vegetation.

Cation Exchange in Temperate Forests

Calcium, potassium, sodium, and magnesium are not chemically incorporated into organic compounds, although they are abundant as dissolved *cations* in cellular and extracellular fluids. The cycles of cations in the ecosystem are only loosely associated with the assimilation and release of energy, but they are tremendously important to cell function.

Cation cycles have been studied most intensively in temperate forests, within which the ions exhibit remarkable mobility. The forest is an open

FIGURE 11-7   Rain gauges installed in a ponderosa pine stand in California to intercept precipitation falling through the canopy of the forest and running down the trunks of trees. Analysis of the nutrient content of water collected in sampling programs like this helps determine the overall cation budget of the forest and the specific routes of mineral cycles.

FIGURE 11-8 A stream gauge at the lower end of a watershed at the Coweeta Hydrological Laboratory, North Carolina. The V-shaped notch regulates the flow of water through the weir in such a way that the flow rate is proportional to the water level in the basin.

system, one which freely exchanges minerals with other parts of the environment. Leached cations are washed out of the system through runoff, both in streams and groundwater. Minerals enter the system in precipitation, in wind-born dust and organic debris, and by weathering of the parent rock over which the forest lies.

Detailed cation budgets have been obtained for small watersheds by measuring the inputs in rainwater collected at various locations in the watershed area (Figure 11-7) and the outputs in water leaving the watershed by way of the stream that drains it (Figure 11-8). Care must be taken to select a watershed with impervious bedrock to eliminate the problem of groundwater movement into and out of the watershed.

Several watersheds in the Hubbard Brook Experimental Forest, New Hampshire, have been studied intensively during the past decade to establish patterns of water and nutrient cycles and the effects of disturbances, primarily clear-cutting, on these cycles. The annual distribution of precipitation and runoff for the Hubbard Brook Forest (Figure 11-9) shows a fairly uniform distribution of rainfall during the year, typical of moist temperate locations. Precipitation exceeds runoff during the cold winter months because of snow accumulation. This pattern is reversed in spring as melting snow swells the streams. The difference between precipitation and runoff during the summer months, seen in Figure 11-9, is accounted for by the evaporation and transpiration of water from the watershed.

Cation budgets have been calculated for the entire watershed from the concentrations of the minerals in precipitation and stream flow (Figure 11-10). Only potassium exhibited a net gain in the watershed. Net losses of other cations would have to be balanced by weathering of the bedrock for the system to be in a steady state. The cation budgets of watersheds around the world vary tremendously with total precipitation, soil acidity, and the relative abundance of minerals in rainfall and the soil. Where sodium is abundant in the soil or in rainwater, as in many coastal areas, sodium output in stream flow is great. Similar patterns apply to other cation budgets.

The uptake of cations by plants closely parallels their availability in the soil and hence their general mobility in the ecosystem (Table 11-3). Calcium is most rapidly cycled, magnesium least. In general, uptake by vegetation is one to ten times the annual loss of cations in stream flow; therefore the average transit time of cations in the ecosystem is probably one to ten years. Considering the great mobility of cations in the soil, the long transit times of cations indicate that plants rapidly assimilate free ions in the soil before they are leached out of the system by runoff and groundwater. The role of vegetation in nutrient cycling is dramatized by experiments in which entire

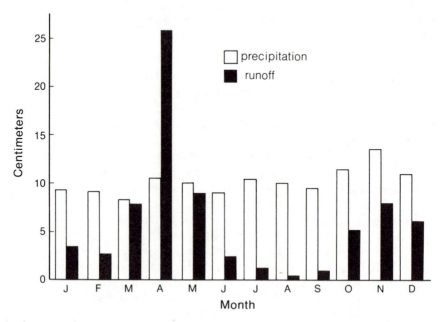

FIGURE 11-9 Average annual distribution of precipitation and runoff for the Hubbard Brook Experimental Forest, New Hampshire, 1955–1969.

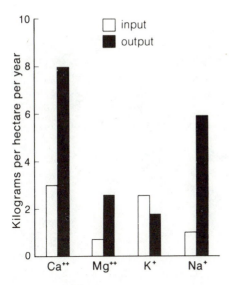

FIGURE 11-10 Cation budgets for the Hubbard Brook Experimental Forest, New Hampshire, during 1963 –1964, expressed as kg ha^{-1} yr^{-1}

watersheds are denuded of trees and shrubs (Figure 11-11). Clear-cutting of small watersheds as in the Hubbard Brook Forest increased stream flow several times as a result of the removal of transpiring leaf surface; losses of cations increased three to twenty times over comparable undisturbed systems. The nitrogen budget of the cut-over watershed sustained the most striking change. Plants assimilate available soil nitrogen so rapidly that the forest usually gains nitrogen at the rate of 1 to 3 kilograms per hectare per year (kg ha^{-1} yr^{-1}, a hectare = 2.47 acres). In the clear-cut watershed, net loss of nitrogen as nitrate (NO_3) soared to 54 kg ha^{-1} yr^{-1}, which is comparable to the annual turnover of nitrogen by vegetation. Precipitation brought only 7 kg ha^{-1} yr^{-1} of nitrogen into the system and, thus, the loss of nitrate represented nitrification of organic nitrogen sources at the normal annual rate by soil microorganisms without simultaneous rapid uptake by

TABLE 11-3 Summary of cation budgets for representative temperate forest ecosystems.

	RANGE OF VALUES (kg ha^{-1} yr^{-1})*			
	Precipitation input	Stream outflow	Net loss	Uptake by vegetation
Calcium	2–8	8–26	3–18	25–201
Potassium	1–8	2–13	–1–5	5–99
Magnesium	1–11	3–13	2–4	2–24
Sodium	1–58	6–62	4–21	—

* Kilograms per hectare per year. A hectare is equal to 10,000 square meters, or 2.47 acres.

FIGURE 11-11   Clear-cut watershed at the Coweeta Hydrological Laboratory, North Carolina, employed in studies of evapotranspiration and runoff in forest ecosystems.

plants. These experiments demonstrate the important role of vegetation in maintaining the fertility of the soil, and they further emphasize that the physical and biological components of the ecosystem cannot exist apart.

Nutrient Cycling in Tropical Forests

Because of the year-round warm temperatures of tropical climates, leaf litter decomposes rapidly throughout the year and does not accumulate in tropical forests to the extent that it does in temperate forests. Warm temperatures and abundant rainfall additionally result in rapid leaching of nutrients from the soil by ground water. The combination of rapid decomposition and rapid leaching is thought to cause many tropical soils to have lower nutrient contents than temperate soils. The rapid mobilization of minerals in tropical soils is balanced in part by rapid uptake by plants, whose roots actively assimilate water and nutrients throughout the year. As a result, minerals are cycled through the soil of tropical forests quickly but are retained by the vegetation.

Of the total biomass of forests, including dead branches and leaves, litter on the forest floor comprises an average of about 20 per cent in temperate coniferous forests, 5 per cent in tropical hardwood forests, and 1 to 2 per cent in tropical rain forests. Because measured litter fall in forests consists mostly of dead leaves, forests may be more meaningfully compared by considering the ratio of accumulated litter on the forest floor to the biomass of living leaves. In temperate forests this ratio is between 5:1 and 10:1, whereas in tropical forests it is less than 1:1. These data are consistent with the notion that dead organic material is decomposed rapidly in the tropics and does not form a substantial nutrient reservoir, as it does in temperate regions.

In northern coniferous forests more than 50 per cent of the organic carbon occurs in the soil and litter, whereas in tropical rain forests less than 25 per cent occurs in soil and litter. Most of the carbon in tropical forests is accumulated in the woody parts of trees. This pattern prevails only for carbon, however, and does not necessarily reflect the distribution of other elements. Nonliving woody materials in the roots, trunks, and branches of trees contain abundant carbon, but relatively little of other elements such as nitrogen and phosphorus. Furthermore, because carbon enters the terrestrial ecosystem from the air by way of photosynthesis, its distribution in the ecosystem must be considered apart from mineral nutrients that plants obtain from the soil.

The distribution of potassium, phosphorus, and nitrogen in two temperate forests and one tropical forest are compared in Table 11-4. Two points seem to be clear. First, the accumulation of nutrients in vegetation, on a weight for weight basis, is slightly greater in the tropical forest. For example, the total dry weight of living vegetation in the Belgian ash-oak forest exceeds that of the tropical deciduous forest in Ghana by 14 per cent, but the accumulation of the three elements per gram of dried vegetation is 32 to 38 per cent lower in the Belgian forest than in the Ghana forest. Second, levels of some nutrient elements in tropical forest soils can be as high as, or higher than, in temperate soils, although the level of phosphorus in the one tropical locality was two orders of magnitude less than that in the temperate localities.

The forest vegetation itself plays a major role in maintaining the fertility of tropical soils against the leaching effects of rainfall. In very sandy regions, where the weathering of the substrate produces poor soils with little nutrient content, the fertility of the soil is built up by nitrogen fixation and by plants trapping and retaining nutrients imported by rainfall. The importance of the forest vegetation in maintaining nutrient concentrations is strikingly

TABLE 11-4 The distribution of mineral nutrients in the soil and vegetation components of representative temperate and tropical forest ecosystems.

Forest (and locality)	Biomass (tons ha^{-1})	NUTRIENTS (kg ha^{-1})		
		Potassium	Phosphorus	Nitrogen
Ash and oak (Belgium)	380			
Living		624	95	1,260
Soil		767	2,200	14,000
Soil/living		1.2	23.1	11.1
Oak and beech (Belgium)	156			
Living		342	44	533
Soil		157	900	4,500
Soil/living		0.5	20.5	8.4
Tropical deciduous (Ghana)	333			
Living		808	124	1,794
Soil (30 cm)		649	13	4,587
Soil/living		0.8	0.1	2.0

demonstrated by the fact that when tropical forests are cleared and planted with annual crops, soil fertility declines rapidly unless fertilizers are applied artificially (Table 11-5).

We do not know to what extent soil nutrients limit the productivity of temperate or tropical forests, but forest thinning experiments and experiments in which fertilizers are added to forests suggest that nutrients limit primary production to some extent. Whether tropical soils are poorer than temperate soils is also open to question. The well-known difficulties encountered in applying temperate agricultural methods to tropical regions may be caused more by the agricultural practices themselves than by the

TABLE 11-5 Influence of mulching cut vegetation into the soil and of adding inorganic fertilizers on soil fertility in continuously planted cotton crops. Data are kilograms of harvested cotton per acre.

Year	CLEAN WEEDED		MULCHED	
	Without fertilizer	Fertilized since 1953*	Without fertilizer	Fertilized since 1953*
1947–48	1,032	—	1,127	—
1953–54	200	440	1,117	1,434
1955–56	186	797	1,464	1,977
1956–57	124	706	986	1,344

* Fertilizer applied: 150 kilograms per hectare (kg ha^{-1}) bicalcium phosphate, 250 kg ha^{-1} sodium nitrate, and 50 kg ha^{-1} potassium sulphate.

intrinsic fertility of the soil. After all, no other ecosystem can match the productivity of a mature tropical forest.

In this chapter, we have traced the cycles of several elements through the ecosystem. The pattern of movement depends partly on the chemical properties of the element and partly on how it is used by living organisms. The cycles of the various elements are similar only to the extent that the decomposition activities of organisms return them to forms that can be assimilated by plants. The rate of flow of a nutrient at each step in its cycle is determined by the ecology of the organisms involved in the step. If a substance is utilized slowly by organisms, it tends to accumulate and form a large reservoir behind the bottleneck in the cycle. Because minerals are cycled, a disruption at any one step in the cycle can be reinforced through successive cycles, leading to instability. For this reason, those who apply ecology to solve environmental problems probably should pay more attention to nutrient budgets than to energy budgets of ecosystems. The disturbance of normal ecosystem function leading to such problems as eutrophication and nutrient depletion can probably be traced to disruption of a critical step in one or more nutrient cycles.

In the preceding chapters, we examined some large-scale measurements of ecosystem function: primary production, energy flow, and nutrient cycling. These properties of structure and function are the sum of the activities of individuals and populations that make up the community. In the remainder of this book we shall consider structure and function in these smaller components of the ecosystem, beginning with organisms, then populations, and, finally, biological communities.

12 | Regulation and Homeostasis

CHANGE PERVADES an organism's surroundings—the annual cycle of the seasons, daily periods of light and dark, and frequent unpredictable turns of climate. The survival of each organism depends on its ability to cope with change in the environment. As we look around us, we are constantly aware of organisms, including ourselves, responding to change. Responses include both the use of the physiological apparatus and changes in the apparatus itself. When we step from a warm room into the outdoors on a cold day, we shiver to generate heat. A few weeks on the beach and our skin darkens to block damaging radiation from the sun. When we shiver, we make use of a muscle response that is always present and available on short notice. Morphological changes, like the production of pigment granules in the skin in response to sunlight, alter the physiological apparatus itself. Such changes require more time. Most responses involving the structure and function of an organism are reversible, as they must be, to follow the ups and downs of the environment. But plants and animals also exhibit more or less permanent and irreversible developmental responses to the particular environments in which they live.

All these responses have evolved to maintain the internal conditions of the organism at some optimum level for its proper functioning (*homeostasis*). But what determines the best internal condition? What is the proper rate of functioning for an organism? And what is the most effective mechanism of response to environmental change? Ultimately, "effectiveness" must be the net influence of a response on the fitness of the individual—the number of descendants it leaves. But the path of causation between shivering on a cold day and number of great-grandchildren is difficult to follow. Homeostatic responses must be treated, like a problem in economics, in terms of costs and benefits. In the case of shivering, the benefit is clearly measured in terms of survival during the next few minutes or hours by an organism that must maintain a high body temperature to function properly. The cost is measured in terms of energy released to produce body heat, which in turn may deplete the fat reserves of the animal and render its life precarious in the face of a sudden food shortage. One way to reduce the cost of temperature regulation is to lower the regulated temperature, just as we turn down the thermostat in our houses and office buildings to save fuel. But turning down the fire of an organism's life also reduces its rate of activity and hence its ability to gather food and escape predators. In the context of the many factors affecting costs and benefits of a particular response the optimum becomes a subtle concept. In this chapter, we shall explore some facets of homeostasis to understand the different ways in which animals and plants cope with the physical factors of their environments.

The Biological Optimum

The molecules responsible for biological function are extremely sensitive to changes in conditions of temperature, pH, and salt concentration. Under high temperatures, the structure of enzyme molecules is altered, and their functions inhibited; exposed to intense light, photosynthetic pigments in plants may break down; with variation salt concentration, the configuration of protein molecules may change. All of these effects have serious consequences for the functioning of the organism.

We do not need to look far to find examples of the influence of environment on the activity and well-being of the organism. In many cases, the effects of varied conditions not only are important to the animal or plant in question but also have economic consequences for man. It is not surprising, therefore, that many of our best-studied neighbors in the natural world are either part of our diet or pests on our crops and livestock. The oyster is such a neighbor.

Oysters pass their larval stages in the brackish waters of small bays and estuaries. The oyster larvae grow most rapidly when the concentration of salt in the water remains between 1.5 and 1.8 per cent, about halfway between fresh and salt water. High salinity depresses growth slightly. Lower salinity slows growth markedly and causes death: at 1 per cent salinity, 90 to 95 per cent of the larvae die within two weeks; at 0.25 per cent salinity, growth ceases and all larvae die within one week.

In the example of the oyster, our measure of environment was that of the surrounding water, not the conditions within the cells of the larval oyster. It is not immediately apparent whether salinity directly determines the salt concentrations within the oyster larva or secondarily influences the cost to the oyster of maintaining its cells at an optimum salt concentration. In either case, salinity weighs heavily on survival.

For some properties, the external environment provides a reasonable measure of internal conditions. Temperature is such a property for small cold-blooded organisms. Almost any measure of activity, such as the swim-

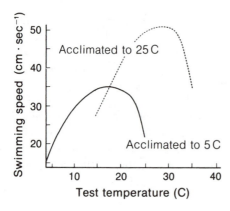

FIGURE 12-1   Swimming speed of the goldfish (*Carassius auratus*) as a function of temperature. Separate curves are graphed for individuals normally maintained at 5 C and 25 C.

ming speed of goldfish portrayed in Figure 12-1, will exhibit a marked temperature dependence, often with a peak rate within a narrow range of optimum temperature.

When the conditions in the surrounding environment differ from the optimum for cellular processes, organisms face a choice between impairment and reduction of cellular function, on one hand, and paying the metabolic price to maintain the proper internal conditions, on the other. The price of maintenance may be exacted either in terms of energetic cost, as in heat production to maintain body temperature, or in terms of constraints on organism function, as in the adaptations required to conserve body water in a dry environment.

Homeostasis and Negative Feedback

Homeostasis refers to the ability of the organism to maintain constant internal conditions in the face of a different and usually varying external environment. All organisms exhibit homeostasis to some degree, although the occurrence and effectiveness of homeostatic mechanisms varies.

Regulation of body temperature is one of the most restricted forms of homeostasis, being fully developed only in the so-called warm-blooded animals: the birds and mammals. Most mammals, including man, closely regulate the temperature of their bodies around 37 C, even though the temperature of the surrounding air may vary from −50 to +50 C. Such close regulation guarantees that biochemical processes within cells can occur under constant temperature (homeothermic) conditions, the cost of homeothermy to the organism notwithstanding. Over the same temperature range within which mammals and birds maintain constant body temperatures, the internal environment of cold-blooded (poikilothermic) organisms, such as frogs and grasshoppers, would conform to external temperature. Of course, frogs cannot possibly function at either high or low temperature extremes, so they are active within a narrow part of the range of the environmental conditions over which mammals and birds are active.

How is body temperature regulated? Most homeotherms have a sensitive thermostat in their brain. This thermostat responds to changes in the temperature of the blood by secreting hormones into the bloodstream to slow down or accelerate the generation of heat in body tissues. In addition, most homeotherms partly regulate body temperature by altering gains and losses of heat from the environment. For example, humans put on heavy clothes in cold weather and avoid standing in the sun when it is hot; birds fluff up their feathers to provide greater insulation against the cold.

Because the so-called cold-blooded organisms cannot generate heat without increasing their rate of movement, many adjust their heat balance behaviorally by altering their exposure to the sun. This may involve such simple behavior as moving into or out of the shade, or orienting the body with respect to the sun. Horned lizards can increase the profile of their bodies exposed to the sun by lying flat against the ground or decrease their exposure by standing erect upon their legs. By lying flat against the ground, horned lizards are also able to gain heat from the sun-warmed surface. Such behavior, widespread among reptiles, effectively regulates body temperature within a narrow range elevated considerably above the temperature of the surrounding air. But this strategy relies on the availability of direct sunlight and, of course, cannot be practiced at night or by inhabitants of continually shaded habitats.

Regardless of the particular mechanism of regulation, all homeostases exhibit properties of a negative feedback system. If we walk from a dark room into bright sunlight, the pupils of our eyes rapidly contract, which restricts the amount of light entering the eye. A sudden exposure to heat brings on sweating, which increases evaporative heat loss from the skin and helps to maintain body temperature at its normal level. Behavior, too, can be homeostatic because it serves either to modify the environment or to change the individual's relationship to its environment. Putting on an overcoat and avoiding the hot sun serve a regulatory purpose, the maintenance of body temperature.

Homeostatic responses act to maintain the internal constancy of the organism. To achieve this goal, responses are controlled in the manner of a thermostat. If a room becomes too hot, a temperature-sensitive switch turns off the heater; if the temperature drops too low, the temperature-sensitive switch turns on the heater. This pattern of response is called *negative feedback*, meaning that if external influences alter a system from its norm, or desired state, internal response mechanisms act to restore that state. A man driving a car down a straight road embodies a negative feedback system. If a sudden gust of wind forces the car to veer to the right, the driver immediately responds by turning the steering wheel to the left, and the car returns to its normal path.

The Cost of Homeostasis

To maintain internal conditions significantly different from the external environment requires work and the expenditure of energy. This fact is entirely compatible with what we know about the laws of diffusion and heat transfer: substances and energy flow more rapidly across large gradients than they do across small ones. The greater the difference between an organism and its environment, the greater the energy cost of homeostasis.

The metabolic cost of homeostasis may be demonstrated by examining temperature regulation by homeotherms in cold environments. Birds and mammals maintain constant high body temperatures, generally between 35 and 45 C, depending on the species. At progressively lower air temperatures, the gradient between the internal and external environments increases, and the rate at which heat is lost across the body surface increases proportionately. An animal that maintains its body temperature at 40 C loses heat twice as fast at an ambient (that is, surrounding) temperature of 20 C (a gradient of 20 C) as at an ambient temperature of 30 C (a gradient of 10 C). To maintain a constant body temperature, an organism must replace heat that is lost by releasing heat energy metabolically. Thus, the rate of metabolism required to maintain body temperature increases in direct proportion to the difference between body and ambient temperature (Figure 12-2).

If the only function of metabolism were temperature regulation, metabolic rate would be zero when body temperature equaled ambient temperature and there would be no net movement of heat between the organism and its environment. But organisms release energy to maintain functions not related to temperature regulation—such as heartbeat, breathing, muscle tone, and kidney function—regardless of the ambient temperature. The metabolic rate of an organism that is resting quietly, not having recently eaten, is called *basal metabolism* and is the lowest level of energy release under normal conditions. Even at this basal level, the rate of energy release is sufficient to maintain body temperature when ambient temperatures exceed a certain level, the *lower critical temperature* (T_C). At lower temperatures, metabolism must increase to maintain a constant body temperature.

An organism's ability to maintain a high body temperature while exposed to very low ambient temperatures is limited over the short term by its physiological capacity to generate heat, and over the long term by its ability to gather food to supply the energy for metabolism. The maximum rate at which an organism can perform work is generally no more than ten to fifteen times its basal metabolism. When the environment becomes so cold that heat loss exceeds the organism's ability to generate heat (point *c* of Figure 12-2), body temperature begins to drop, a condition that is fatal to most homeotherms. The lowest temperatures that homeotherms can survive for long periods are usually determined by their ability to gather food (point *b*)

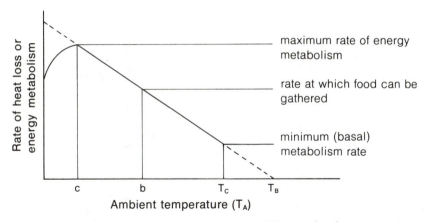

FIGURE 12-2 Relationship between energy metabolism and ambient temperature for a homeothermic bird or mammal whose body temperature is maintained at T_B. T_C is the lower critical temperature, below which metabolism must increase to maintain body temperature. Points c and b are the lower lethal temperature and the lowest temperature at which the organism can maintain itself indefinitely.

rather than their ability to metabolize the energy in food. Animals can quite literally starve to death at low temperatures because they release energy for heat production more rapidly than they can find food to supply their metabolic requirement. At warmer temperatures, surplus energy is available for such functions as reproduction. Points c and b in Figure 12-2 might appropriately be called the critical physiological temperature and the critical ecological temperature. Below c, organisms die. Between c and b, organisms survive but only for relatively brief periods and with a negative energy balance. Above b, energy balance is positive and not only can the individual survive indefinitely, but also can engage in other energy-consuming activities besides foraging. This picture is, of course, simplified. It would be greatly complicated if we properly accounted for the energy expended to gather food, or if we acknowledged that the availability of food changes as ambient temperature and other climate conditions change.

When homeostasis costs more than the organism can afford to spend, certain economy measures are available. For example, the temperature of portions of the body may be lowered, thereby reducing the difference between air and body. Because the legs and feet of birds are usually not insulated by feathers, they would be a major avenue of heat loss in cold regions if they were not maintained at a lower temperature than the rest of the body (Figure 12-3). This is accomplished in gulls by a countercurrent heat-exchange arrangement in which the arteries carrying warm blood to the feet are cooled by passage close to the veins that carry cold blood back from the feet. In this way, heat is transferred to venous blood and transported back into the body rather than lost to the environment.

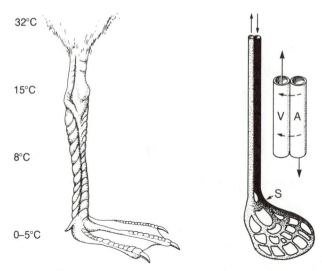

FIGURE 12-3 Skin temperatures of the leg and foot of a gull standing on ice. Countercurrent heat exchange between arterial blood (A) and venous blood (V) is diagrammed at right. Arrows indicate direction of blood flow and heat transfer (dashed arrows). A shunt at point S allows the gull to constrict the blood vessels in its feet, thereby reducing blood flow and heat loss further, without having to increase its blood pressure.

Because of their small size, hummingbirds have a large surface area relative to their weight and consequently lose heat rapidly compared with their ability to produce heat. As a result, hummingbirds need very high metabolic rates to maintain their at-rest body temperature near 40 C. Species that inhabit cool climates would risk starving to death overnight if they did not become *torpid,* that is, lower their body temperatures and enter into an inactive state resembling hibernation. The West Indian hummingbird, *Eulampis jugularis,* lowers its temperature to 18 to 20 C when resting at night. It does not cease to regulate body temperature, but merely changes the setting on its thermostat to reduce the difference between ambient and body temperature.

Regulators and Conformers

Animals that maintain constant internal environments are usually referred to as *regulators*; those which allow their internal environments to follow external changes are called *conformers* (Figure 12-4). Few organisms are ideal conformers or regulators. Frogs regulate the salt concentration of their blood but conform to external temperature. Even warm-blooded animals are partial temperature conformers: In cold weather, our hands, feet, nose, and ears (in other words, our exposed extremities) become noticeably cool.

Organisms sometimes regulate their internal environments over mod-

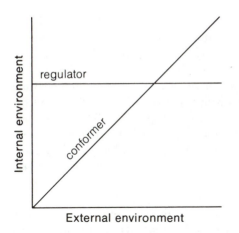

FIGURE 12-4 Relationship between internal and external environments in idealized regulating and conforming organisms. Regulators maintain constant internal environments with homeostatic mechanisms, whereas conformers allow their internal environments to follow changes in the external environment. The difference between the curves represents the gradient that regulators maintain between internal and external environments.

erate ranges of external conditions, but conform under extremes. Small aquatic amphipods of the genus *Gammarus* regulate the salt concentrations of their body fluids when they are placed in water with less concentrated salt than their blood, but not when they are placed in water with more concentrated salt (Figure 12-5). The fresh-water species *G. fasciatus* regulates the salt concentration of its blood at a lower level than the salt-water species *G. oceanicus*, and thus begins to conform to concentrated salt solutions at a lower level. In their natural habitat, however, neither the fresh-water species nor the salt-water species encounters salt more concentrated than that in their blood. Among animals that inhabit salt lakes and brine pools, however, the salt concentrations in the blood are actively kept below that of the surrounding water. The brine shrimp *Artemia*, for example, maintains the salt concentration of its blood below 3 per cent even when placed in a 30 per cent salt solution.

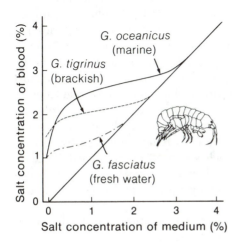

FIGURE 12-5 Salt concentration in the blood of three gammarid crustaceans from different habitats as a function of the salt concentration of their external environment. The normal salt concentration of sea water is 3.5 per cent. The inset shows *G. fasciatus*.

The extent of temperature regulation among homeotherms exhibits a level of variability similar to that of the effectiveness of osmoregulating mechanisms. Most birds maintain body temperatures between 40 and 43 C, and mammals keep theirs between 36 and 39 C depending on the species, although some forms are poorer regulators. The body temperature of the opossum, a marsupial, varies widely between 29 and 39 C, and the temperatures of sloths and armadillos may drop below 30 C. Although temperature regulation by these primitive mammals is irregular (hetero-thermic), temperature may be precisely regulated at a high level by preg-nant females to provide the embryo the constant environment it requires for proper and rapid development.

Metabolic generation of heat for temperature regulation is not restricted to birds and mammals. Pythons maintain high body temperatures while incubating eggs. Some large fish, such as the tuna, maintain temperatures up to 40 C in the center of their muscle masses while they are swimming. Countercurrent heat exchange between vessels carrying blood from the muscles to the skin and back keeps the outer layers of the tuna's body from warming up and thereby reduces heat loss. A preflight warm-up period is frequently encountered in large moths. Even among plants, temperature regulation has been discovered in the floral structures of philodendron and skunk cabbage.

Clearly, organisms other than birds and mammals are physiologically capable of generating heat to maintain elevated body temperatures, and many do so under certain conditions. Why, then, is the distribution of homeothermy throughout the animal and plant kingdom so limited? Part of the answer certainly lies in a consideration of body size. Birds and mammals are relatively large compared to representatives of other groups. As body size increases, volume increases relatively more rapidly than surface area, across which heat is lost. In general, the lower the ratio of surface to volume, the more comprehensive and precise regulation can be made.

Although body size may explain why mammals are homeotherms and insects generally are not—large moths that exhibit preflight warm-up ap-proach the size of small mammals—large fish and reptiles also have not made the adaptive shift to homeothermy, at least not in a big way. For most fish, the high rate of heat loss in the aquatic environment, owing to the thermal conductance of water, and the low availability of oxygen must preclude the high metabolic rates necessary for temperature regulation in all but the most active species. The metabolic rate of resting birds and mammals may be ten times that of fish, amphibians, and reptiles of similar size. Marine mammals can maintain high body temperatures partly because they breathe air, which provides a rich source of oxygen, partly because they have either fur or blubber for insulation, and partly because they can keep their body surface cold by countercurrent heat exchange and thereby reduce heat loss.

Homeothermy implies a constant body temperature; but, furthermore, most homeotherms maintain their body temperatures considerably above that of the surrounding air. Logic tells us that if the energetic cost of temperature regulation is to be minimized, the regulated temperature should be close to the average ambient temperature. Why are the body temperatures of birds and mammals so high and so completely independent of the environment? The advantages of high body temperature are poorly understood, but could be several. First, increased temperature raises the level of sustained activity that is possible and may increase alertness, both contributing to capture of food and avoidance of predators. Second, elevated body temperature reduces the need for frequent use of evaporative cooling to dissipate body heat. For terrestrial organisms, potential savings in water may justify the energetic cost of elevated temperature. Third, the large gradient in temperature generally maintained between body and environ- ment may allow more rapid control over rate of heat loss in response to changes in rate of activity than is possible when a small gradient is main- tained. This in turn would allow body temperature to be regulated more closely. These considerations illuminate only one facet of the optimization of physiological functions, a general problem in the relationship between organism and environment that has barely been touched.

What is clear, however, is that organisms have developed a variety of physiological mechanisms, morphological devices, and behavioral ploys to lessen the tension in their relationship to the physical environment. This versatility may be seen particularly well in the adaptations of desert birds and mammals to their stressful environment.

Temperature Regulation and Water Balance in Hot Deserts

Heat stress is one of the most critical factors to an organism's survival. Warm-blooded animals maintain their body temperatures only a few degrees below the upper lethal maximum. In cool environments, animals can quickly dissipate excess heat, generated by activity or absorbed from the sun, by conduction and radiation to their surroundings. But when air temperature exceeds body temperature, evaporative water loss must become the primary route of heat dissipation, and desert animals cannot afford to use scarce water to regulate their temperatures. Inactivity, use of cool microclimates, and seasonal migrations provide escape from heat stress for desert animals, but they also limit the animal's ability to exploit the desert environment (Figure 12-6).

Temperature regulation and water balance are closely linked. Where fresh water is scarce, organisms have a wide variety of behavioral, morpho- logical, and physiological adaptations for conserving water and using it efficiently to dissipate heat. The daily activity patterns of desert animals are closely tied to problems of temperature regulation. Many small mammals, such as kangaroo rats, which eat only dry seeds, appear aboveground only at night when the desert is cool. Because they avoid hot temperatures,

FIGURE 12-6 A jackrabbit seeking refuge from the hot sun of southern Arizona in the shade of a mesquite tree. The large ears and long legs of desert jackrabbits effectively radiate heat from the body when the temperature of the surroundings is lower than body temperature.

kangaroo rats can survive in the desert without having to drink. Their only source of water is the small amount in the seeds they consume and that produced during the metabolism of foodstuffs. (Remember that both carbon dioxide and water are by-products of oxidative respiration.) Ground squirrels, in sharp contrast, remain active during the day, but they conserve water by allowing their body temperatures to rise when they are aboveground. Before their body temperatures become dangerously high, ground squirrels return to their cool burrows where they dissipate their heat load by convection rather than evaporative heat loss. By alternately appearing aboveground and retreating to their burrows, ground squirrels extend their activity into the heat of the day and pay a relatively small price in water loss.

Among vertebrates, birds are perhaps the most successful inhabitants of the desert. They remain active in the heat long after other animals have sought refuge. The success of birds derives from their low excretory water loss (nitrogen is excreted as crystallized uric acid rather than urea) and from feeding on insects, from which they obtain some free water. Even some seed-eating birds can persist without water in the desert, provided they avoid both full sun and shade temperatures above 35 C.

The behavior of the cactus wren, a desert insectivore (Figure 12-7), shows that it, too, must respect the physiological demands of the hot desert climate. In cool air, wrens lose two to three milliliters (ml) of water each day in the air they exhale. Water loss increases rapidly above 30 to 35 C, to over 20 ml per day at 45 C; active birds might use five times that much water to dissipate their heat load. (The wren's body contains about 25 ml of water.) In the cool temperatures of the early morning, wrens forage throughout most of the environment, actively searching for food among foliage and on the ground. As the day brings warmer temperatures, wrens select cooler parts of their habitat, particularly the shade of small trees and large shrubs, always managing to avoid feeding where the temperature of the microhabitat exceeds 35 C. When the minimum temperature in the environment rises above 35 C, the wrens become less active. They even feed their young less frequently during hot periods.

The cactus wren apparently does not obtain enough water in its diet to allow it to dissipate excess heat through evaporation of water; one almost never observes cactus wrens panting. The behavior of several seed-eating birds, such as the English sparrow and the house finch, contrasts sharply with that of the cactus wren. These species occur only where there is water, and because they can drink freely they can remain active even during the

FIGURE 12-7 The cactus wren, a conspicuous resident of deserts in the southwestern United States and northern Mexico.

hottest parts of the day, when temperatures may reach 60 C on the ground. They pant at a furious rate under such conditions.

Many desert birds build enclosed nests or place their nests in holes in the stems of large cacti, where the young are protected from the sun and from extremes of temperature. The cactus wren builds an untidy nest, resembling a bulky ball of grass, with a side entrance. Once a pair of wrens have built their nest, they cannot change its position or orientation. For a month and a half, from the beginning of egg laying until the young fledge, the nest must provide a suitable environment day and night, in hot and cool weather. Cactus wrens usually nest several times during the period of March through September. Early nests are oriented with their entrance facing away from the direction of the cold winds of early spring; during the hot summer months, nests are oriented to face prevailing afternoon breezes, which circulate air through the nest and facilitate heat loss (Figure 12-8). Nest orientation is an important component of nesting success, particularly during the hot summer months. Nests oriented properly for the season are consistently more successful, 82 per cent, than nests facing the wrong direction, 45 per cent.

Cactus wrens are so conservative about their use of water that they do not even let the water in the feces of their young go to waste. The fecal sacs, which adults remove from the nest during cool parts of the year, are left in the nest during the hot weather. The evaporation of water from the fecal sacs presumably helps to cool the air in the nest, and the increased humidity reduces respiratory water loss. The re-use of fecal and excretory water is fairly common in desert organisms. In the desert iguana, secretions from salt glands, located in the orbits of the eyes, run down to small pits at the entrance of the nasal passages, where the water evaporates into the inhaled air and reduces respiratory water loss. Some storks have been observed to defecate on their legs during periods of hot weather, thus benefiting from evaporative cooling of their fecal water. Adult roadrunners eat the fecal sacs of their young; the kidneys of adults apparently have greater urine-concentrating abilities than the kidneys of the young, so the adults can extract water from the ingested fecal sacs. These examples merely emphasize the diversity of adaptations—morphological, physiological, and behavioral—that are focused on the solution of a single problem: heat dissipation and water balance in desert organisms.

The secret of life is the distinctiveness of living systems compared to the physical environment and their ability to utilize energy to maintain this difference in an open dynamic state. The energetic cost of maintaining organisms depends in part on how different they are from the surrounding physical environment and how their behavior differs from the passive existence of inanimate objects. The success of adaptations depends on minimizing this cost relative to the rate of energy intake. We have seen the many ways in which organisms can adjust their adaptations to the physical environment. Foremost among these are reducing the surface area of the

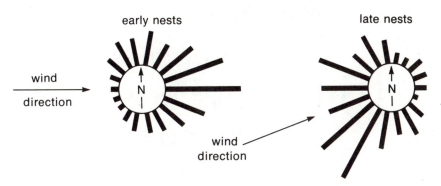

FIGURE 12-8 Orientation of nest entrances of the verdin (*Auriparus flaviceps*) during the early (cool) and late (hot) part of the breeding season, based on 182 and 206 nests, respectively, from near Las Vegas, Nevada and Tucson, Arizona.

body across which exchange with the environment occurs, seeking out those portions of the habitat in which the costs of regulation are minimum, utilizing physical mechanisms of regulation, such as absorption of solar radiation and, conversely, convective cooling, where possible, and employing conservation and recycling, particularly of water.

Homeostasis is made all the more difficult and costly by the inconstancy of the physical world both in time and space. Homeostatic mechanisms that suffice during the summer may not be adequate during winter. Rapid changes in the environment are countered by rapidly adjusting the rate of physiological processes, but slower changes allow time for altering both the homeostatic mechanism itself and the interface between the organism and its environment. In the next chapter, we shall examine the mechanisms by which animals and plants adjust to spatial and temporal variation in their environments.

13 | Organisms in Heterogeneous Environments

VARIATION IN THE ENVIRONMENT is a fact of life for all plants and animals, except perhaps for inhabitants of the abyssal depths of the oceans. Inasmuch as the adaptations of an organism are best suited for particular conditions, variation poses a critical problem to most forms of life. Those changes that occur rapidly—from within a few seconds to within a few hours or days at most—are accommodated by the homeostatic responses described in the last chapter. Such responses primarily involve change in the rate of physiological processes, or they may be behavioral reactions. They hold in common the property that they are rapidly reversible; responses to short-term variation must be as labile as the environment. But other changes in an organism's surroundings happen more slowly and are more persistent. Seasonal changes, in particular, permit a long response time for the organism, which is thereby freed to undergo substantial morphological and physiological change.

The type of response adopted by the organism is dictated by the rate and amount of environmental change. In this chapter, we shall explore some ways plants and animals adjust to long-term, persistent changes in the environment. Furthermore, for organisms that are immobile—plants and many sessile and slow-moving animals—spatial variation has many of the implications of long-term temporal variation.

The Environment in Time and Space

Daily, seasonal, and tidal cycles produce highly predictable changes in the environment. Beyond these regular cycles are the unpredictable changes that cause climate to vary from seasonal norms and the changes that result from biological cycles, such as fluctuations in the abundance of prey populations and developmental phases of hosts or parasites. These changes frequently happen on short notice or without warning, and the organism must therefore include their expectation in its bag of response tricks.

Spatial variation in the environment poses different problems for animals and plants. Where animals are faced with different types of habitats or places within habitats, they can choose among them. And, of course, animals must express habitat preferences, for one cannot be everywhere at once. In contrast to animals, plants are, with few exceptions, stuck where they land as seeds. They must make do with the particular conditions of their surroundings, or they die.

British plant ecologist John Harper and his co-workers have shown that for proper germination, seeds require quite specific combinations of light, temperature, and moisture, which vary even among closely related species. Irregularities in the surfaces of natural soils provide the variety of conditions needed to allow the germination of many species, but Harper dramatized the differences between species by creating an artificially heterogeneous soil environment. Three species of plantains, common lawn and roadside weeds in the genus *Plantago*, sown in seed beds responded differently to the variations in environment produced by slight depressions, by squares of glass placed on the soil surface, and by vertical walls of glass or wood (Figure 13-1). In fact, relatively few seeds of any species germinated on the smooth areas of soil surface that had not been disturbed experimentally.

Temporal fluctuations sometimes interact within the three-dimensional structure of the habitat to produce more complex patterns of variation. The edge of the sea is alternately covered by water and exposed to the air by a twice-daily cycle of tides (Figure 13-2). Over the region between the highest and lowest tides, called the *intertidal zone,* the proportion of time during which the surface is covered by seawater changes continuously from 0 to 100 per cent. Above the intertidal zone, conditions are essentially terrestrial; below, they are marine. Within the zone, animals and plants are exposed to the influences of both realms to a varying degree. Organisms that live high in the intertidal zone must tolerate occasional submergence, but they must also be able to withstand the desiccating influence of air and the wide variations in temperature so typical of terrestrial environments. Near the high-tide mark, one finds only barnacles and periwinkles, which have door-like coverings to the openings of their shells and can close themselves off completely from the air when the tide is out. The highest seaweeds, *Ulva* and *Fucus,* begin to appear a foot or so lower. Along the west coast of the United States, mussel beds with associated species of gooseneck barnacles, limpets, chitons, snails, starfish, and worms occupy a narrow band in the middle intertidal region. The seaweed *Laminaria* forms a dark brown mantle over even richer and more diverse communities at the lower edge of the intertidal zone, where forms less tolerant of terrestrial conditions, such as sea anemones, sea urchins, nudibranchs, tunicates, sponges, and myriad crabs and amphipods abound.

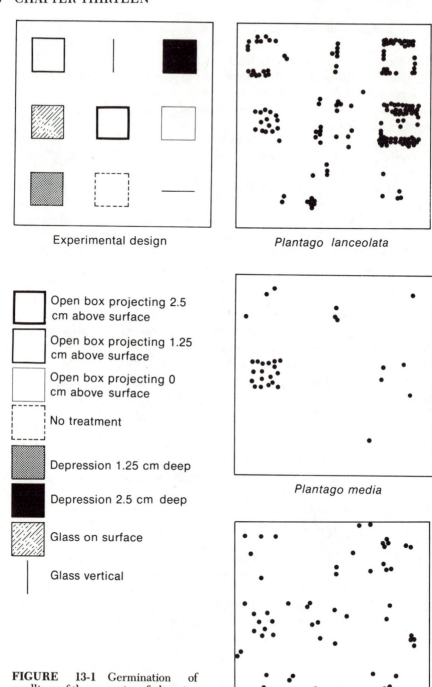

Experimental design

Plantago lanceolata

Open box projecting 2.5 cm above surface

Open box projecting 1.25 cm above surface

Open box projecting 0 cm above surface

No treatment

Depression 1.25 cm deep

Depression 2.5 cm deep

Glass on surface

Glass vertical

Plantago media

FIGURE 13-1 Germination of seedlings of three species of plantains (genus *Plantago*) with respect to artificially produced variation in the soil surface.

Plantago major

FIGURE 13-2   A portion of the coast of Kent Island, New Brunswick, at the mouth of the Bay of Fundy during high tide (left) and during low tide (right). The daily tidal range in this area is about 25 feet.

The Graininess of the Environment

Fluctuations that are of long duration require a different response from those that are of short duration. The choices presented to organisms by spatial variation depend on the distance between patches of different habitat type. But what is long and brief, or near and far, depends on the life span and response time of the organism and on its mobility. The upper and lower surfaces of a leaf do not matter much to us as we push aside the branch of a tree, but they are different worlds to the aphid that sits on the undersurface of a leaf and sucks plant juices. A tropical storm or passing cold front means little more to us than perhaps the spoiling of plans for a picnic, but such events may completely encompass the adult life of a mayfly that is ready to mate and lay eggs.

These different perspectives on variation in the environment are summarized by the concept of environmental *grain*. Imagine for a moment that variation consists of patches of uniform conditions that are distributed like a mosaic in time and space, resembling a patchwork quilt or the pattern made by alternating fields of different crops. The patches have different sizes, and patterns of different kinds of patches can be superimposed. For example, patches of air temperature tend to be much larger than patches of soil moisture because topographic influences subdivide the soil moisture landscape more finely. The concept of grain relates the size of the patch to the activity space of the organism. We define a *coarse-grained* environment as one in which the patches are relatively so large that the individual can choose among them. In a *fine-grained* environment, the patches are so small that the individual cannot usefully distinguish among them, and the environment appears essentially uniform. To anyone other than a trained botanist, a field appears to be a uniform carpet of plants—a patch that we distinguish from patches of forest, marsh, beach, and so on. But to the caterpillar, a single plant within the field may be its home for the duration of larval life. Even grasshoppers, which can fly from plant to plant, choose carefully among the various species in the field, for some provide better feeding than others. Therefore, the patches that are individual plants in a field are fine-grained to us, but coarse grained to small insects. Needless to say, grain also depends on the kind of activity in which one is involved. If we set out in the field to pick flowers of a particular kind, we would perceive individual plants as coarse-grained patches.

Patches may be thought of as occurring in time as well as in space. At one extreme, conditions that fluctuate through a daily cycle or over a shorter period may be thought of as fine-grained to most organisms, inasmuch as animals and plants do not have time to undergo major adjustments in response. At the other extreme, seasonal changes and longer cycles are decidedly coarse-grained for most organisms. Given persistent changes of several months' duration, plants and animals can complete morphological changes, such as dropping their leaves and increasing the thickness of their fur, and corresponding changes in physiological mechanisms, as in the case of animals that undertake migrations to areas with more favorable conditions.

The Activity Space of an Organism

Animals actively move among patches of conditions in a coarse-grained environment as the environment changes temporally. Conditions within each spatial patch change over diurnal and seasonal cycles, and by moving among them, animals can remain within ranges of conditions close to their optimum. Although plants cannot pick themselves up and move off to another patch, most regulate their activity levels according to the suitability of conditions at a particular time. Simply by closing the stomata on their leaves, plants are able to shut themselves off from some unfavorable conditions.

As the sun changes its position each day and throughout the year, the thermal environment of organisms in exposed habitats changes drastically. The surface of the desert, cool in the early morning, becomes a furnace at noon, and drives animals to shaded sites or to their burrows (Figure 13-3). As the conditions within different patches change, the activity space also changes.

The diurnal behavior cycle of lizards is geared to the varying temperatures of habitat patches. Although lizards do not generate heat metabolically for temperature regulation, they do take advantage of solar radiation and warm surfaces to maintain their body temperatures within the optimum range. At night, these sources of heat are not available, and the lizard's body temperature gradually drops to that of the surrounding air. The mallee dragon (*Amphibolurus fordi*), an agamid lizard of Australia, is most active when its body temperature is between 33 and 39 C. In the early morning, before its body temperature has risen above 25 C and when its movement

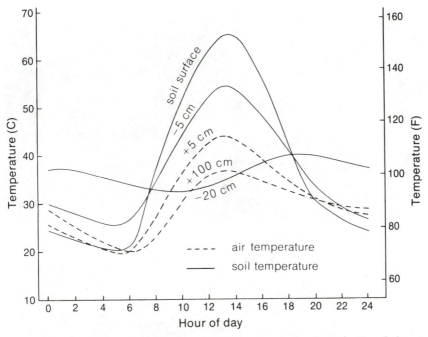

FIGURE 13-3 Daily fluctuation in the temperature of a desert soil and the air above it. Measurements indicate the depth below the surface (solid lines) and height above it (dashed lines).

is still sluggish, the mallee dragon basks within large protective clumps of the grass *Triodia* (Figure 13-4). In fact, *Amphibolurus* is so dependent upon these grass clumps that it occurs only where *Triodia* is established. When the temperature of an individual rises above 25 C, it moves out of the *Triodia* clump and basks in the sunshine nearby, with its head and body in direct contact with the ground surface from which additional heat is absorbed. When body temperature enters the normal activity range (33–39 C), *Amphibolurus* ventures farther from *Triodia* clumps to forage, its head and body normally raised above the ground as it moves. When body temperature exceeds 39 C, the lizards move less rapidly and seek the shade of small *Triodia* clumps; above 41 C, they re-enter large *Triodia* clumps, at whose centers they find cooler temperatures and deeper shade. The lizards may also pant to dissipate heat by evaporative cooling. If heat stress is not avoided, *Amphibolurus* loses locomotor ability above 44 C, and will die if body temperature exceeds 46 C.

On a typical summer day, during which air temperature varies from about 23 C at dawn to 34 C at midday, the mallee dragon does not begin to forage until about 8:30 A.M. By 11:30 A.M., it has become too hot for normal activity and most individuals seek shade and inactivity. By 2:30 P.M., the air has cooled off enough so that foraging is resumed until 6 P.M.,

FIGURE 13-4 The mallee dragon at different points of its activity cycle: top, early morning basking in *Triodia* grass clump; middle, midmorning basking on ground (note body flattened against surface to increase exposed profile and contact with warm soil); bottom, normal foraging attitude.

after which time the lizards retreat back into *Triodia* clumps and their bodies rapidly cool. If they remained in the open after this time of day, they would easily be caught by warm-blooded predators.

The desert iguana (*Dipsosaurus dorsalis*) of the southwestern United States faces a more severe environment, with greater annual fluctuation, than the mallee dragon. In summer, air temperatures can reach 45 C; in winter, temperatures frequently drop below freezing. During mid-July, the thermal environment changes so rapidly between extremes that the desert iguana can be normally active within its preferred body temperature range of 38 to 43 C for about 45 minutes in the middle of the morning and a similar period in the early evening (Figure 13-5). During the remainder of the day, lizards seek the shade of plants or the coolness of their burrows, where the temperature rarely rises above the preferred range (for example, see Figure 13-3). At night, the lizards enter their burrows, partly to escape predation and partly because at dawn the burrow is warmer than the surface, hence the early morning warm-up period for the lizard is reduced.

Whereas the activity period of the desert iguana in summer is restricted to two brief bouts separated by a long period of inactivity to avoid midday heat stress, spring is more favorable. In May, the thermal environment

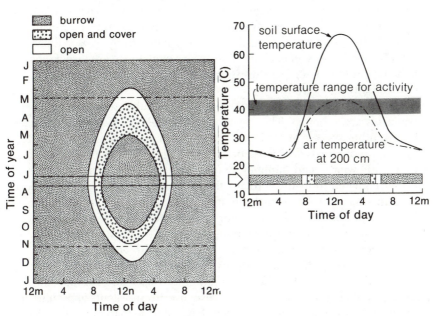

FIGURE 13-5   Seasonal activity space of the desert iguana in southern California. At left, the daily activity budget is portrayed for an entire seasonal cycle. At right, the activity budget for July 15 is shown with the time course of environmental temperature.

does not exceed the preferred range of *Dipsosaurus*, and individuals remain active aboveground from 9 A.M. to 5 P.M., occasionally seeking the shade of plants to cool off. In winter, cold temperatures restrict the activity of *Dipsosaurus* to brief periods in the middle of the day when body temperature warms up enough so that individuals can come above ground and forage. Between early December and the end of February, most days are so cold that the desert iguana remains inactive in its burrow.

Acclimation

The time course of a response to changing conditions must be shorter than the period of environmental change. Otherwise, today's response may be effective only for yesterday's conditions. For each type of response, some types of environmental fluctuations are coarse-grained, others are fine-grained. Acclimation is the modification of an organism's morphology or physiology in response to long-term environmental change. Many birds have a heavier plumage, with greater insulating properties, during the cold winter months than during hot summer months. These species replace their body feathers only twice each year, and each plumage must be suited to the average conditions of the environment between each molt. The willow ptarmigan, a ground-feeding arctic bird, sheds its lightweight brown summer plumage in the fall for a thick white winter plumage, which provides both insulation and camouflage against a background of snow. With increased insulation, ptarmigans require less energy expenditure to maintain their body temperatures during the winter (Figure 13-6). Seasonal change in plumage thickness effectively shifts the regulatory response range to match the prevalent temperature range of the season. In winter, maintenance of body temperature at −40 C requires the same expenditure of energy as it does at −10 C in summer (a conceivable temperature in the ptarmigan's arctic home). The metabolic response curves of the ptarmigan and most other species of homeotherms would seem to suggest that winter-acclimated individuals are energetically most efficient over all temperatures within both winter and summer ranges. But heat-conserving devices of winter-acclimated individuals retard the dissipation of heat and would lead to heat prostration with normal activity in summer. Adjusting insulation to enhance heat conservation in winter and to facilitate heat dissipation in summer maintains constant body temperature at the least possible cost.

Temperature-conforming animals and plants also acclimate to seasonal changes in their environment. By switching between enzymes and other biochemical systems with different temperature optima, cold-blooded animals adjust their tolerance ranges in response to prevalent environmental conditions. Experiments with lobsters have shown that the upper lethal temperature increases as the temperature to which the lobsters are acclimated increases. Lobsters kept at 5 C die when exposed to 26 C, but tolerate 28 C when acclimated to 15 C, and tolerate 31 C when acclimated

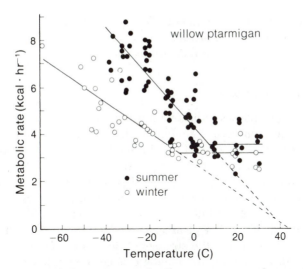

FIGURE 13-6   Metabolic responses of willow ptarmigan acclimatized to summer and winter temperatures. Winter-acclimated birds have thicker plumages providing better insulation than summer birds. Hence their metabolic rates are lower at any given temperature, and their critical temperature is also lower.

to 25 C. Acclimation does not, however, allow an organism to respond infinitely to environmental change. Regardless of their previous temperature experience, lobsters cannot be physiologically acclimated to withstand temperatures above 31 C, because their physiological systems are adapted to much colder environments and the capacity of lobsters to acclimate to extreme temperature is limited.

Although acclimation can produce remarkable adjustment of a species to a wide range of conditions, it nonetheless operates within somewhat limited ranges. If we consider swimming speed in fish as an index to activity level, its relationship to water temperature shows at once the capabilities and limitations of acclimation. Goldfish swim most rapidly when they are acclimated to 25 C and placed in water between 25 and 30 C, conditions that closely resemble their natural habitat (see Figure 12-1). Lowering the acclimation temperature to 15 C increases the swimming speed at 15 C but reduces it at 25 C. (Increased tolerance of one extreme often brings reduced tolerance of the other.) Reducing acclimation temperature further, to 5 C, well beyond the normal lower range of temperatures experienced by gold-fish in nature, does not increase swimming speed at 5 C, but does reduce swimming speed at moderate temperature.

Acclimatory changes in physiology enable a basically warm-water fish, like the goldfish, to better tolerate temperatures near the upper and lower ends of the normal temperature range in its environment. When brook trout, a cold-water species, are tested over a similar range of temperatures, they perform best when acclimated to 15 C, approximately the summer temperature of their environment. The difference between goldfish and trout represents the accumulation of evolutionary modifications to the different temperature ranges of their respective habitats.

Developmental Responses

Developmental responses indicate a flexibility of growth and development processes that are sensitive to environmental variation. These responses are generally not reversible. Once fixed during development, they remain unchanged for the remainder of the organism's life. Because developmental responses have distinct limitations for short-term environmental changes (long response time and irreversibility), plants and animals are likely to exhibit developmental flexibility only in environments with persistent variation for the individual. For plant species whose seeds may settle in many different kinds of habitats, the strategy of developmental flexibility makes good sense. Also, when environmental changes occur slowly compared to the life span of an organism, developmental responses may be the most appropriate type of adjustment. For example, an organism that lives much less than a year can adjust itself to seasonal conditions by developmental response.

A striking example of developmental flexibility occurs in several species of aquatic plants whose morphology varies according to whether the plant grows on the land, partly submerged in water, or completely underwater. Leaves of arrowleaf plants that grow underwater are not structurally rigid, and because they lack a waterproof waxy cuticle, they may absorb nutrients directly from the water into the leaves (Figure 13-7). If these underwater leaves are removed from the water, they collapse. The aerial leaves of the plant are broad, more rigid than underwater leaves, and are covered with a thick cuticle to decrease transpiration. The stiff aerial leaf would probably snap off if it were exposed to water currents. An arrowleaf that grows partly submerged has both kinds of leaves. Terrestrial individuals have large root systems compared with those of aquatic individuals. The latter can absorb nutrients through their leaves, whereas terrestrial plants must obtain all their nutrients from the soil by way of their root system. These differences in growth form result from developmental responses of the arrowleaf to the environmental conditions of each habitat. In the aquatic environment, the development of thick cell walls in leaf tissue is suppressed; in well-drained soil, the growth of roots is more pronounced.

Light intensity is another factor that can influence the course of development in plants. Loblolly pine seedlings grown in shade have smaller root systems and more foliage than seedlings grown in full sunlight. Because the shaded environment has lower water stress, shade-grown seedlings can allocate more of their production to stem and needles. Sun-grown seedlings must grow more extensive root systems to obtain water. The larger proportion of foliage of the shade-grown seedlings results in a higher rate of

FIGURE 13-7   Variation in the morphology of the arrowleaf plant when it grows on land (left), partially submerged in water (center), and fully submerged (right).

photosynthesis per plant under given light conditions, particularly when the light intensity is low (Table 13-1). The different developmental responses of shade-grown and sun-grown seedlings adapt each to the different conditions of light intensity and moisture stress in its environment.

A fascinating and complicated case of developmental response involves the wing development of water striders, fresh-water bugs of the genus

TABLE 13-1   Proportions and rates of photosynthesis of loblolly pine seedlings grown under shade and full sunlight.

	Shade-grown	Sun-grown
Per cent of dry weight in*		
roots	35	52
needles	47	37
stem	18	11
Photosynthetic rate ($mgCO_2$ hr^{-1} gm^{-1})†		
low light intensity	1.9	1.0
moderate light intensity	4.6	4.0
high light intensity	7.2	6.6

* Six-month-old seedlings.
† Four-month-old seedlings. Light intensities were 500, 1,500, and 4,500 foot candles.

Gerris. European species fall into four categories of wing length depending on the type of habitat in which they live (Table 13-2). At one extreme, species inhabiting large permanent lakes have short wings or are wingless and do not disperse between lakes. At the other extreme, species inhabiting temporary ponds are usually long-winged and disperse to find suitable habitats for breeding each year. Between these extremes, species that inhabit small ponds, which are more or less persistent from year to year but tend to dry up during the summer, frequently exhibit both long- and short-winged forms.

The life cycle of most *Gerris* species in central Europe, England, and the southern parts of Scandinavia includes two generations per year. The first, or summer, generation hatches during the spring, reproduces during the summer, and then dies. The second, hatched from eggs laid by females of the summer generation, develops to the adult stage during late summer, then overwinters before breeding the following year in early spring. In species that inhabit seasonal ponds, the summer generation is dimorphic, having both long- and short-winged forms (Figure 13-8). All the individuals in the winter generation are long-winged and able to fly. They leave the pond in late summer and move into nearby woodlands where they over-winter. In the spring, they return to small bodies of water to lay eggs.

Dimorphism in the summer generation reflects two extreme strategies. The long-winged forms are capable of flying to other habitats if the pond in which they developed happens to dry up, especially if this happens early in the season. The short-winged forms bank on the persistence of the habitat into late summer, and they convert nutrients that would have become wing and flight muscles in long-winged forms into eggs. Thus short-winged individuals tend to be more fecund, but their progeny are sometimes destroyed when a pond dries up earlier than usual.

All members of the winter generation have long wings. From this we may deduce that seasonal dimorphism is not genetic but rather is a devel-

TABLE 13-2 Wing lengths of water striders (*Gerris*) found in habitats with different levels of permanence and predictability.

Characteristic of habitat	Characteristic wing length*	Mechanism of determination
Permanent	short	genetic
Fairly persistent but unpredictable	both short and long	genetic dimorphism
Seasonally temporary	seasonally dimorphic (summer)	developmental switch
Very unpredictable	long	genetic

* Short wings are not functional and prevent dispersal.

FIGURE 13-8   Alary polymorphism in the water strider *Gerris odontogaster*. The long-winged form is on the left and the short-winged form is on the right.

opmental response determined primarily by length of day. If the daylight increases continually during larval development and exceeds eighteen hours (in southern Finland) during the last larval stage prior to change into the adult form, the individual will have short wings. If the daylength begins to decrease before the end of larval development (as it would if the development period extended beyond June 21), a long-winged adult is produced. Thus summer generation individuals become short- or long-winged depending upon when they hatch and upon their rate of larval development. The switch between wing length determination is also influenced by temperature. High temperatures, which could lead to early drying of ponds, favor the development of long-winged forms.

Animal Migration

Under conditions of extreme drought or cold, physical conditions may become sufficiently stressful, and food sufficiently difficult to find, that plants and animals can no longer maintain normal activity. Faced with these conditions, some organisms leave the environment, seeking more favorable conditions elsewhere, while others enter a dormant state, sealing themselves off from the rigors of the environment.

Migrations are widespread in nature, particularly among flying animals: birds, bats, and some insects. Many of them perform impressive feats of long-distance navigation each year (Figure 13-9). Shorebirds, like the golden plover, breed in the Canadian Arctic and spend their winters as far south as Patagonia at the southern tip of South America. They make yearly round trips of up to 25,000 miles. The arctic tern probably holds the record for long-distance migration with a yearly round trip of 36,000 miles between its North Atlantic breeding grounds and its Antarctic wintering grounds. A few insects, like the monarch butterfly, perform impressive migratory movements each year, but most species of insects overwinter in a dormant state as eggs or pupae.

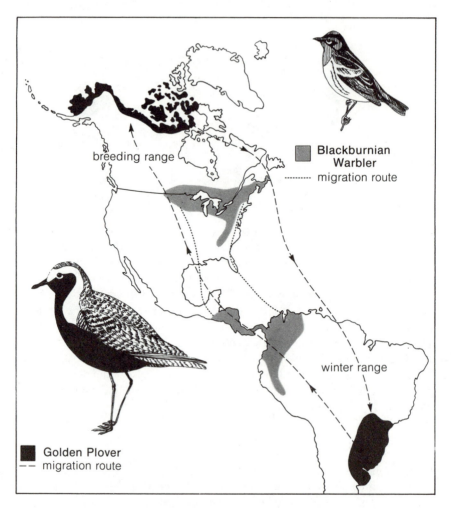

FIGURE 13-9 Breeding and wintering ranges of the golden plover (left) and Blackburnian warbler (right). Migration routes are indicated by dashed lines.

Each fall, hundreds of species of land birds move out of temperate and arctic North America in anticipation of cold winter weather and dwindling supplies of their invertebrate food. Similarly, montane birds make altitudinal migrations of several thousand feet, the mountain quail making its annual trek on foot. Because the Southern Hemisphere winter is less harsh than the Northern Hemisphere winter, South American species have less need to escape the southern winter to take advantage of the temperate summer occurring to the north.

Mammals, other than bats, do not have the migratory abilities of birds and some insects, but some mammals do exhibit impressive seasonal movements. The barren-ground caribou of northern Canada migrates from its summer home on the tundra into the spruce forest for the winter because its food (lichens and mosses) is covered by snow on the tundra but remains accessible in the protected spruce forest habitat.

Some marine organisms also undertake large-scale migrations to reach spawning grounds, to follow a food supply, or to keep within suitable temperature ranges for development. The migration of salmon from the ocean to their spawning grounds at the headwaters of rivers and the reverse migration of adult fresh-water eels to their breeding grounds in the Sargasso Sea are striking examples. Lobsters undertake less conspicuous annual migrations up to several hundred kilometers off Long Island and Massachusetts, from deep waters in summer to shallow in winter, always staying within zones of cool temperature. Predatory gastropods and some sea urchins are known to undertake seasonal movements into shallow waters for feeding.

Some populations exhibit irregular or sporadic movements that are tied to food scarcity during particular years rather than to seasonal conditions. The occasional failure of cone crops in coniferous forests of Canada and the mountains of the western United States forces large numbers of birds that rely on seeds to move to lower elevations or latitudes. Birds of prey that normally feed on rodents disperse widely when their prey populations decline sharply. Snowy owls move southward from their arctic hunting grounds into the northern United States and adjacent Canada every four years on the average, corresponding to a periodic scarcity of lemmings. In desert areas, irregular rainfall forces animals such as the budgereegah (an Australian parakeet) into a nomadic existence in a continual search for areas where rain has recently fallen. Even insects are subject to irregular movements. Outbreaks of migratory locusts, from areas of high local density where food has been depleted, can reach immense proportions and cause extensive crop damage over wide areas (Figure 13-10). The behavioral traits that lead to population irruption are actually a developmental response to population density. When the locusts grow up in sparse populations, they develop into a solitary phase as adults. In dense populations, frequent contact with other individuals stimulates a developmental switch to the gregarious phase, which can often lead to population outbreaks following local depletion of food resources.

Dormancy and Storage

When the environment becomes so extreme that normal life processes can no longer function, or that the maintenance of normal activity would quickly lead to starvation or desiccation, plants and animals incapable of migration must enter into a physiologically dormant state. For many small invertebrates and cold-blooded vertebrates, freezing temperatures directly curtail activity and lead to dormancy. Many mammals enter into hibernation be-

FIGURE 13-10   A dense swarm of migratory locusts in Somalia, Africa, in 1962.

cause of a lack of food, rather than because of physiological inability to cope with the physical environment.

Lack of water is the key factor in the adoption of the deciduous habit by plants. Many tropical and subtropical trees shed their leaves during seasonal periods of drought. Temperate and arctic broad-leaved trees shed their leaves in the fall to avoid desiccation. Moisture frozen in the soil is unavailable to plants; if these trees kept their leaves through the winter, transpiration of water from the leaves would wilt them as quickly as if the tree were cut through its trunk.

Insects enter into a resting state known as *diapause*, in which water is chemically bound or reduced to prevent freezing and metabolism drops to near nought. In summer diapause, drought-resistant insects either allow their bodies to dry out and tolerate desiccation or secrete an impermeable outer covering to prevent drying. Plant seeds and the spores of bacteria and fungi have similar dormancy mechanisms. Regardless of the mechanism, dormancy serves the single purpose of shutting off the organism from the environment and reducing exchange between the two. In this way animals and plants ride out unfavorable conditions and await better ones before reassuming an active and interactive state.

Although homeostasis helps maintain function in the face of a changing physical environment, environmental changes often plunge organisms from feast into famine. When the environment becomes barely tolerable and small fluctuations in food or water supply can mean disaster, many plants and animals store food and water reserves. Desert cacti absorb water during rainy periods and store it in their succulent stems. Cacti use these reserves during the long dry intervals between desert rains. Many temperate and arctic animals store fat during periods of mild weather in winter as a reserve of energy for periods when heavy snow covers food sources. Tropical animals sometimes store fat prior to the onset of seasonal dry periods. Instead of accumulating body fat, many winter-active mammals (beavers and squirrels) and birds (acorn woodpeckers and jays) cache food underground or under the bark of trees. During winter months, piñon jays of the western United States normally feed on insect grubs in the soil, but they cannot do so when snow covers the ground. In the fall, the jays harvest the vast crops of piñon pine nuts and bury them in caches near the base of trees. The nut stores are usually placed at the south side of a tree's trunk, and after a snowfall, the snow first melts on the south side, thus exposing the cache.

Fat is often stored in anticipation of greatly increased energy demands—as before long migrations or periods of reproduction. Many species of migrant birds store large quantities of fat, often half their normal weight, particularly if they undertake long flights over water. The side-blotched lizard stores fat during the fall and winter to provide energy for egg formation in the spring. The fat bodies of lizards may also serve as an insurance policy against poor conditions, for some species accumulate larger fat stores where winters are harsh than they do in milder regions.

All deciduous plants store materials during the summer and early fall to provide energy and nutrients needed for flowering and the early growth of leaves in the spring. Just before trees begin to leaf out, sap rises in the trunk to the tips of the branches, carrying sugars and other nutrients.

Many plants store nutritive materials in their roots to allow recovery after their shoots have been destroyed by fire or by defoliating insects. Every few years, in the northeastern United States, tent caterpillars defoliate black cherry trees in the early spring. The cherry trees respond by putting out new sets of leaves, drawing on untapped reserves of nutrients in their roots. Where fires frequently sweep through habitats—as in the chaparral of southern California—many plants store food reserves in fire-resistant root crowns, which sprout and send up new shoots shortly after a fire has passed (Figure 13-11). Root sprouting promotes the recovery of vegetation and stabilization of the soil more rapidly than would be possible if the vegetation could grow back only from seed. The seeds of many annual plants are also fire-resistant, and so get an early foothold in burned-over areas, growing up and producing seed before shrubby vegetation crowds them out.

FIGURE 13-11   Root-crown sprouting by chamise following a fire in the chaparral habitat of southern California. The upper photograph was taken on May 4, 1939, six months after the burn. The lower photograph, showing extensive regeneration, was taken on July 16, 1940.

In this chapter, we have seen how animals and plants respond to variation in their environments to maintain their activity at an optimum level. But in spite of the battery of homeostatic defenses available to the organism, its activity space within the entire range of conditions possible is greatly limited. This restriction takes on an additional geographical perspective when we consider the distribution of organisms and the factors that delimit geographical ranges in the next chapter.

14 | Environment, Adaptation, and the Distribution of Organisms

NO SINGLE TYPE of plant or animal can tolerate all the conditions found on earth. Each thrives within relatively narrow ranges of temperature, precipitation, soil conditions, and other environmental factors. The geographical range of any population cannot exceed the geographical distribution of suitable environmental conditions. Moreover, the preferences and tolerances of each species differ, and so, although the distributions of species broadly overlap one another, no two are found under exactly the same range of conditions.

This chapter is concerned with the factors that determine the distributions of species' populations. On a local scale, environment plays an overwhelming role in distribution, confining populations to regions and habitats within which the species thrives and perpetuates itself. On a global scale, geographical barriers and historical accidents of distribution assume prominent roles in determining the presence or absence of a particular species. One can find localities in Asia, North America, and Europe with closely matched climate and soils; but although the vegetation of each locality superficially resembles the vegetation of the other two, most of the species differ. The plants of each region are adapted to tolerate similar environments, but barriers to dispersal—oceans, deserts, and mountain ranges—restrict their distributions. That species often can flourish beyond their natural geographical ranges is demonstrated by the success of dandelions, Norway maples, starlings, and honeybees, all of which were introduced from Europe to the United States, sometimes intentionally and sometimes accidentally, by man.

Climate and the Geographical Distributions of Plants

The range of the sugar maple, a common forest tree in the northeastern United States and southern Canada, is limited by cold winter temperatures to the north and hot summer temperatures to the south (Figure 14-1). Sugar maples cannot tolerate average monthly temperatures above 75 to 80 F (24 to 27 C) or below about 0 F (−18 C). The western limit of the sugar maple, determined by dryness, coincides closely with the western limit of forest-type vegetation, in general. Because temperature and rainfall interact to determine availability of moisture, sugar maples tolerate lower annual rainfall at the northern edge of their range (about 20 inches, or 50 cm) than at the southern edge (about 40 inches, or 100 cm). To the east, the distribution of the sugar maple is limited abruptly by the Atlantic Ocean.

Within its geographical range, sugar maple is more abundant in northern forests, where it sometimes forms single-species stands, than in the more diverse forests in the south. Sugar maples occur most frequently on moist, podsolized soils that are slightly acid.

The range of the sugar maple overlaps three other tree-sized species of maples: black, red, and silver (Figure 14-2). The range of the black maple falls almost entirely within the distributional limits of the sugar maple, indicating similar but narrower tolerance of temperature and rainfall extremes. In fact, the two maples are so similar that foresters only recently have recognized them as different species, by slight differences in the shape of the fruits and by the presence or absence of hairs on the underside of the leaves. The northern limit of red maple nearly coincides with that of the sugar maple, but red maple extends south to the Gulf Coast and appears to be less tolerant of drought in the midwestern United States. Silver maple also extends beyond the southern limit of sugar maple, but unlike the red

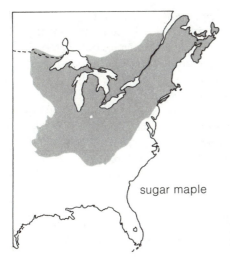
sugar maple

FIGURE 14-1 The range of the sugar maple in eastern North America.

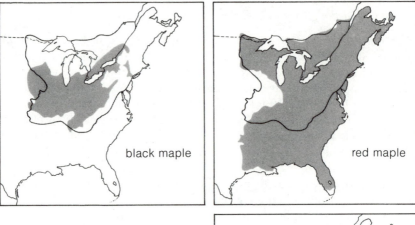

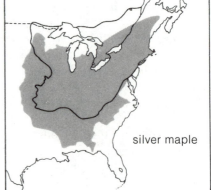

FIGURE 14-2 The ranges of black, red, and silver maples of eastern North America. The range of the sugar maple is outlined on each map to show the area of overlap.

maple, it is found farther to the west, extending along stream valleys where soils are moist.

Where their ranges overlap, maples exhibit distinct preferences for local environmental conditions created by differences in soil and topography. Black maple frequently occurs together with sugar maple, but prefers drier, better-drained soils with higher (therefore less acidic) calcium content. Silver maple is widely distributed, but prefers the moist, well-drained soils of the Ohio and Mississippi river basins. Red maple is peculiar in favoring either very wet, swampy conditions (it is often called swamp maple) or dry, poorly developed soils. Whether these extremes have some other soil factor in common that the red maple likes or whether red maple consists of two distinct physiological types, each with difference preferences, is not known.

The Local Distribution of Plants

The effects of different factors on the distribution of plants are manifested on different scales of distance. Climate, topography, soil chemistry, and soil

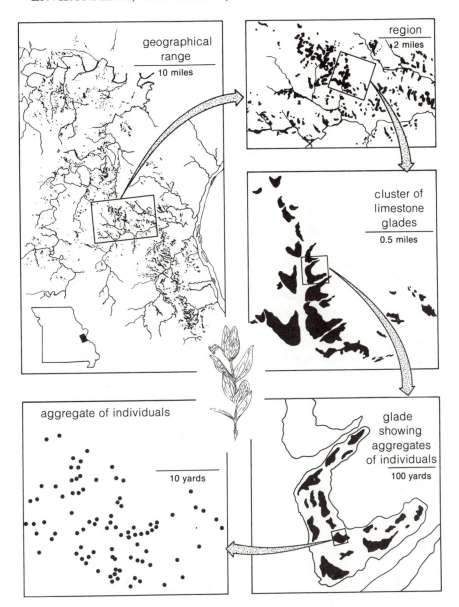

FIGURE 14-3 Hierarchy of distribution of *Clematis fremontii*, variety *riehlii*.

texture exert progressively finer influence on geographical distribution. For example, the perennial shrub *Clematis fremontii*, variety *riehlii*, exhibits a hierarchy of distribution patterns revealed by examining its distribution on different geographical scales (Figure 14-3). Climate and perhaps related species in other areas restrict this species of *Clematis* to a small part of the midwestern United States. Variety *riehlii* is found only in Jefferson County, Missouri. Within its geographical range, *Clematis fremontii* is restricted to dry rocky soils on outcroppings of dolomite, which are distributed with respect to mountain and stream systems. Small variations in relief and soil quality further restrict the distribution of *Clematis* within each dolomite glade to sites with suitable conditions of moisture, nutrients, and soil structure. Local aggregations occurring on each of these sites consist of many, more or less evenly distributed individuals.

Elevation, slope, exposure, and underlying bedrock—factors that greatly influence the plant environment—vary most in mountainous regions. The heterogeneity of such regions breaks the geographcal ranges of species into isolated areas with suitable combinations of moisture, soil structure, light, temperature, and available nutrients. Ecologists frequently turn to the varied habitats of mountains to study plant distribution. Along the coast of northern California, mountains create conditions for a variety of plant communities ranging from dry coastal chaparral to tall forests of Douglas fir and redwood. Of all environmental influences on the distribution of forest trees, moisture is the most important. When localities are ranked on scales of available moisture and exchangeable calcium, the distribution of each species among the localities exhibits a distinct optimum (Figure 14-4). The coast redwood dominates the central portion of the moisture gradient and frequently forms pure stands. Cedar and Douglas fir, and two broad-leaved evergreen species with small thick leaves—manzanita and madrone—are found at the drier end of the moisture gradient. Three deciduous species—alder, big-leaf maple, and black cottonwood—occupy the moister end.

The distribution of species along the moisture gradient may coincide with distribution along an available nutrient gradient. For example, the dry soils in which cedar thrives are also poor in nutrients. Low exchangeable calcium indicates that the soil has few cation-exchange sites or that these are largely occupied by hydrogen H^+ ions; in either case, the soil is relatively infertile. Because moisture and nutrient availability are so highly correlated, however, we cannot determine whether water, nutrients, or a third factor related to both, exerts the greatest influence on the distribution of cedar. In fact, cedars are virtually restricted to serpentine barrens in northern California. There may also be instances of distribution with no apparent relationship to nutrients and moisture. Soil nutrients vary widely over the range of soil moisture conditions that favor the redwood, which occurs in all but the most impoverished soils—the serpentine barrens.

The ecological distributions of some species apparently affect others. Madrone and Douglas fir overlap each other broadly along the moisture gradient, but occur apart on the nutrient gradient. Furthermore, madrone occupies the central part of the soil nutrient gradient, with Douglas fir forming two peaks of abundance, one on poorer soils and the other on richer soils. The separate peaks of abundance may represent distinct subpopulations of Douglas fir, each with different tolerance ranges.

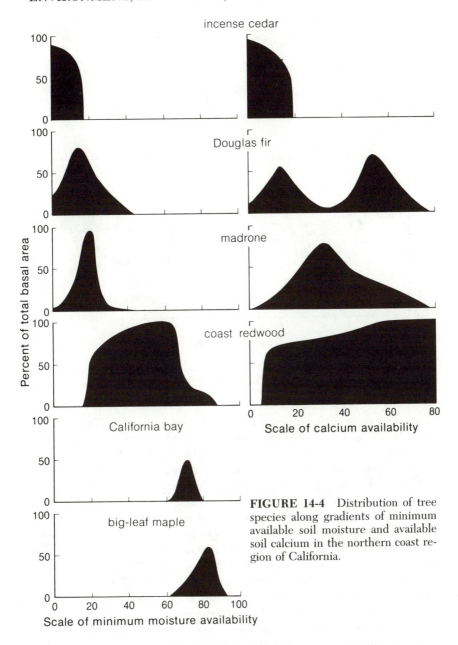

FIGURE 14-4 Distribution of tree species along gradients of minimum available soil moisture and available soil calcium in the northern coast region of California.

Environmental variables do not change independently of each other. Increasing soil moisture usually alters the status of available nutrients in the soil. Variation in the amount and source of organic matter in the soil creates parallel gradients of acidity, soil moisture, available nitrogen, and so on. Because these variables interact, one should examine the distribution of

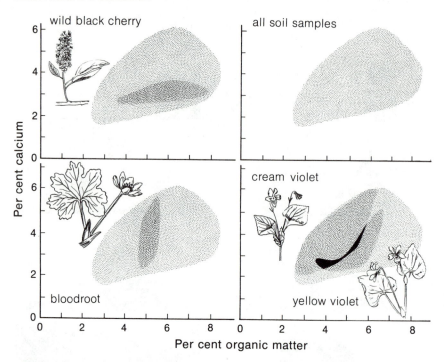

FIGURE 14-5   The occurrence of four forest-floor plants with respect to the calcium and organic-matter contents of the soil in woodlands of eastern Indiana.

plants with respect to all the variables at the same time. In Figure 14-5, for example, distributions of some forest-floor shrubs, seedlings, and herbs in woodlands of eastern Indiana are related to levels of organic matter and calcium in the soil. These soils contain between 2 and 8 per cent organic matter and between 2 and 6 per cent exchangeable calcium. Furthermore, levels of calcium and organic matter are interrelated: soils rich in organic humus tend to be rich in inorganic nutrients. Within the range of soil conditions found in these woods, each species shows different preferences. Black cherry is found only within a narrow range of calcium but is tolerant of variation in the percentage of organic matter. Bloodroot is narrowly restricted by the per cent of organic matter in the soil but is insensitive to variation in calcium. The distributions of yellow violets and cream violets extend more broadly over levels of organic matter and calcium in the soil, but they do not overlap. Cream violets prefer relatively higher calcium and lower organic-matter content than yellow violets; where one is found, the other is usually absent. In nature, the distributions of species on all scales of geographical distance are not only related to physical and chemical properties of the environment but are influenced by the presence of other species as well.

The Suitability of the Environment for Populations

For each organism there exists some combination of environmental conditions that is optimum for its growth, maintenance, and reproduction. To either side of the optimum, biological activity falls off until the organism ceases altogether to be supported by the environment. We see this pattern over and over again, whether we examine the dependence of photosynthetic rate on leaf temperature, the distribution of plants along moisture and nutrient gradients, the vertical range of seaweeds and marine snails within the intertidal zone, or the size of prey captured by predatory animals. Within the range of environmental conditions that will adequately support an individual, the perpetuation of a population is confined more narrowly. The successful population requires resources for growth and reproduction in addition to those required for individual maintenance.

We may visualize the general relationship of an organism to its environment on a graph on which we relate rate of biological activity (by whatever measure we choose) to a gradient of environmental conditions. This relationship is portrayed in Figure 14-6 as a bell-shaped curve showing a distinct optimum over the middle of the gradient. In nature, these curves can be quite asymmetrical. Oyster development, for example, proceeds best in moderate salinities but is depressed more by low salinity than high salinity (at least up to the maximum for seawater). Optimum oxygen concentration for brook trout occurs toward the upper end of the oxygen gradient in lakes and streams.

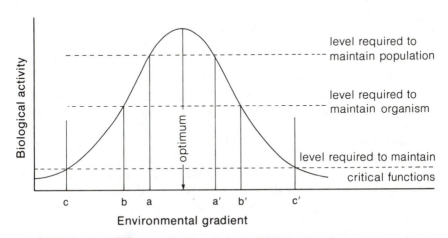

FIGURE 14-6 The general relationship of biological activity to a gradient of environmental conditions. Levels of activity required to maintain critical biological function, the organism, and the population, determine the lethal extremes (c and c'), the limits of persistence for organisms (b and b'), and populations (a and a').

Regardless of the shape of the activity curve, the ecological distribution of a species along the gradient is governed by three levels of tolerance. First, extreme conditions may totally disrupt critical biological functions, resulting in rapid death. Such lethal conditions occur beyond points c and c' in Figure 14-6. Along a temperature gradient, for example, temperatures below freezing and above 50 C are lethal for most organisms.

Second, organisms must sustain a certain level of activity to maintain themselves in a steady state for long periods. Within points b and b' on the environmental gradient, the organism can exist indefinitely. Outside these limits, the organism's activity level is too low to be self-maintaining, and the organism can venture beyond these limits only briefly.

Third, populations can maintain their size only if reproduction balances death. Reproduction requires resources, hence biological activity, over and above the level needed for self-maintenance. Populations persist, therefore, within a narrower range of conditions than the individual can tolerate. Individuals may live in environments that are inadequate for population maintenance, but their numbers can be maintained only by immigration from populations in more suitable habitats where reproduction exceeds death.

Because organisms can live only under characteristic combinations of favorable environmental conditions, one should be able to determine the suitability of a particular place for any species. For example, optimum climates for the Mediterranean fruit fly lie between 16 and 32 C, and 65 to 75 per cent relative humidity. Populations thrive under these conditions provided, of course, that food is available. Outside this optimum, fruit flies can maintain populations under conditions between about 10 and 35 C, and 60 to 90 per cent relative humidity; adult individuals can persist at temperatures as low as 2 C, and relative humidity as low as 40 per cent; more extreme conditions are usually lethal (Figure 14-7). The Mediterranean fruit fly is a major agricultural pest, but populations reach outbreak proportions only where conditions are within the biological optimum most of the year and rarely exceed the limits of tolerance. Thus Tel Aviv, Israel, where mean monthly temperature varies between 7 and 31 C, and humidity ranges between 59 and 73 per cent, is frequently plagued by the fly and requires extensive control measures, although conditions favor outbreaks more in some years than others. The climate of Paris, France, is generally too cold for fly populations to reach damaging levels, and Phoenix, Arizona, is too dry. The climates of Honolulu, Hawaii, and Miami, Florida, where the pest has been accidentally introduced, are quite suitable for rapid population growth.

As we saw in the last chapter, seasonal changes in climate usually exceed the optimum range of conditions of most species. Consequently, population growth is favored during restricted periods, and populations exhibit seasonal cycles of abundance and activity. Temperature governs the seasons in most

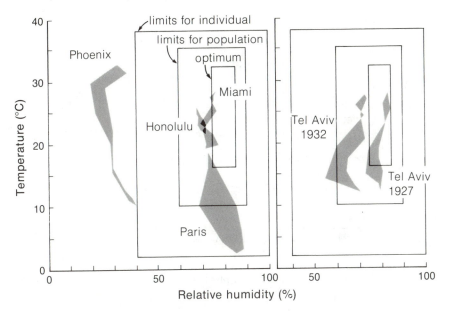

FIGURE 14-7 The seasonal course of air temperature and relative humidity at selected locations in relation to conditions favorable for the Mediterranean fruit fly. The inner rectangle encloses conditions optimum for growth, the middle rectangle encloses conditions suitable for development, and the outer rectangle delimits the extreme tolerance range.

temperate and arctic regions, restricting the growing season to the warm summer months.

In the tropics, rainfall sets the seasonal pace of activity. Arboreal mosquitoes are abundant in Panama only during the rainy months of May through December. Most mosquito populations persist through the dry season as desiccation-resistant eggs. Species of arboreal mosquitoes whose eggs cannot withstand drying must find permanent sources of water in which to lay their eggs; adults do not live long enough to span the entire dry season.

Seasonal changes in the food value of plants sometimes determine shifts in diet and the timing of reproduction in herbivores. In the chaparral regions of California, winter rainfall and warming temperatures in the early spring stimulate plant growth, greatly increasing the abundance and protein content of food plants for deer (Figure 14-8). Range quality declines during the dry summer months. Deer require 13 per cent protein in their diet for optimum growth and reproduction; 7 per cent protein barely provides a maintenance diet. Chamise is a common chaparral shrub used extensively by deer, but it does not provide a maintenance diet during most of the year and cannot adequately support a deer population without supplemental

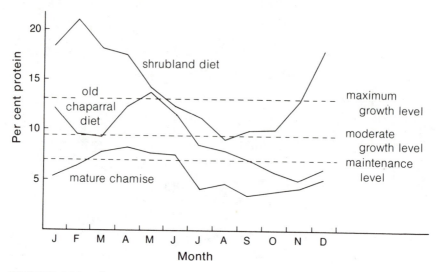

FIGURE 14-8   The protein content of plants used by deer in Lake County, California.

foods. Mature chaparral vegetation, including all species of plants used by deer, provides an adequate maintenance level of protein during most of the year and supports moderate reproduction and growth in early spring.

Heterogeneity of soils in the coast ranges of northern California has resulted in striking cases of limited plant distribution. In particular, several species of pines (*Pinus*) and cypresses (*Cupressus*) are restricted to serpentine soils (e.g., Sargent cypress), while others are found only on extremely acid soils (e.g., Bishop pine, lodgepole pine, pygmy cypress). When grown on soils from different localities, seedlings of these endemics often do best when planted in soil resembling that on which the population is normally found. Thus, lodgepole pine grows only in acid soils and Sargent cypress, the serpentine endemic, grows somewhat better on serpentine soil than on "normal" soil, and not at all on acid soil (Figure 14-9). Not all endemics perform best on their home soil. When given the chance in an experimental garden, pygmy cypress, normally restricted to acid soils, grows much better on "normal" and serpentine soil. Wherefrom it derives its tolerance for serpentine soils is not known. Nor is it known why pygmy cypress is excluded from soils on which its seedlings grow vigorously. Clearly, edaphic conditions are not the only ones limiting the distributions of these species.

Adaptation and the Environment

It is no accident that different species find different portions of the environmental spectrum suitable for maintaining their populations. The adaptations of an organism—its form, physiology, and behavior—cannot easily be separated from the environment in which it lives. Organism and environment go hand in hand. Insect larvae from stagnant aquatic environments in ditches and sloughs can survive longer without oxygen than related species from well-aerated streams and rivers; species of marine snails that occur high in the intertidal zone, where they are frequently exposed to air, can withstand a greater degree of desiccation than species from lower levels.

In general, the forms of organisms are closely linked to their ways of life and to the environmental conditions in which they live. Compare the leaves of deciduous forest trees with those of desert species. The former are typically broad and thin, providing a large surface area for light absorption and for water loss. Desert trees have small, finely divided leaves—or some-

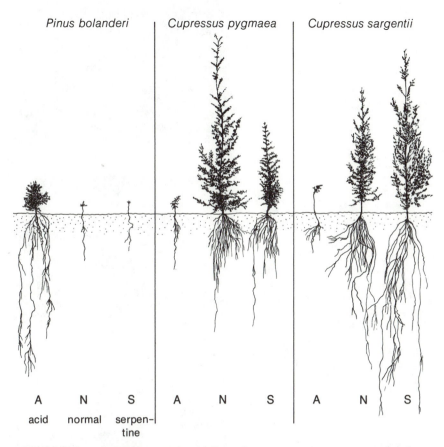

FIGURE 14-9 Seedling growth of lodgepole pine, pygmy cypress, and Sargent cypress in acid (left), normal (center), and serpentine (right) soils. Acidity (pH) of the three soils was 4.5, 6.8, and 7.1; exchangeable calcium (millequivalents per 100g) was 0.67, 12.90, and 3.40; exchangeable magnesium (me per 100g) was 0.05, 8.9, and 17.0.

times none at all (Figure 14-10). Leaves heat up in the desert sun. Structures lose heat by convection most rapidly at their edges, where wind currents disrupt insulating boundary layers of still air. The more edges, the cooler the leaf and the lower the water loss. Moreover, small size means a large portion of each leaf is given over to its edge. One may find many exceptions to the relationship between leaf size and moisture availability, but the pattern holds for most species. Even on a single plant, leaves exposed to full sun are likely to show adaptations for dissipating heat and conserving water that shade leaves lack. In the white oak, sun leaves are more deeply lobed than shade leaves and hence have more edge per unit of surface area (Figure 14-11).

FIGURE 14-10 Leaves of some desert plants from Arizona. Mesquite leaves (top) are subdivided into numerous small leaflets, which facilitate the dissipation of heat when the leaves are exposed to sunlight. The paloverde carries this adaptation even farther (center); its leaves are tiny and the thick stems, which contain chlorophyll, are responsible for much of the plant's photosynthesis (hence the name paloverde, which is Spanish for green stick). Cacti rely entirely on their stems for photosynthesis; their leaves are modified into thorns for protection. Unlike most desert plants, limberbush (bottom) has broad succulent leaves, but limberbush plants leaf out only for a few weeks during the summer rainy season in the Sonoran Desert. Photographs are about one-half size.

sun   shade

FIGURE 14-11 Silhouettes of sun and shade leaves of white oak.

The vertical distribution of species of algae within the intertidal range is closely paralleled by their rates of photosynthesis in water and when exposed to air (Table 14-1). Along the coast of central California, several common seaweeds have vertical ranges of 0.5 to 2.5 feet within the approximately 8-foot intertidal range. Height above sea level determines the proportion of time exposed to air on an annual basis. Exposure is about 10 per cent at 0 feet, 40 per cent at 3 feet and 90 per cent at 6 feet. Algae from the lower part of the tide range (*Prionitus* and *Ulva*) have a reduced photosynthetic rate while exposed; algae from the middle and upper parts of the tide range have elevated rates of photosynthesis in air. The relative rates of photosynthesis while submerged and exposed depend on tolerance of des-

TABLE 14-1 Characteristics of intertidal species of algae from various height zones.

| | SPECIES | | | | |
Characteristic	*Prionitis*	*Ulva*	*Iridaea*	*Porphyra*	*Fucus*
Mean height above sea level (ft)	−1.0	+0.5	+1.0	+3.0	+3.0
Time exposed to air (per cent)	5	15	25	40	40
Photosynthesis rate ($mgC\ dm^{-1}\ hr^{-1}$)*					
in air	0.74	1.00	1.08	2.52	3.13
in water	0.81	1.37	0.35	0.90	0.49
air to water ratio	0.9	0.7	2.9	2.8	6.6
Rate of water loss ($\%\ hr^{-1}$)	25	40	25	13	11
Photosynthesis drops to one-half submerged rate at					
water loss (per cent)	10	24	32	47	60
hours	0.4	0.6	1.3	3.6	5.3

* dm = decimeter; $1\ dm^2 = 100\ cm^2$

iccation. In algae from the lower portion of the intertidal, photosynthesis, after exposure to air, drops to one-half the value of submerged plants in 0.4 to 0.6 hours after a 10 to 24 per cent loss of water (at a rate of 25 to 40 per cent per hour). In algae from the middle and upper intertidal zone, photosynthesis first increases upon exposure. It does not drop to one-half the value while submerged for 1.3 to 5.3 hours after a 32 to 60 per cent loss of water (at a rate of 11 to 25 per cent per hour). Thus, the adaptations of these algae—low rates of water loss and relative insensitivity to desiccation—suit them to the long periods of exposure to air experienced in the upper parts of the tidal range.

Not only do species in different environments have different adaptations, but species living together are also adapted to utilize different parts of the environment. That is, they have different activity spaces within the same habitat.

The water relations of coastal sage and chaparral plants in southern California demonstrate divergent courses of adaptation. Chaparral habitats generally occur at higher elevation than the coastal sage habitats, and thus are cooler and moister. Both vegetation types are exposed to prolonged summer drought, but water deficiency is greater in the sage habitat. Plants of the coastal sage habitat are typically shallow-rooted with small, delicate deciduous leaves (Figure 14-12). Chaparral species have deep roots, often extending through tiny cracks and fissures far into the bedrock. Their leaves, typically thick, with a waxy outer covering (cuticle) that reduces water loss, are persistent. The more delicate leaves of coastal sage species are often dropped during the summer drought period. Leaf morphology influences photosynthetic rate in conjunction with its influence on transpiration (Table 14-2). The thin leaves of coastal sage species lose water rapidly, but also carry on photosynthesis rapidly when water is available to replace transpiration losses. When leaves are clipped from plants and placed in a chamber where transpiration and photosynthesis can be measured, both functions

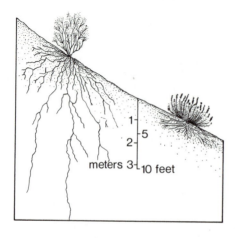

FIGURE 14-12 Profiles of the root systems of chamise, a chaparral species (left), and black sage, a member of the coastal sage community.

TABLE 14-2   Characteristics of chaparral (left) and coastal sage (right) vegetation in southern California.

Characteristics	VEGETATION TYPE	
	Chaparral	Coastal sage
Roots	Deep	Shallow
Leaves	Evergreen	Summer deciduous
Average leaf duration (months)	12	6
Average leaf size (cm^2)	12.6	4.5
Leaf weight (g dry wt dm^{-2})*	1.8	1.0
Maximum transpiration (g H_2O dm^{-2} hr^{-1})	0.34	0.94
Maximum photosynthetic rate (mg C dm^{-2} hr^{-1})	3.9	8.3
Relative annual CO_2 fixation	49.8	46.8

* dm = decimeter; 1 dm^2 = 100 cm^2

decline as the leaves dry out and their stomata close to prevent further water loss. Coastal sage species, such as the black sage, have high photosynthetic and transpiration rates at the beginning of the transpiration experiment, but shut down quickly owing to rapid water loss (Figure 14-13). Chaparral species such as the toyon (rose family) have maximum photosynthetic rates which are only one-fourth to one-third those of coastal species, but they resist desiccation and continue to be active under drying conditions for longer periods. In this property, the chaparral species resemble the

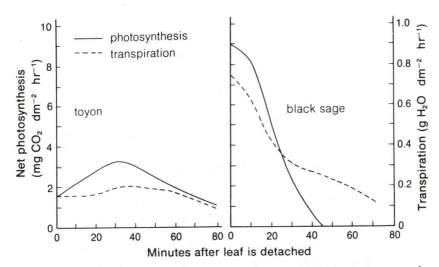

FIGURE 14-13   Time-course curves for photosynthesis and transpiration under standard drying conditions for a chaparral species (left) and a coastal sage species (right). Note that transpiration continues well after photosynthesis has been shut off; hence leaf dormancy is an ineffective long-term solution to drought.

desiccation-resistant algae of the mid-intertidal zone mentioned earlier. For the latter, however, drought comes twice each day.

When chaparral and coastal sage species grow together near the overlapping edges of each other's range, they exploit different parts of the environment: deep perennial sources of water and shallow ephemeral sources of water. In spite of these differences and the corresponding adaptations of leaf morphology and drought response, the annual productivity of both types of species is about the same where there are intermediate levels of water availability. In drier habitats, the prolonged seasonal absence of deep water tips the balance in favor of the adaptations found in deciduous coastal sage vegetation. Increasing availability of deep water at higher elevations favors evergreen chaparral vegetation.

Ecotypes

Botanists have long recognized that a species, when grown in different habitats, may exhibit various forms corresponding to the conditions under which it is grown. In the hawk-weed *Hieracium*, for example, woodland plants generally have an erect habit, those from sandy fields are prostrate, and those from sand dunes are intermediate in form. Leaves of the woodland ecotype are broadest, those of the dunes ecotype are narrowest, and those of sandy fields, intermediate. Plants from sandy fields are covered with fine hairs, a trait the others lack. When individual plants of the same species are taken from different habitats and grown in a garden under identical conditions, differences in growth form often persist, generation after generation. About sixty years ago, the Swedish botanist Göte Turesson collected seeds of several species of plants that occurred in a wide variety of habitats and grew them in his garden. He found that even when grown under identical conditions, many of the plants exhibited different forms depending upon the habitats from which they were originally taken. Turesson called these forms *ecotypes*, a name that persists to the present, and suggested that ecotypes represented genetically differentiated strains of a population that are restricted to specific habitats. Because Turesson grew these plants under identical conditions, he realized that the differences between the ecotypes must have a genetic basis, and they must have resulted from evolutionary differentiation within the species according to habitat.

Jens Clausen and co-workers David Keck and William Hiesey conducted similar experiments, published in 1948, on an introduced species of yarrow, *Achillea millefolium*, in California. *Achillea*, a member of the sunflower family, grows in a wide variety of habitats ranging from sea level to more than 10,000 feet elevation. Clausen collected seed from plants at various points along the altitude gradient and planted them at Stanford, California, near sea level. Although the plants were grown under identical conditions for several generations, individuals from montane populations retained their distinctively small size and low seed production (Figure 14-14), thereby demonstrating ecotypic differentiation within the population. Such regional

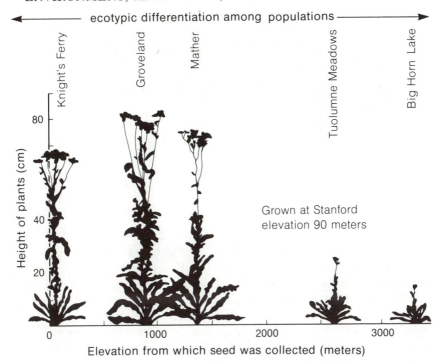

FIGURE 14-14 Ecotypic differentiation in populations of the yarrow, *Achillea millefolium*, demonstrated by raising plants derived from different elevations under identical conditions in the same garden.

and habitat differences in adaptations undoubtedly broaden the ecological tolerance ranges of many species by dividing the species into smaller sub-populations, each differently adapted to local environmental conditions.

The physical environment clearly plays an immense role in shaping the adaptations of plants and animals, and in determining their distributions. In the last two chapters, we have seen how the activity space of an organism is shaped by the interaction of adaptation and environment and how the concept of activity space can be extended on a geographical scale to encompass the distribution of the species.

Whether a population can persist under a given set of environmental conditions or not depends on the balance between births and deaths of individuals. Like the activity spaces of organisms, population processes reflect the interaction between the environment, including both physical and biological factors, and the intrinsic nature of the population, particularly its capacity to reproduce. In the following chapters, we shall examine the balance of birth rates and death rates in population dynamics, and how this balance is affected by the activities of other populations—whether they are predators, prey, or competitors for resources.

15 | Population Growth and Regulation

ALTHOUGH IT IS the individual that reproduces and the individual that dies, the ecological effect of these events can be appreciated only by examining the entire population. This chapter is about populations—their growth and regulation. Populations are open, dynamical systems into which individuals enter by birth and leave by death. Even though the size of a population may remain nearly constant over a long period, its composition is in continual flux. Change in population size, brought about when births or deaths get the upper hand, is more generally the rule in natural populations. But the single great fact about populations is that while numbers of births and deaths may fluctuate wildly in response to varying environmental influences, the numbers of births and deaths very nearly balance over the long run. That is, in spite of their ups and downs, changes in population size also appear to be under the influence of regulatory processes.

Population Structure in Space

Populations are composed of many individuals spread out over the geographical range of the species. But individuals are not usually distributed evenly. Ecologists recognize three patterns of dispersion in populations: *clumped* (underdispersed), *random*, and *spaced* (overdispersed) (Figure 15-1). These types of distribution patterns are caused by different types of individual behavior. On one hand, clumping results from the attraction of individuals to the same place, either an intrinsically suitable environment or a gathering spot for such social functions as mating. On the other hand, an evenly spaced distribution usually signifies antagonistic interactions between individuals. In the absence of mutual attraction or social dispersion, individuals are distributed at random, without regard to the position of other individuals in the population.

Variation in the suitability of the environment causes individuals to congregate where the environment is most favorable. Although the sugar maple occurs widely throughout the eastern United States and Canada,

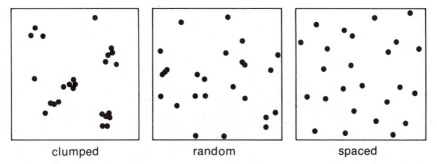

clumped   random   spaced

FIGURE 15-1 Diagramatic representation of the distribution of individuals in clumped, random, and evenly spaced populations.

maple trees are most numerous on soils that are slightly acid; they may be missing altogether from some hardwood forests within their geographical range. On a smaller distance scale, the salamanders in a maple forest congregate under fallen logs where the humidity of the environment falls within their range of tolerance.

Within suitable parts of the geographical range, behavioral interactions between individuals may space members of the population at regular distances. Amidst the pandemonium of seabird breeding colonies, nests are separated just far enough to prevent the adults from pecking each other while sitting on their nests (Figure 15-2). Most land birds, many fish and arthropods, and some reptiles and mammals defend exclusive *territories* within which they search for food and raise their offspring. All antagonistic behavior, including territorial defense, tends to space individuals evenly over suitable habitats. Plant populations exhibit even spacing of individuals owing to mutual shading and root competition.

When we view the spatial distribution of a population at a moment in time, the positions of individuals are fixed as if in a photograph. Of course, most animals continually move. Even plants disperse during reproductive periods either as pollen grains or seeds; phytoplankton float freely in the water. Dispersal is an adaptation of the organism, allowing it to find a suitable habitat to settle in. But dispersal has an important population function as well. Spatial and temporal variation in the suitability of a habitat result in population growth in some areas and population decline in others. Movement of individuals between areas enables a population to respond to local variation in the environment more rapidly than if local reproduction and death were the only avenues of response open to populations.

Age Structure and Sex Ratio

All individuals are not alike. Demographers, people who study the structure and dynamics of populations, distinguish individuals according to sex and age. Because death rates and fecundity rates vary by sex and age, the future

FIGURE 15-2 A nesting colony of Peruvian boobies on an island near the coast of Peru. The densely packed birds space their nests more or less evenly, the distance being determined by behavioral interactions between individuals. Along the entire length of the Peruvian coast, however, sea bird populations are clumped, during the breeding season, on a few offshore islands with suitable nesting sites.

growth or decline of a population is influenced by its age and sex structure. For example, the growth potential of a human population with many females between 15 and 35 years of age is greater than a population consisting mostly of old men and preadolescents.

The growth rate of a population—the net result of birth and death rates—also influences the age structure of a population. The size of the human population of Sweden has been relatively constant for many decades. As a result, the proportion of the population in each age class (the *age structure* of the population) parallels the expected percentage survival of newborn babies to each age (the *survivorship curve*) (Figure 15-3). Consider for a moment a population with a constant number of births and an equal number of deaths each year. Because births exactly balance deaths, the size of the hypothetical population is also constant. If, in such a population, 95 per cent of newborn babies lived to 45 years of age, the population would contain 95 per cent as many 45-year-olds as newborn. Similarly if 35 per cent of babies lived to be 80, the population would contain 35 per cent as many 80-year-olds as newborn. In Sweden, the correspondence between age structure and survivorship is reasonably close, although the relative dearth of old individuals betrays a slight though persistent excess of births over deaths. In 1965, the population of Sweden was increasing at a rate of 0.6 per cent per year.

In sharp contrast, the population of Costa Rica had a growth rate of about 4.1 per cent in 1963. A high birth rate swelled the young age classes, resulting in a bottom-heavy structure. Although a male baby born in 1963 had an 80 per cent probability of surviving to 45 years of age, the number of 40- to 45-year-old men in the population was only one-fifth the number of 1- to 5-year-old male children. The discrepancy between the size of the age class and survivorship is caused by the smaller number of babies producing the 40 to 45 year age class compared to the 1 to 5 year class. Costa Rican men 40 to 45 years old in 1963 were born between 1918 and 1923, when the population was much smaller.

Any variation in the death rate or number of young born creates irreg-

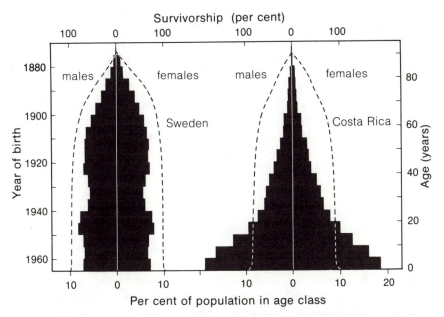

FIGURE 15-3 Population age-structure and survivorship, calculated separately for males and females, in Sweden (1965) and Costa Rica (1963). Because Sweden's population has grown slowly, its age structure resembles the survivorship curve. Declining birth rates during the Depression, and the baby boom which followed World War II, are responsible for irregularities in the age-structure of Sweden's population. Costa Rica's rapid population growth has resulted in a bottom-heavy age structure.

ularities in the age structure of a population. An exceptionally productive year might send a large year-class through the age structure of the population in such a manner that 1-year-olds would predominate the first year, 2-year-olds the following year, and so on. Populations in which the number of young born or their survival during the first year varies greatly from year to year exhibit irregular age structure. Environmental changes that increase adult death rate usually affect all ages equally; although population size may decline, age structure does not change greatly. Subsequent population growth will, however, result in a relative increase in the younger age classes.

Seasonal patterns of reproduction and mortality in short-lived animals create cyclic changes in age structure. Honeybee populations expand rapidly in the spring and assume the bottom-heavy age distribution characteristic of growing populations. As reproduction declines toward the end of the summer, the younger age classes dwindle in number, and finally disappear altogether after reproduction ceases. During winter, surviving honeybees move through progressively older age classes.

The ratio of males to females (the *sex ratio*) in most animal populations is about 1:1. The sex ratio changes with age if births and deaths favor one sex over the other. For example, more males are born into human populations, but females live longer. The sex ratios of some parasitic species and social insects are grossly biased in favor of females. Males have no role in reproduction beyond fertilizing the eggs, and one male can go a long way toward that end. Some fish, lizards, and aquatic invertebrates have abandoned sex altogether: females produce eggs that do not require fertilization to develop. Plants are a demographic nightmare. The flowers of most species have both male and female sexual organs. In species with separate sexes, sex ratios vary from place to place within a population and rarely are evenly balanced. Until recently, population biologists have paid little attention to plants and correspondingly little is known about their population structure.

The Life Table

The *life table* is a summary of the vital statistics of a population: the *survivorship* to, and *fecundity* of, each age class. Survivorship to a given age is usually denoted by the symbol l_x, where x is the age specified; fecundity at a given age is denoted by b_x. These statistics are usually computed only for females because the fecundity of males is extremely difficult to determine in populations where mating occurs promiscuously.

The construction of a life table for a population requires that the ages of individuals be known, a straightforward problem for captive or laboratory populations and for (or perhaps including) human populations, but more difficult for natural populations. Age-specific survival rates are normally estimated from the survival of individuals marked at birth or from the deaths of individuals of known age, but both methods have serious limitations. The first requires that individuals do not emigrate from the area in which they were marked; otherwise they would erroneously be counted as dead. The second method requires that large numbers of individuals be marked to obtain sufficient recoveries of dead individuals. For example, of 180,718 robins banded as adults in the United States between 1946 and 1965, only 2,444 were ever retrapped or found dead in subsequent years; of 7,604 brown pelicans banded as nestlings, only 375 were ever seen again.

When the ages of individuals can be determined by their appearance, without knowledge of date of birth, the construction of a survivorship table becomes a relatively simple task. Among plants, for example, the age of temperate forest trees can be determined from growth rings in their wood. Among animals, the age of fish can be determined from growth rings in their scales or ear bones; the age of mountain sheep can be estimated from the size of their horns.

Estimated age of death, judged by the size of horns on skulls, was used to calculate survivorship for the Dall mountain sheep (Figure 15-4) in Mount McKinley National Park, Alaska. In all, 608 skeletal remains of sheep were found: 121 of the sheep were judged to have been less than 1-year-old at death; 7 between 1 and 2 years, 8 between 2 and 3 years; and so on, as shown in Table 15-1. We may construct a survivorship table using the following reasoning: all 608 dead sheep must have been alive at birth; all but 121 that died during the first year must have been alive at the age of 1 year (608 − 121 = 487); all but 128 (121 dying during the first year and

FIGURE 15-4 Dall mountain sheep in Mount McKinley National Park, Alaska. The size of the horns increases with age.

TABLE 15-1   Survivorship table for the Dall mountain sheep constructed from the age of death of 608 sheep in Mount McKinley National Park, Alaska.

Age interval (years)	Number dying during age interval	Number surviving at beginning of age interval	Number surviving as a fraction of newborn (survivorship)
0–1	121	608	1.000
1–2	7	487	0.801
2–3	8	480	0.789
3–4	7	472	0.776
4–5	18	465	0.764
5–6	28	447	0.734
6–7	29	419	0.688
7–8	42	390	0.640
8–9	80	348	0.571
9–10	114	268	0.439
10–11	95	154	0.252
11–12	55	59	0.096
12–13	2	4	0.006
13–14	2	2	0.003
14–15	0	0	0.000

7 dying during the second) must have been alive at the end of the second year (607 − 128 = 480); and so on, until the oldest sheep died during their fourteenth year. Survivorship (right-hand column in Table 15-1) was calculated by converting the number of sheep alive at the beginning of each interval to a decimal fraction of those alive at birth. Thus, the 390 sheep alive at the beginning of the seventh year represented 64.0 per cent (decimal fraction 0.640) of the original newborn in the sample.

Survivorship provides half the life table. One must also know the age-specific fecundity to fully understand the dynamics of the population. Field biologists have devised techniques for estimating fecundity from embryo counts in mammals, nest checks in birds, ratios of juveniles to adults in many kinds of animals, and direct counts of eggs in amphibians, insects, marine invertebrates, and others. As in the matter of death, the age at which a female produces offspring is all-important. An average litter size of 5 young at age 3 contributes to population growth less than the same size litter born to females at age 2 simply because more females live to age 2 than to age 3. To calculate the total production of young by a population, the average fecundity of each age group must be multiplied by the number of adults in that age group, and the products summed over all ages.

Calculating Population Growth Rate from the Life Table

The growth rate of a population depends on two calculations from the life table: the net reproductive rate and the mean generation time. The *net reproductive rate* is the expected number of female offspring to which a newborn female will give birth in her lifetime. Of course, some individuals die before they attain reproductive age and thus leave no offspring; exceptionally long-lived females give birth to many more young than the population average. But it stands to reason that if females produce more than one female progeny on the average, the population will grow; if females fail to replace themselves on the average, the population will decline.

The rate at which a population grows or declines also depends on the *mean generation time*, which is the average age at which females produce offspring. The earlier young are born, the earlier they, in turn, have offspring and the more rapid population growth will be.

The net reproductive rate, denoted here by the letter R, is calculated by adding the expected production of offspring by a female at each age. In the example presented in Table 15-2, each female could expect to produce 0.5 young, on the average, at age 1 (average fecundity of 1.0 times expected survivorship of 0.5), 1.6 young at age 2 (0.40 times 4.0), and so on, totalling an expected net reproductive rate of 3.1. In other words, at the time of death, females in the population would have left an average of 3.1 female offspring.

The mean generation time (T) is calculated by finding the average age at which a female gives birth to her offspring. In the present example, 0.5 young are produced at age 1, 1.6 young at age 2, 0.8 young at age 3, and 0.2 young at age 4. We calculate the average age by adding the products of each age and the number of offspring produced at the same age, and dividing the sum by the total number of offspring produced. In the example in Table 15-2, the sum of the products is 6.9. Dividing this figure by the net reproductive rate (3.1), we obtain a mean generation time of 2.22 years. Our calculations show that a group of newborn individuals will themselves produce 3.1 times as many newborns in an average of 2.22 years. It is

TABLE 15-2 Life table of a hypothetical population, demonstrating the calculation of the net reproductive rate and mean generation time.

Age (x)	Survivorship (l_x)	Fecundity (b_x)	Expected offspring $(l_x b_x)$	Product of age and expected offspring $(x l_x b_x)$
0	1.00	0.0	0.0	0.0
1	0.50	1.0	0.5	0.5
2	0.40	4.0	1.6	3.2
3	0.20	4.0	0.8	2.4
4	0.10	2.0	0.2	0.8
5	0.00	0.0	0.0	0.0
	Net reproductive rate =		3.1	
	Total weighted age =			6.9

difficult to compare population growth rates expressed on a generation basis. For example, is a factor of 3.1 every 2.22 years greater or less than a factor of 1.4 every 1.36 years? To avoid this problem, population growth rates are usually expressed on an annual basis, and are specified by the Greek letter lambda (λ). The population described above would increase by a factor of 1.6 each year.*

The Growth Potential of Populations

The age-specific survivorship and fecundity of a population are used to calculate its rate of growth. Birth and death rates reflect the interaction of the individual with its environment—its ability to channel resources into reproduction and its ability to avoid predation, starvation, or death by exposure. The mathematics of demography translates the individual's activities into the dynamics of the population. If the environment of the individual is unchanging, the growth rate of the population should not vary. If a population increases by a factor of 1.6 each year, 100 individuals become 160 after 1 year, 256 after 2 years, 410 after 3 years, progressing in a geometric fashion (Figure 15-5).

Populations increase by multiplication (geometric growth) rather than by simple addition (arithmetic growth) because young born into the population themselves grow up to give birth. Population increase thus resembles compound interest on a bank account in which the earned interest (newborn young) is periodically added to the principal (reproducing adults).

The geometric growth rates of populations under conditions that are ideal for growth reveal a basic fact: populations have a tremendous intrinsic capacity for growth. Population increase, if unrestricted by limiting resources or predators, would be sufficient to raise the numbers of most kinds of animals to astronomical figures in a very short time. Charles Darwin appreciated this growth potential more than a century ago, when he wrote in *On the Origin of Species* that "There is no exception to the rule that every organic being naturally increases at so high a rate that, if not destroyed, the earth would soon be covered by the progeny of a single pair."

We can best appreciate the capacity of a population for growth by following its rapid increase when introduced to a new region with a suitable environment. The number of colonists is at first so low that crowding and

---

* The annual growth rate of a population (λ) is calculated by the formula

$$\lambda = R^{1/T}$$

where R is the net reproductive rate and T, the mean generation time. In the example presented in Table 15-2 ($R = 3.1$ and $T = 2.22$), the annual growth rate is

$$\lambda = 3.1^{1/2.22}$$
$$= 3.1^{0.45}$$
$$= 1.6$$

In a strict sense, the population would grow somewhat faster than $\lambda = 1.6$ because early born progeny will themselves give birth to offspring before their parents will have died.

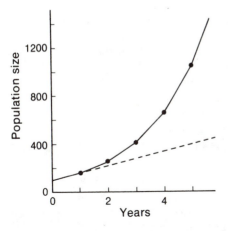

FIGURE 15-5 Growth of a population whose annual rate of increase (λ) is 1.6. The initial population size (time = 0) was arbitrarily set at 100 individuals. The dashed line represents the population size if growth had continued by adding a constant amount (60 individuals) to the population (arithmetic growth) rather than multiplying the population by a constant factor (1.6 times each year) (geometric growth).

depletion of resources do not hinder population growth. When domestic sheep were introduced to Tasmania, a large island off the coast of Australia, the population increased from less than 200,000 in 1820 to more than 2,000,000 in 1850. The ten-fold increase in 30 years is equivalent to an annual rate of increase of 8 per cent ($\lambda = 1.08$). Two male and six female ring-necked pheasants introduced to Protection Island, Washington, in 1937 increased to 1,325 adults within five years. The 166-fold increase represents a 180 per cent annual rate of increase ($\lambda = 2.80$). Even such an unlikely creature as the elephant seal, whose population had been all but obliterated by hunting during the nineteenth century, increased from 20 individuals in 1890 to 30,000 in 1970 ($\lambda = 1.096$). If we are unimpressed, we should consider that another century of unrestrained growth would find 27,000,000 elephant seals crowding surfers and sunbathers off southern California beaches. Before the end of the next century, the shorelines of the Pacific Ocean would give lodging to one trillion elephant seals.

Elephant seal populations do not hold any growth records. Life tables of populations maintained under optimum conditions in the laboratory have shown that potential annual growth rates (λ) may be as great as 24 for the field vole, ten billion (10^{10}) for flour beetles, and 10^{30} for the water flea *Daphnia*. Rapid growth rates are more conveniently expressed in terms of the time required for the population to double in number. Corresponding doubling times are 7.6 years for the elephant seal, 8 months for the pheasant, 80 days for the vole, 10 days for the flour beetle, and less than 3 days for the water flea. Populations of microorganisms (bacteria and viruses) and many unicellular plants and animals can double in a day or a few hours.

Geometric and Exponential Growth

When a population grows at a constant rate, the increase in number of individuals over a period of time t can be predicted from the number of individuals at the beginning of the period $N(0)$ and the rate of growth λ.

We put a zero in parentheses following the N to signify the initial population, when time is arbitrarily set equal to zero, i.e., the present. Now, population size at any time t in the future may be calculated from the equation

$$N(t) = N(0)\lambda^t \tag{15-1}$$

For example, beginning with 100 individuals, a population with an annual growth rate λ of 1.6 increases to 1050 after 5 years (see Figure 15-5). We obtain this value by substituting $\lambda = 1.6$, $t = 5$, and $N(0) = 100$ into equation 15-1.

λ describes the geometric growth rate of a population. As we have seen, it is the factor by which a population changes over a specified interval of time, such as the period between annual counts. It is also a biologically realistic description of growth in a population with periodic reproduction, such as we see in the short summer breeding seasons of many temperate-zone plants and animals. Other populations grow continuously, without seasonal fluctuations. The human population and those of many animals under domestication are familiar examples. In addition, the growth of populations of animals or plants with short lifespans may be approximately continuous during the growing season, as may those of long-lived individuals when viewed over periods of many years.

In continuously growing populations, we may wish to describe the instantaneous growth rate of the population, which is the slope of the growth curve at a given point, $dN(t)/dt$. The notation may be read as "the rate of change in population size N at time t with respect to time." The instantaneous growth rate is defined in terms of the population size $N(t)$ and the exponential growth rate of the population r according to the expression

$$dN(t)/dt = rN(t) \tag{15-2}$$

The exponential growth rate is analogous to the constantly increasing interest on a bank account that is continually compounded. When we say that a population of a million individuals grows at an exponential rate of 5 per cent per day ($r = 0.05$), its size increases at a rate of 50,000 individuals per day only for an instant, because an instant later some individuals will have been added to the population, and they also contribute to its increase. That is, $N(t)$ grows continuously and so, therefore, does $dN(t)/dt$. In order to find the relationship between population size and time, it is necessary to integrate equation 15-2. The resulting expression

$$N(t) = N(0)e^{rt} \tag{15-3}$$

for exponential growth is similar to equation 15-1 for geometric growth. As you can see, λ in equation 15-1 is equivalent to e^r in equation 15-3. Conversely, $r = \log_e \lambda$. Geometric and exponential growth rates are readily exchanged in equations for population growth. For a population remaining constant in size, $\lambda = 1$ and $r = 0$; for a growing population, λ exceeds 1 and r exceeds 0; for a declining population, λ exceeds 0 but is less than 1, and r is negative. In the development of population biology as a scientific discipline, r has become more generally used than the geometric growth rate to express population increase.

The Regulation of Population Size

Under the most favorable conditions, populations occasionally attain their maximum potential rates of exponential increase. If these rates were to prevail for long periods, populations, even of elephant seals, would increase so as to cover the earth. Fortunately, ecological factors bearing upon the population act to bring its growth under control.

The number of individuals in a population is limited by the availability of resources. Barnacle populations cannot increase beyond the point that all the available rock surface is covered. The number of pairs of titmice in a forest cannot exceed the number of available nesting sites. Predators cannot become so numerous that they reduce their prey below the level required for their own maintenance.

As populations increase, the physiological effects of crowding and depletion of resources begin to change the life table by decreasing birth rate or increasing death rate, or both, with the result that population growth rate declines (Figure 15-6). If the density of a population is low, relative to the abundance of its resources, the birth rate will exceed the death rate, and the population will increase. The growth rate of the population (number of individuals added to or removed from the population) is equal to the birth rate minus the death rate. As the population increases, deaths increase proportionately faster than number of individuals in the population, and the number of births per individual decreases. If population density exceeds the level that the environment can support, the death rate will exceed the birth rate and the population will decline. For each set of environmental conditions, a level of population density exists at which birth and death rates exactly balance each other and the population neither grows nor declines. This point of population equilibrium is referred to as the *carrying capacity* of the environment (K). Whenever the population falls below or exceeds K, population responses cause population size to return to the carrying capacity. At low population densities (point B in Figure 15-6), population growth rate is positive; at high densities (point A), growth rate is negative. Hence K represents a stable equilibrium.

Laboratory populations maintained at different densities, but under identical conditions, demonstrate the response of birth rate and death rate to density. In a series of experiments on the water flea, *Daphnia pulex*, initial population densities varied between 1 and 32 individuals per milliliter of water. The populations were grown in small beakers with cultures of

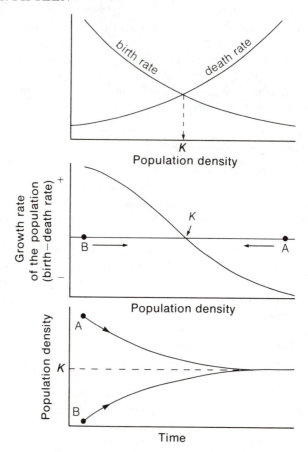

FIGURE 15-6 Diagramatic representation of birth and death rates in a regulated population (top), the resulting growth rate of the population with respect to density (middle), and the approach of population size to equilibrium (*K*) with time (bottom). A and B represent the densities of two populations whose birth and death rates are not in equilibrium.

green algae provided for food. Survivorship and fecundity of females were noted for two months after the beginning of the experiment and the data were used to construct life tables for each population density, shown graphically in Figure 15-7. Fecundity decreased markedly with increasing population density. Survivorship actually increased at densities up to eight individuals per milliliter; high fecundity apparently reduces survivorship, indicating a trade-off between reproductive effort and adult survival. At densities of eight or more individuals per milliliter the body growth of individual water fleas was stunted, suggesting that depletion of food resources ultimately limited birth rates and survivorship in the dense cultures.

Population growth rate (λ), calculated from life table data for the water fleas, decreases with increasing density and falls below 1.0 (equilibrium

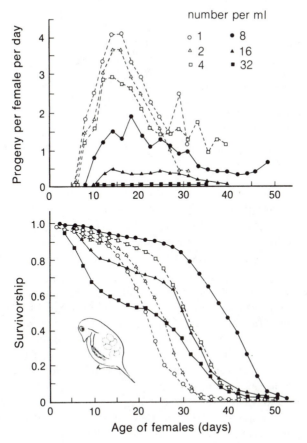

FIGURE 15-7 Fecundity and survivorship of females in laboratory populations of the water flea *Daphnia pulex* maintained at different densities.

population size) at a density of about 20 individuals per milliliter (Figure 15-8). Under the conditions of temperature, light, water quality, and food availability provided in the laboratory, water flea populations reach a stable equilibrium of 20 individuals per milliliter, representing the carrying capacity of the environment, regardless of the initial density of the culture.

The Carrying Capacity of the Environment

Populations are regulated by the balance between two opposing forces: the inherent growth potential of the population and the limits to population growth imposed by the environment. The ability of the environment to support a species varies with climate and resource availability, and so then does the equilibrium population density of the species.

Population density is limited by two kinds of resources. On one hand,

FIGURE 15-8 Values of λ calculated from the life table data portrayed in Figure 15-7. Population
growth rate decreases as a function of
density. A density of 20 individuals
per milliliter represents the apparent
carrying capacity of the environment,
at which point λ is equal to 1.0 and
the population just maintains itself.

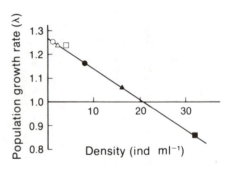

nonrenewable resources such as space or nesting sites can be completely
utilized by a population, thereby placing an abrupt upper limit on population size. On the other hand, *renewable resources* such as food, water, and
light are supplied continually to populations. The resource demands of a
large population can reduce renewable resources to levels at which they are
so difficult to locate or assimilate that they will not support further population growth, but renewable resources are never completely exhausted.
Renewable resources are maintained at an equilibrium level by the balance
between exploitation and production. When population size equals the
carrying capacity of the environment, the resource requirements of the
population just match the renewal rate of its resources. If the population
exceeds its carrying capacity, exploitation exceeds renewal, resources are
depleted, individuals starve, and the population begins to decline. Conversely, if the rate of resource production increases, the environment can
support a larger population. Dense populations are usually found in the
most productive habitats, as shown by the relationship between the population density of birds and primary plant productivity in a variety of habitats
(Table 15-3).

   Populations of barnacles and mussels on rocky coasts are limited by a
nonrenewable resource—space. Individuals become so crowded that larvae

TABLE 15-3 Censuses of birds in six habitats listed in increasing order of primary
production.

Habitat	Locality	Total breeding pairs per 100 acres	Net primary production (g m^{-2} yr^{-1})
Desert	Mexico	22	70
Prairie	Saskatchewan	92	300
Chaparral	California	190	500
Pine forest	Colorado	290	800
Hardwood forest	West Virginia	320	1,000
Flood-plain forest	Maryland	581	1,500

FIGURE 15-9 A crowded population of mussels.

have no place to settle (Figure 15-9). Such high densities occur because barnacles and mussels are restricted to flat surfaces at the edge of their food supply, plankton obtained directly from sea water washing over the animals. As a result, these filter feeders have little effect on the abundance of their food, and therefore food does not limit the size of their populations. Other intertidal creatures that feed on algae that grows directly on the rock surface can exhaust their food supplies. Limpets and herbivorous snails are much more sparsely distributed in the intertidal region than are barnacles.

The influence of resource productivity on the carrying capacity of the environment for mule deer populations has been demonstrated by range management experiments. Chaparral habitats in northern California were turned into shrublands by mechanical thinning, controlled burning, and seeding with herbaceous plants. As a result of stimulated plant growth, experimental shrublands had 2.8 times as much edible foliage in woody species and 17 times as much edible foliage in herbaceous species as unaltered chaparral control areas. Deer in shrublands were able to select higher quality foods than in chaparral habitats as well (14 versus 9 per cent protein content). As a result of increased food production, populations of deer in artificially maintained shrublands were more than twice as dense as populations in chaparral; individuals were 5 to 10 per cent heavier in shrubland habitats.

The direct physiological effects of food availability on reproduction were shown in a study of white-tailed deer in New York. Per cent of females pregnant and number of embryos per doe (incidence of twins) was directly related to range conditions (Table 15-4). The opening of one Adirondack study area to hunting incidentally demonstrated the relationship between food availability and reproduction. Before hunting was allowed, the over-crowded deer population in the DeBar Mountain region had a pregnancy rate of 57 per cent, with 0.7 surviving embryos per pregnant female. After the population had been reduced by two years of hunting, pregnancy rates rose to nearly 100 per cent with 1.8 embryos per female; nearby areas inaccessible to hunters exhibited no increase in reproductive performance.

The Description of Regulated Population Growth

In a population whose density is determined by its resources, population growth rate is inversely related to the density of the population (Figure 15-6). Equation 15-3, which describes the exponential increase of populations with constant growth rates, clearly is not applicable. Instead, we must make the exponential rate of increase a function of population size, that is $r(N)$, so that $dN(t)/dt = r(N)N(t)$. A simple relationship for $r(N)$ that has much experimental and empirical justification is a linear decrease in r as N increases, hence

$$r(N) = r_o(1 - N/K) \qquad (15\text{-}4)$$

In this expression, the influence of population size on growth rate is adjusted in relation to the carrying capacity of the environment K. When N is very small relative to K, $(1 - N/K)$ is close to 1, and $r(N)$ is close to r_o, the maximum potential rate of exponential increase. As population size approaches the carrying capacity, N/K approaches 1 and the term $(1 - N/K)$ approaches zero. When N equals K, $r(N)$ equals 0. Of course, when N exceeds K, $r(N)$ is negative and the population declines. The expression $r(N)$ balances two opposing forces: the biotic potential of the individual for reproduction (r_o) and the detrimental effects of crowding on that potential $(1 - N/K)$.

TABLE 15-4   Pregnancy rate and number of embryos per pregnant female in white-tailed deer populations in New York. The localities are listed in decreasing order of range condition.

Region	Per cent of females pregnant	Embryos per pregnant female
Western (best range)	94	1.7
Catskill periphery	92	1.5
Catskill center	87	1.4
Adirondack periphery	86	1.3
Adirondack center (worst range)	79	1.1

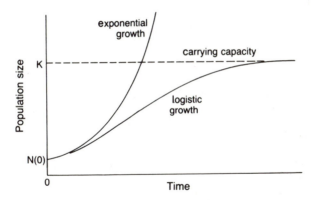

FIGURE 15-10 Representation of unrestricted exponential population growth, and logistic growth with an upper bound to population size imposed by the carrying capacity (K). Initial population size (time = 0) is $N(0)$.

In a population regulated according to equation 15–4 the instantaneous growth rate is described by

$$dN(t)/dt = r_oN(t)[1 - N(t)K]$$ (15-5)

Like equation 15-2, this expression may be integrated to obtain a slightly more complicated equation describing the increase in population size over time. It is written

$$N(t) = \frac{K}{1 + \left[\dfrac{K - N(0)}{N(0)} \right] e^{-rt}}$$ (15-6)

and is usually called the *logistic* equation. Starting with a small population, the form of the logistic equation is sigmoid, or S-shaped. Initially, the population increases at an accelerating rate, as in unrestricted exponential growth. But as external factors limiting population size are felt more strongly, the growth curve bends over and finally levels off at the carrying capacity (Figure 15-10). Of course, when the population exceeds the carrying capacity, its growth rate is negative and it decreases until it levels off at K.

Fluctuations in Natural Populations

Natural environments are rarely so constant as the laboratory. Variation in climate and food availability, through their influence on survivorship and fecundity, continually change the direction and rate of growth of natural populations. The degree of variation in the size of a population depends

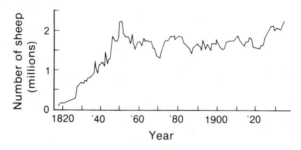

FIGURE 15-11 Number of sheep on the island of Tasmania since their introduction in the early 1800s.

partly on the magnitude of environmental fluctuations and partly on the inherent stability of the population. Populations of large, long-lived, slowly reproducing plants and animals are comparatively insensitive to changing environments because of their inherent homeostatic capacities. For example, after sheep became established on Tasmania, their population varied irregularly between 1,230,000 and 2,250,000 individuals over nearly a century (Figure 15-11). Short-lived organisms with high reproductive capacities are more sensitive to short-term fluctuations in the environment; their populations frequently increase and decrease by factors of hundreds or thousands over days or weeks (Figure 15-12). Because sheep raised for wool production live for five to ten years, the population at any one time consists of young born over a relatively long period, thereby evening out the influence of short-term variations in birth rate on population size. The lifespan of the single-celled algae that constitute the phytoplankton of a lake is measured in days; the individuals in an algae population are replaced rapidly and the population is thus vulnerable to the capriciousness of the environment.

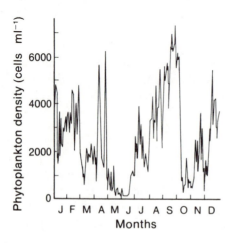

FIGURE 15-12 Variation in the number of phytoplankton in samples of water from Lake Erie during 1962.

In seasonal environments, reproduction occurs only when climate and resources combine to produce favorable conditions. Seasonal changes in temperature, moisture, and nutrients also affect mortality, either directly or indirectly. When lifespan is so short that many generations occur within a year, seasonality can greatly influence population size. Near Adelaide, South Australia, populations of the tiny insect pest *Thrips imaginis*, which infests roses and other cultivated plants, undergo regular cycles of increase during seasonally favorable conditions, followed by rapid decline when the environment becomes either too dry to too cold to support population growth (Figure 15-13). Adelaide has a Mediterranean climate: winters are cool and rainy, summers are hot and dry. The winter months are generally too cold to sustain growth of thrips populations; summers are too dry. The spring (October through December in the Southern Hemisphere) brings an ideal combination of moisture, warmth, and plant flowering for population growth.

Thrips subsist mainly on plant pollen, whose abundance varies seasonally with the production of flowers. The climate of Adelaide is mild enough that some flowers are always available and thrips are active all year. During the winter, however, the depressing effect of cool temperature on development rate and fecundity (Table 15-5) brings a marked decline in the number of thrips. Population growth is further checked by high mortality during the immature stages; the period from egg to adult is so long during the winter that most flowers wither and fall off before the thrips reach maturity. The warm weather of spring increases the net reproductive rate of the thrips while shortening the mean generation time. Under these conditions, populations rapidly increase to infestation levels. Increasing mortality caused by ensuing summer drought halts population growth and results in rapid population decline in late November or early December.

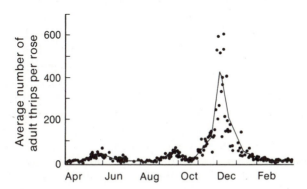

FIGURE 15-13 Number of *Thrips imaginis* per rose from April 1932 through March 1933 near Adelaide, Australia. Dots indicate daily records; the curve is a moving average for successive 15-day intervals.

TABLE 15-5 Influence of temperature on development rate, lifespan, and fecundity of *Thrips imaginis*.

	TEMPERATURE (C)	
	8 to 12	23 to 25
Length of adult life (days)	250	46
Total eggs laid per female*	192	252
Daily egg production	1.4	5.6
Development period (days)	44	9

* With pollen (a protein source) in their diet. If adults are raised without pollen, their egg production falls to 20 eggs at 24 C, and life span increases to 77 days.

Density-Dependence and Density-Independence

Many of the environmental factors that influence the growth rate of thrips populations do so independently of population density. Temperature interacts with development rate to influence mean generation time whether thrips are numerous or few. Hot, dry air robs thrips of their body moisture regardless of their numbers. Factors whose influence on population growth rate is independent of population size are called *density-independent factors*. Although such factors can cause great fluctuation in the numbers of individuals in populations, their action does not lead to a stable equilibrium.

The influence of *density-dependent factors* on population size changes with the density of individuals (see Figure 15-6). These factors somehow depend on the interaction of individuals. A dense population depletes its food supply, utilizes suitable breeding and resting places, attracts predators, and hastens the spread of disease.

Density-dependence is often associated with biotic factors and density-independence with physical factors, although this distinction does not always pertain. For example, the killing effects of a frost (a "density-independent" factor) depend on whether the individual can find a sheltered refuge. If the environment contains a limited number of well-protected living places, the influence of such climatic factors may well be "density-dependent." A severe frost does not affect all individuals in a population equally; some survive, others die. Climate and other physical factors also indirectly exert a density-dependent influence on population size through their effect on food supply.

The rapid increase of thrips populations in the spring, as well as their decline in the fall, seemingly results from changes in density-independent influences on population size. Thrips rarely infest every flower and they consume only a small portion of the food available to them, even when populations are dense. Population increase is fostered in the spring by favorable conditions of temperature and moisture. Density-independent factors overwhelmingly influence the size of the thrips populations, but they cannot regulate population size over long periods. In fact, the peak annual population of thrips varies little from year to year compared to seasonal variation in population size. It would be quite unlikely that the product of all the daily values of λ for the thrips population would be very close to 1.0 (representing an unvarying long-term population trend) just by chance unless density-dependent factors operate at some season. These presumed factors have not actually been identified for the thrips population, but research on Canadian insect pests, particularly the spruce sawfly, has identified the factors responsible for regulation of some insect populations.

Population Dynamics of the Spruce Sawfly

The spruce sawfly was first introduced to Quebec from Europe in about 1930. By 1938, it had spread throughout most of Quebec, the maritime provinces of Canada, and New England, causing widespread damage, including complete defoliation of spruce forests. Most sawflies are female—fewer than one in a thousand are male—and reproduction is predominately asexual. Females lay eggs singly in slits cut in spruce needles; the larvae feed on the foliage after they hatch (Figure 15-14). After the larvae are fully grown, they drop to the ground and burrow a few inches into moss, where they spin a tough cocoon around themselves and metamorphose into adults. The sawfly population goes through several generations each summer, but toward fall, pupae (the developmental stage between larval and adult forms) enter into a diapause state in their cocoons to spend the winter. The Canadian government introduced several parasites in an attempt to control the sawfly, but eventually a virus, accidentally introduced with one of the parasites, proved to be the most effective control agent.

Canadian foresters and entomologists initiated a census program during the height of the sawfly outbreak in the late 1930s. The program was

FIGURE 15-14 The larch sawfly, a close relative of the European spruce sawfly, feeding on needles of western larch.

continued for more than 20 years. Larvae were sampled by spreading large canvas sheets on the ground under a tree and vigorously shaking the limbs directly above it. The dislodged larvae were collected and reared individually in vials to determine rates of disease and parasitism in the population. A census of cocoons was taken each year. Wooden trays were filled with moss gathered from an area free of sawflies. The trays were then set out in an infested forest during July, before the first generation larvae began to drop to the forest floor, to serve the larvae in place of naturally occurring moss. The following May, after most sawflies had emerged from their cocoons, the trays were brought back to the laboratory and the cocoons classified as either (a) sound but unemerged, (b) emerged normally, (c) parasitized, (d) preyed upon by wireworms, or (e) preyed upon by small mammals, mostly rodents. Each type of predator and parasite leaves a distinguishing mark on the tough and leathery cocoons.

The size of the sawfly population varied by factors typically as great as 10 and even as great as 100 over a period of a few years (Figure 15-15). Rates of parasitism and disease also varied greatly. Statistical analysis of population trends in relation to biotic and climatic factors identified factors responsible for fluctuations in population size; the analysis also distinguished density-dependent from density-independent factors. When the annual growth rate of the population (λ) was plotted on a graph with respect to population size, growth rate was found to decrease significantly with increasing density—an indication of population regulation by density-dependent factors.

Values of λ were at first calculated as the ratio of one year's population of larvae to the previous year's population of larvae. To determine whether larval parasites and disease organisms contributed to density-dependent population regulation, λ was recalculated as the ratio of one year's newly-hatched larvae to the previous year's emerged adults. By using the previous year's adult stage as the basis of comparison, larval mortality would be eliminated from the relationship between population growth rate and density. If λ still exhibited density-dependence, one would have to look elsewhere for the action of density-dependent factors. If the density dependence of λ disappeared by eliminating larval mortality from the analysis, one could pinpoint the regulating factor to the larval stage. Similar analyses were used to test each mortality factor for density-dependent influence.

The sawfly study provided useful insight into the regulation of insect populations. First, although disease and weather were important components of mortality, neither exerted a stabilizing effect on population size. (Disease was, however, largely responsible for bringing serious outbreaks under control.) Second, parasites were found to be the only factor that acted in a density-dependent manner during the larval period. Third, additional unidentified density-dependent factors must have acted at other stages of the life history because larval mortality could not account for all the regulatory influences.

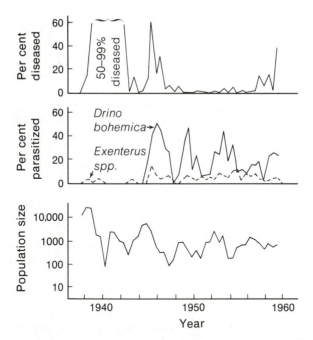

FIGURE 15-15 Relative population size, disease rate, and parasitism rate of European sawfly larvae in a black spruce forest. Note the apparent long-term constancy underlying the short-term fluctuations in population size.

Entomologists have reached similar conclusions from studying other insect populations. For example, whereas climate accounts for 80 per cent of variation in numbers of the diamondback moth, a pest on cabbage, the long-term regulation of population size resides with parasites that strike late in the larval period. In other studies, the critical stage at which mortality occurred, as well as the mortality factor itself, varied from species to species. Only rarely did density-dependent factors cause most of the short-term variation in population size.

Population Cycles

Some populations of long-lived organisms, particularly birds and mammals, fluctuate with great regularity; peaks and troughs in their numbers occur at intervals of anywhere from three to ten years. The most remarkable of the cycles are exhibited by some mammal populations of the New World arctic, where the regularity of population fluctuations is precise enough to predict population size several years in advance. For example, lynx populations of Canada have a cycle of approximately ten years (Figure 15-16), closely following cyclic changes in the population of their principle prey, the snowshoe hare (Figure 15-17).

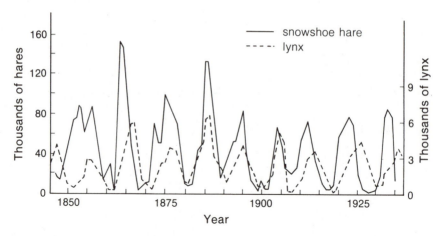

FIGURE 15-16 Population cycles of the lynx and the snowshoe hare in the Hudson Bay region of Canada, as indicated by fur returns from trappers to the Hudson's Bay Company.

Ecologists are not completely agreed on the cause of population cycles. The populations vary much too regularly for the cycles to be caused solely by variation in the environment. The most reasonable explanation ties the cycles to inherent population properties of predator-prey interactions. In one version of this hypothesis, an increase in the hare population is followed by an increase in the lynx population, which eventually becomes so dense

FIGURE 15-17 A snowshoe hare in its winter coat. The hares are brown in summer.

that the hare population can no longer sustain the predation rate and declines. Lynx follow suit as their food supply diminishes. When the lynx become scarce, the hare population can increase and the cycle begins again. This version probably does not adequately explain the lynx and hare population cycles because there are too many bits of contrary evidence. First, the reproductive potential of the hare is so much greater than that of the lynx that the lynx population could not increase fast enough to exterminate the hares unless some other factor, perhaps insufficiency of food, slowed the growth rate of the hare population. Second, lynx population peaks occasionally coincide with or precede hare population peaks rather than following them by a year or so. Third, on some islands from which lynx are absent, snowshoe hare populations fluctuate just as much as they do on the mainland. Perhaps the population cycles are caused by periodic decline in quality or quantity of plants on which the hares feed, which in turn causes a decline in the hare population (and the lynx population), and subsequently allows the plants a chance to recover from overeating by hares.

Attempts to dissect the population cycles of arctic mammals into their working parts by field observation and experimentation have met with little success. Lemmings have been studied intensively at Barrow, Alaska, where population density varies in a three- to five-year cycle by factors of 100 or more between trough and peak years. Predators undoubtedly exert a strong depressing influence on population growth because most of the increase in number of lemmings occurs during the winter beneath a protective mantle of snow. Nevertheless, the reduction of lemming population size in summer by predators is small compared to longer-term changes in population size. The only habitat characteristic that appears to parallel the lemming cycle at Barrow is the quality (not quantity) of the vegetation (Table 15-6). During one peak year, vegetation contained 22 per cent protein, but only 14 per cent during the next population trough. If lemming populations and plant quality have parallel cycles, the two may be interrelated: the total available nutrients in the tundra ecosystem may be of such small quantity that they

TABLE 15-6 Features of a lemming population cycle, including vegetation characteristics, at Barrow, Alaska.

| | YEAR | | | |
	1960	1961	1962	1963
Relative peak density	125	0.5	1 to 10	50
Male body weight (g)	92	47	69	59
Litter size	7.6	7.0	7.3	6.7
Breeding season (days)	58	80	73	83
Green vegetation (lbs/A^{-1})	111	278	115	149
Per cent protein	22	14	17	19
Protein in plants (lbs/A^{-1})	24	40	20	28

are mostly transferred to the bodies of lemmings during peak years, thereby depressing plant growth and nutrient quality. Nutrients would then be restored to the cycle only after large numbers of lemmings die and their remains decompose.

Time Lags and Population Cycles

We expect density-dependent factors to restore population size to an equilibrium level, and yet cyclic populations never reach a single equilibrium point, but fluctuate around that point. If we can determine why density-dependent responses fail to damp population cycles, we may come closer to understanding predator-prey cycles. Experiments by Australian ecologist A. J. Nicholson on population regulation in the sheep blowfly demonstrate that if the action of density-dependent factors is delayed, population cycles may occur.

The life cycle of the fly includes four stages: egg, larva, pupa, and adult. In laboratory cages both the larvae and the adults are fed liver. When larvae were fed 50 grams of liver each day, and adults received unlimited sugar solution to supplement their diet, the number of adults in the population fluctuated in a regular cycle from a maximum of about 4,000 per cage to a minimum of zero (at which point all the individuals in the population were either eggs, larvae, or pupae). The period of the oscillation varied between 30 and 40 days, which is the maximum lifespan of individual flies (Figure 15-18).

Regular fluctuations in the blowfly population were caused by a time lag in the response of fecundity and mortality to the density of adults. Large adult populations laid many eggs, but most of the larvae hatched from those eggs starved or failed to pupate owing to an inadequate amount of food; few became adults. Because large adult populations produced few progeny, and because adults live a maximum of four weeks, peak populations declined rapidly. During adult population lows, few eggs were laid and most of the larvae survived. The population then increased to peak levels again. Fluctuations in population numbers were not damped because the effects of adult population size were felt during the larval period of the *next* generation. These effects were expressed in adult populations two to four weeks after the eggs were laid when the progeny were adults of the next generation.

The time lag in the blowfly population could be eliminated by restricting the amount of food available to adults. Since adults must consume protein to produce eggs, limiting their food curtails egg production to a level determined by the availability of liver rather than by the number of adults in the population. When adult blowflies were fed only one gram of liver per cage per day, and the larvae were fed a special food preparation that the adults could not eat, the recruitment of new adults into the population was determined at the egg-laying stage by the influence of food on fecundity

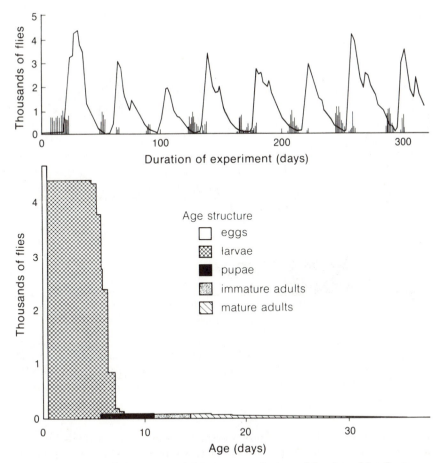

FIGURE 15-18   Fluctuations in laboratory populations of the sheep blowfly. Larvae were provided 50 grams of liver daily; adults were given unlimited supplies of food and water. The continuous line represents the number of adult blowflies in the population cage. The vertical lines represent the number of adults that eventually emerged from eggs laid on the days indicated by the lines. The average age structure of the population, representing the average survivorship curve, is shown below.

(nearly all the larvae survived); as a result, fluctuations in the population all but disappeared (Figure 15-19).

Behavioral Aspects of Population Regulation

Social behavior among animals ranks an individual according to its position on a scale of social dominance or according to the quality of territory it procures. Strong territorial systems or hierarchical systems of social dominance can prevent may individuals in a population from reproducing. Environmental fluctuations often affect the proportion of individuals reproduc-

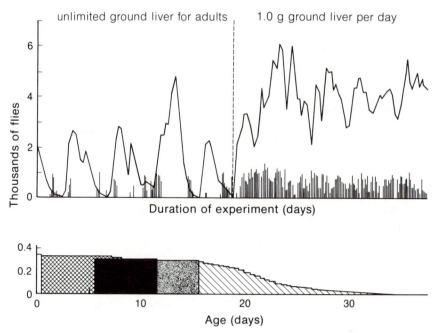

FIGURE 15-19   Effect on fluctuations in a laboratory population of sheep blowflies of limiting the amount of liver available to adults. The experiment resembled that depicted in Figure 15-18 in all other respects. Average age structure is shown for the latter half of the experiment.

ing in addition to the fecundity or survivorship of individuals that do reproduce. The attainment of sexual maturity by mammals often depends on the successful procurement of breeding territories. Within social groups, individuals vie for position in the dominance hierarchy and for the opportunity to reproduce, which depends on high social status (Figure 15-20).

By limiting the occupation of suitable habitats, territoriality can exert a strong stabilizing effect on population size. Social behavior immediately adjusts the density of organisms to year-to-year fluctuations in their resources, and thereby eliminates the time lag from the density-dependent response. Mammals that maintain breeding territories in the summer and hibernate through the winter—chipmunks, squirrels, bears, raccoons—do not exhibit well-defined population fluctuations. Mammals and birds that do show cyclic populations are typically nonterritorial for a large part of the year.

Individuals excluded from holding territories in optimum habitats either establish territories in poor habitats where they cannot breed or wander through the population searching for the occasional opening resulting from the death of a territory holder. Experimental removal of territorial birds, fish, and insects has demonstrated the presence of large reserves of non-

FIGURE 15-20 Two bull elks fighting among a herd in Wyoming. The outcome of the encounter will affect each adversary's social position and his prospects of reproducing.

breeding individuals that quickly occupy vacated habitats. In addition to limiting reproduction, failure to procure a territory exposes the individual to increased risk of mortality through starvation, disease, exposure, or predation.

Social Pathology

Working with artificial laboratory populations of mammals, many physiologists have observed decreased survival and extreme impairment of reproduction at high population density, even when food, water, and nesting sites are provided in excess of requirements. Social stress leads to a variety of abnormal physiological symptoms, collectively referred to as the *general adaptive syndrome*, which includes curtailment of growth and reproduction, delay of sexual maturity, increased mortality of embryos, inadequate lactation, and increased susceptibility to disease.

When house mouse populations are confined with unlimited food and nesting boxes, mice continue to reproduce regardless of the density of the population, but the survival of embryos and young drops so low that population growth ceases and population size may even decline (Figure 15-21). As density increases, individuals fight more frequently and the number of wounded or diseased adults, mostly males, rises.

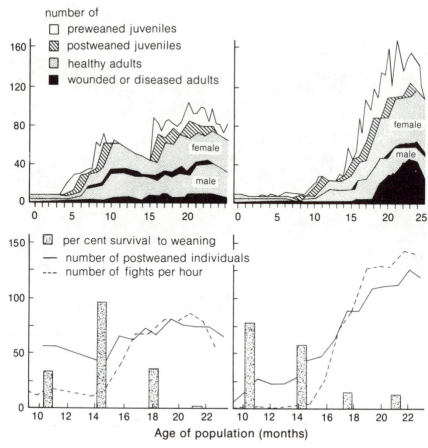

FIGURE 15-21 Growth of two confined house mouse populations supplied with unlimited food, showing the age structure, level of aggression, survival of young to weaning, and the incidence of diseased or wounded adults.

Social pathology has been suggested as a mechanism of population regulation in natural populations of mammals and as a cause of population cycles, but concepts of social stress developed for confined populations do not apply widely in nature. First, few natural populations achieve the densities that can be maintained in the laboratory. The densities of natural populations are kept low by emigration of socially subordinate individuals from regions of high population density. Second, where resources can support dense populations, evolutionary responses in social behavior will favor those individuals capable of tolerating extreme crowding. We see this in many colonial organisms that reproduce normally at densities far beyond levels that would be tolerated by species which maintain large breeding territories. That reproductive behavior can evolve to function normally at high population densities can best be appreciated by a visit to a colony of

FIGURE 15-22 A dense breeding colony of royal terns on the coast of North Carolina. The social behavior of the terns has evolved to allow tolerance of high densities.

prairie dogs or sea birds, where young are successfully raised amidst utter bedlam (Figure 15-22).

We have examined the balance of forces that regulate population size. Some forces are density-dependent. The available resources in its environment limit the inherent capacity of a population to grow and tend to keep the population at a level for which the resouces are adequate. If the availability of resources changes, the size of the population changes in an appropriate manner. Other forces—cold, drought, and the like—are density-independent. They may limit the growth of a population without regard to its density or to its resources.

Changes in population size are compensated by homeostatic population responses, including growth and dispersal. The response of a population to its resources can lead to cyclic fluctuations in population size if the influence of the response is delayed by time lags. Such cyclic fluctuations have great influence on the total stability of the ecosystem and on the efficiency of energy and nutrient flow.

Inseparable from the notion of density-dependent forces and responses of populations to them is the fact that individuals compete with one another for resources. In the next chapter, we shall examine the influence of competition, whether between individuals of the same or of different species, on population processes and on the coexistence of populations of different species within the community.

16 | Competition

THE TENDENCY OF POPULATIONS to increase exponentially expresses the unrestricted biotic potential of the individual. The realization of this potential diminishes as population density increases; this is the essence of population regulation. The growth rate of a population is dependent upon the fecundity and longevity of individuals. The depressing influence of density on growth rate reflects the detrimental effect of competition on the survival and reproductive performance of individuals.

How does competition operate? When the availability of a particular food resource is reduced by the feeding activities of each individual that uses the resource, the ability of other individuals in the population to obtain the resource is reduced. Organisms whose resource requirements are most similar will compete most intensely, and so competition between individuals of the same species is usually more intense than competition between individuals of different species.

The Competitive Exclusion Principle

Field observations and laboratory experiments convey opposite impressions of nature. We frequently observe many ecologically similar species coexisting in nature, clearly using many of the same resources. Closely related species of trees grow in the same habitat, all needing sunlight, water, and soil nutrients. Coastal estuaries and inland marshes harbor a variety of fish-eating birds, including egrets, herons, terns, kingfishers, and grebes (Figure 16-1). As many as six kinds of warblers (small insectivorous birds), having similar morphology and feeding habits, may be found in one locality in forest habitats of the northeastern United States. Eight species of large predatory snails frequent the shallow waters of Florida's Gulf Coast.

By contrast, closely related species rarely coexist in the laboratory. If two species are forced to live off the same resource, inevitably one persists and the other dies out. Reconciling these observations has been a major task for ecologists over the past half century, since the Russian biologist G. F. Gause tried to make two similar species coexist in the laboratory.

Gause established laboratory cultures of closely related species of pro-

277

FIGURE 16-1 Three closely related species of egrets (common, reddish, and snowy) feeding in the Aransas National Wildlife Refuge, Texas.

tozoa, of the genus *Paramecium*, in the same nutritive medium. The species flourished when separate, but in mixed cultures only one species survived (Figure 16-2). Similar experiments with fruit flies, mice, flour beetles, and annual plants have always produced the same result: One species persists and the other dies out, usually after thirty to seventy generations.

The results of laboratory competition experiments led to the formulation of the *competitive exclusion principle*, also called Gause's principle, which states that two species cannot coexist on the same limiting resource. This principle was first appreciated by the American naturalist Joseph Grinnell and the earliest experimental demonstration of the principle was published by the English plant ecologist A. G. Tansley in 1917. But it remained for Gause's extensive laboratory experiments to raise the principle to the level of general recognition by ecologists.

The word "limiting" is included in the statement of the competitive exclusion principle because only resources that limit population growth can provide the basis for competition. Nonlimiting resources, like atmospheric oxygen, are superabundant compared to the needs of organisms, and their use by one organism does not make them less available to others.

How can the principle of competitive exclusion, verified by laboratory

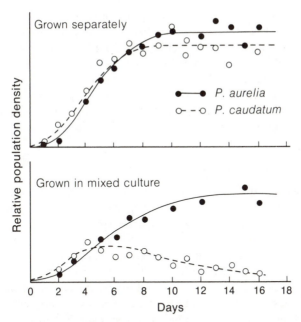

FIGURE 16-2 Increase in populations of two species of *Paramecium* when grown in separate cultures (above) and when grown together (below). Although both species thrive when grown separately, *P. caudatum* cannot survive when grown with *P. aurelia*.

experiments, be reconciled with observations of similar species coexisting in nature? Coexistence suggests that competition between species generally is much weaker in natural communities than in laboratory populations. Species avoid competition by partitioning resources and habitats among themselves, a response that species in simplified laboratory environments cannot make. But the existence of resource partitioning in nature does not mean that competition does not occur. The inhibiting influence of populations upon each other's growth has been demonstrated in many instances by the response of a population to the removal of ecologically similar species from the habitat—for example, in competition for space among barnacles and competition for light and moisture among tropical forest trees. In general, however, when species can avoid extensive overlap with other species in the use of resources, competition can be reduced sufficiently to permit indefinite coexistence. It seems that only in laboratory situations, where species are forced to exploit a single resource, is competition so intense that only one population can persist.

Does competitive exclusion ever occur in nature? Does competition influence the number of species that can coexist in a community? Do the numerous ways in which species exploit the environment, and thus avoid competition, allow any conceivable number of species to be crammed into a community? Or is there some degree of ecological similarity between species that cannot be exceeded without leading to exclusion of one or more of them, and are natural communities saturated with species whose similarities just approach this level? Ecologists have not resolved these problems; the factors that determine the number of species in a community constitute an important area of ecological research. As we shall see, the analysis of competition has provided a useful approach.

Examples of Competitive Exclusion

The species we find coexisting in nature are the successful ones. What of the species that failed? Fossils of extinct species prove that populations have died out. Was their demise caused by superior competitors?

Competitive exclusion is a transient phenomenon. The evidence of exclusion having taken place is lost when the poorer competitor is eliminated. We can observe competitive exclusion in the laboratory because we can mix populations according to our whims and follow the course of their interaction. The closest natural analogy to the laboratory experiment is the accidental or intended introduction of species by man. When new immigrants are superior competitors to resident species, the immigrants can reduce or eliminate local populations. The explosion of rabbit populations following their introduction to Australia worsened range conditions and thereby reduced populations of many native marsupial herbivores, such as kangaroos and wallabies. (Rabbit-control programs were aimed more at preserving the range for another introduced competitor—the sheep—than preserving the native fauna.)

Competition between introduced species has produced some striking demonstrations of the competitive exclusion principle. When many species of parasites are introduced simultaneously to control a weed or insect pest, the control species are brought together in the same locality to prey on, or parasitize, the same resource. We should not be surprised that competitive exclusion has occurred frequently under these conditions. Between 1947 and 1952, the Hawaii agriculture department released 32 potential parasites to combat several species of fruit pests, including the Mediterranean fruit fly. Thirteen of the species became established, but only three kinds of braconid wasps proved to be important parasites of fruit flies. Populations of these species, all closely related members of the genus *Opius*, successively replaced each other from early 1949 to 1951, after which only *Opius oophilus* was commonly found to parasitize fruit flies (Figure 16-3). As each parasite population was replaced by a more successful species, the level of parasitism of fruit flies by wasps also increased, suggesting superior competitive ability on the part of the most recently established wasp population.

A similar pattern of competitive replacement involving wasps that are parasites of scale insects, which are pests of citrus groves, has been thor-

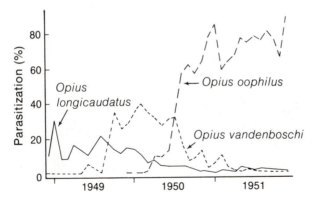

FIGURE 16-3   Successive change in predominance of three species of wasps of the genus *Opius*, parasitic on the Oriental fruit fly.

oughly documented in southern California. With the failure, owing to the evolution of resistance by pests, of chemical pesticides to provide adequate long-lasting control, agricultural biologists turned to the importation of insect parasites and predators. Yellow scales have infested California citrus groves since oranges and lemons were first planted there. In the late 1800s, the red scale was accidentally introduced and has replaced yellow scale almost completely, perhaps itself a case of competitive exclusion. Of the many species introduced in an effort to control citrus scale, tiny parasitic wasps of the genus *Aphytis* (from the Greek *aphyo*, to suck) have been most successful. One species, *A. chrysomphali*, was accidentally introduced from the Mediterranean region and became established by 1900.

The life cycle of *Aphytis* begins when adults lay their eggs under the scaly coverings of hosts. The newly hatched wasp larva uses its mandibles to pierce the body wall of the scale and proceeds to consume nearly all the body contents. After the wasp pupates and emerges as an adult, it continues to feed on scales while producing eggs. Each female can raise 25 to 30 progeny under laboratory conditions, and the development period is so short (egg to adult in fourteen to eighteen days at 27 C or 80 F) that populations may produce eight to nine generations per year in the long growing season of southern California.

In spite of its tremendous population growth potential, *A. chrysomphali* did not effectively control scale insects, particularly not in the dry interior valleys. In 1948, a close relative from southern China, *A. lingnanensis*, was introduced as a control agent. This species increased rapidly and almost completely replaced *A. chrysomphali* within a decade (Figure 16-4). When both species were grown in the laboratory, *A. lingnanensis* was found to have the higher net reproductive rate, whether the two species were placed separately or together in population cages.

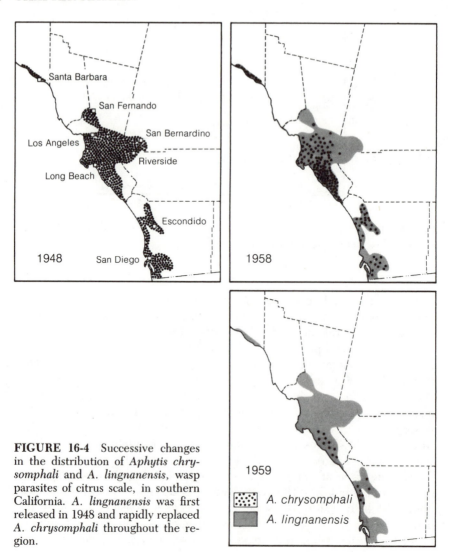

FIGURE 16-4 Successive changes in the distribution of *Aphytis chrysomphali* and *A. lingnanensis*, wasp parasites of citrus scale, in southern California. *A. lingnanensis* was first released in 1948 and rapidly replaced *A. chrysomphali* throughout the region.

Although *A. lingnanensis* had replaced *A. chrysomphali* throughout most of southern California, it still did not provide effective biological control of scale insects in the interior valleys because cold winter temperatures periodically reduced parasite populations. Wasp development slows to a standstill at temperatures below 16 C (60 F), and adults cannot tolerate temperatures below 10 C (50 F). Pupae are most resistant to cold, but average winter pupal mortality was 42 per cent, with extremes of about 80 per cent in the cold interior valleys.

In 1957, a third species of wasp, *A. melinus*, was introduced from areas in India and Pakistan where temperatures range from below freezing in winter to above 40 C in summer. As expected, *A. melinus* spread rapidly throughout the interior valleys of southern California, where temperatures resemble the wasp's native habitat, but did not become established in coastal areas (Figure 16-5).

To determine that competition was responsible for changes in the wasp populations, *A. lingnanensis* and *A. melinus* were cultured under controlled laboratory conditions—27 C (80 F) and 50 per cent relative humidity—which resembled the climate of coastal areas more closely than that of the inland valleys. The wasps were provided with oleander scale grown on lemons as hosts, and new scale-infested lemons were added every seventeen days. Populations of wasps were counted just before new food was added. When cultured separately, both species parasitized about 40 per cent of the scales, and wasp populations averaged 6,400 for *A. lingnanensis* and 8,300 for *A. melinus*. When grown together, *A. melinus* was reduced in four months from 50 per cent to less than 2 per cent of the total wasp population. During the process of competitive exclusion, the combined population of the two species (average 7,400) remained at approximately the same level as that of either species grown separately. We may infer that population size was limited by host availability and that under the conditions of the experiment, *A. lingnanensis* was the superior competitor.

Competition in Plant Populations

Plants differ from animals in two ways that bear on the study of competition. First, few terrestrial species have generation times of less than a year. Plant ecologists, therefore, often cannot continue experiments for a long enough

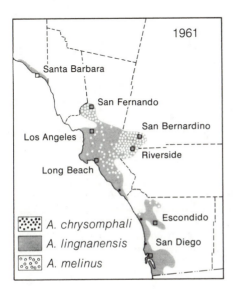

A. chrysomphali

A. lingnanensis

A. melinus

1961

Santa Barbara

San Fernando

San Bernardino

Los Angeles

Riverside

Long Beach

Escondido

San Diego

FIGURE 16-5 Distribution of three species of *Aphytis* in southern California in 1961. *A. melinus* predominates in the interior valley, while *A. lingnanensis* is more abundant near the coast.

period to demonstrate competitive exclusion. Second, plant growth, as well as survival, is greatly affected by the variety of conditions under which the plant may live. In particular, plants grow slowly when crowded and do not attain their full stature, even though they may produce seed. In contrast, animal populations usually respond to crowding with increased mortality and reduced fecundity, not stunted growth.

When horseweed (*Erigeron*) seed is sown at a density of 100,000 (10^5) seeds per square meter (equivalent to about ten seeds in the area of your thumb nail), the young plants compete vigorously. As the seedlings grow, many die and the density of surviving seedlings decreases (Figure 16-6). At the same time, the growth of surviving plants exceeds the decline of the population, and the total weight of the planting increases. Over the entire growing season, the hundredfold decrease in population density is more than balanced by a thousandfold increase in the average weight of each plant.

The relationship between plant weight and density, shown in Figure 16-6, is often called a *self-thinning curve*. When obtained from very dense plantings, self-thinning curves reflect the maximum capacity of the soil to support plant growth. Combinations of density and size lying outside the self-thinning line never occur because they are beyond the carrying capacity of the environment. If horseweed were sown at a density of 10,000 (10^4) seeds per square meter, rapid growth with little early mortality would carry the population vertically on the graph in Figure 16-6, until it reached the

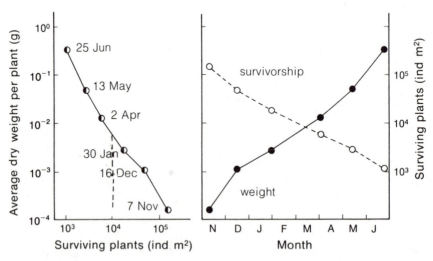

FIGURE 16-6 Progressive change in plant weight and population density in an experimental planting of horseweed sown at a density of 100,000 seeds per square meter (10 cm^{-2}). The relationship between plant density and plant weight as the season progressed is shown at left.

self-thinning line. Sown at densities of 1,000 per square meter, few of the horseweed plants would die before reaching maturity.

Each species has a characteristic thinning curve under particular conditions of soil, light, temperature, and moisture. Regardless of the density of the initial planting, the eventual harvest of mature plant biomass is surprisingly uniform. The self-thinning curve does not, however, tell the full story of within-species competition. When sown at low initial density, most individuals grow vigorously. At high density, only a few individuals reach large size. Other survivors do poorly because they are crowded by individuals that obtained an initial growth advantage owing to a favorable site for germination (Figure 16-7).

Because of the flexible growth response of plants and their long generation times, botanists usually assess competition between plant populations by comparing total plant weight or number of seeds produced in an experimental plot. These indexes were used to measure competition between two closely related species of oats, *Avena barbata* and *A. fatua*. Plants were grown in pots at six densities: 8, 16, 32, 64, 128, and 256 individuals per pot. When the species were grown separately, density had relatively little effect on germination, establishment of seedlings, or survival to maturity. Growth responses were, however, markedly different: At high densities, individual plants attained smaller average weight and height and produced fewer seeds.

To measure the influence of interspecific competition on plant growth, pots were sown at each of several densities with different proportions of the two species. In experiments with seed densities of 128 per pot, for example, a group of pots would be planted with 128 *barbata* and 0 *fatua*, 112 *barbata* and 16 *fatua*, 64 *barbata* and 64 *fatua*, and so on, each totaling 128 seeds.

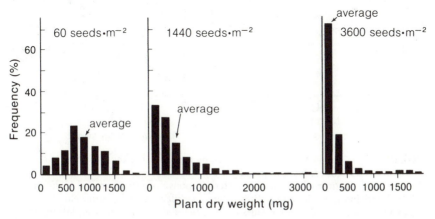

FIGURE 16-7 The distribution of dry weights of individuals in populations of flax plants sown at different densities.

For each species, the pots created conditions ranging from pure intraspecific competition (128 vs. 0), to strong interspecific competition (16 vs. 112).

When grown separately at densities of 128 seeds per pot, both species showed survival rates of about 96 individuals per pot (75 per cent). If the survival of a *fatua* individual were affected by competition with *barbata* individuals no differently than from competition with other *fatua* individuals, we would expect 75 per cent of *fatua* seeds to survive regardless of the ratio of the two species in the pot. Thus 12, 48, and 84 individuals would survive from initial plantings of 16, 64, and 112 seeds. The relationship between the initial planting and the final outcome is often depicted on a graph called a *replacement series diagram* (Figure 16-8). The left-hand figure represents the expected number of *fatua* surviving to maturity if interspecific competition and intraspecific competition had equal influence on survival. A straight line drawn between 96 individuals on the right-hand axis (128 *fatua* seeds per pot) and 0 on the left-hand axis (0 *fatua* seeds per pot) represents 75 per cent survivorship of *fatua* seeds. If, on one hand, interspecific competition depressed the survival of *fatua* seedlings, relative to the effect of intraspecific competition, the outcomes of the experiment in the mixed species pots would fall below the line of equal competitive effect. If, on the other hand, *fatua* grew better under conditions of strong interspecific competition than it did under strong self-inhibition, the points would fall above the line of equal competitive effect.

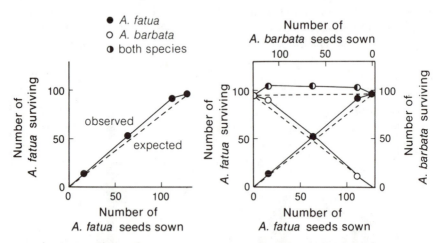

FIGURE 16-8 Replacement series diagrams representing the outcome of competition between *Avena fatua* and *A. barbata* grown at a density of 128 seeds per pot. The left-hand diagram shows the expected response of *fatua* survival if the effects of intra- and interspecific competition did not differ. Actual results closely match this expectation. At right, the results of *A. fatua* and *A. barbata* are plotted on the same diagram. The expected total survivorship of both species is added to the diagram.

Survival of both *fatua* and *barbata* exhibited no differential response to intraspecific and interspecific competition. Weight and seed production did, however, display differences (Figure 16-9). Compared with the outcomes of single-species cultures, *A. fatua* produced more seeds per plant when grown in competition with *barbata*; *A. barbata* produced fewer seeds per plant when grown with *fatua*. Thus *fatua* competed well against *barbata* under the conditions of the experiment, and we would expect *fatua* eventually to exclude *barbata*.

We may visualize the outcome of competition on a ratio diagram (Figure 16-10), in which the ratio of seeds produced by one species to seeds produced by the other is compared with the initial ratio of seeds planted. If

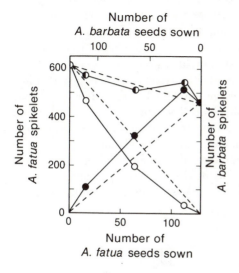

FIGURE 16-9 Replacement series diagram for seed production (measured as number of flower spikelets per pot) in *Avena fatua* and *A. barbata*.

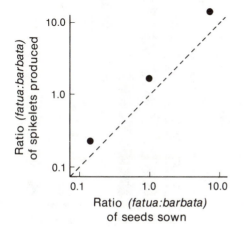

FIGURE 16-10 Ratio diagram of competition between *Avena fatua* and *A. barbata*, representing the outcome of the experiment depicted in Figure 16-9.

the ratio of the species of seed produced from a pot were identical to the ratio of seeds sown, neither species would have a competitive advantage. This situation is represented by the diagonal line in Figure 16-10. If *fatua* were relatively more productive than *barbata*, the outcome of competition experiments would lie in the upper left-hand part of the diagram, representing an increase in the ratio of *fatua* to *barbata*. Conversely, if *barbata* were favored, the outcome would lie in the lower right-hand portion of the diagram. At high initial seed density, all experimental results favored *fatua* over *barbata*. Inasmuch as the proportion of *fatua* in mixed populations would continually increase, *fatua* eventually would exclude *barbata* under the conditions of the experiment.

Plant ecologists have begun to study competition between species by transplanting individuals to natural sites, with or without close competitors, and following their subsequent growth and reproduction. This technique has been applied to the study of competition between two species of *Desmodium*, which are small herbaceous legumes common in oak woodlands in the midwestern United States. Small individuals of each species, *D. glutinosum* and *D. nudiflorum*, were planted either 10 cm from a large individual of the same species, 10 cm from a large individual of the other species, or at least 3 m from any *Desmodium* plant. Total increase in length of all leaves, both old and new, added together provided an index of subsequent growth. The results of the experiment (Figure 16-11) showed that both species grew most rapidly in the absence of individuals of either species. Note that the transplanted individuals of *Desmodium* were placed among numerous unrelated plant species, regardless of their proximity to other

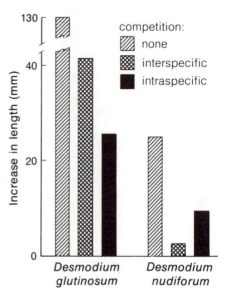

FIGURE 16-11 The growth responses of two species of *Desmodium* when planted near individuals of the same species (black bars), near individuals of the other species (crosshatched bars), and at a distance from individuals of either species (hatched bars).

Desmodium individuals. It was also clear, however, that *D. glutinosum* exerted a far stronger depressing effect on the growth of *D. nudiflorum* than the reverse. All other things being equal, *D. glutinosum* should be able to replace *D. nudiflorum* in the habitats in which the transplant experiments were performed, but leaf growth does not provide a full measure of plant productivity, and the disadvantage of *D. nudiflorum* could be balanced during some other stage of the life cycle.

Experiments with laboratory and artificially manipulated natural populations show that competition is a powerful force in the ecology of populations. But the persistence of closely related species in the same habitats also tells us that competition need not lead to competitive exclusion. To determine the conditions under which competing species can coexist, population biologists have resorted to the analysis of mathematical models of population growth, including the mutual effects of competitors on the growth rates of their populations. Most of these models are built upon a simple mathematical expression for the growth of a single population limited by the availability of its resources.

A Mathematical Model of Competition

In the previous chapter, we described the rate of growth of a population regulated by density-dependent factors with

$$dN/dt = rN(1 - N/K) \tag{16-1}$$

The term $(1 - N/K)$ expresses the effects of competition among individuals in the population for limited resources. When kept in separate culture, populations of *Paramecium* increase to a well-defined carrying capacity after a period of sigmoid growth (see Figure 16-2). When Gause put populations of two species in the same culture, one had a detrimental effect on the growth of the other. This competitive relationship can be expressed mathematically by adding a term to the equation for logistic growth (16-1). We shall use the subscripts 1 and 2 to denote the population sizes, carrying capacities, and potential growth rates of species 1 and 2. Now, for species 1,

$$dN_1/dt = r_1N_1(1 - N_1/K_1 - a_{12}N_2/K_1) \tag{16-2}$$

In this equation, the term $a_{12}N_2/K_1$ represents the effects of interspecific competition, specifically the amount by which individuals of species 2 reduce the growth rate of population 1. Notice that the number of individuals of species 2 is expressed in terms of the carrying capacity for species 1 because it is the latter's resources that species 2 usurps to effect dN_1/dt. Also, the population of species 2 is adjusted by a coefficient of competition a_{12}, which expresses the equivalence of individuals of the two species with respect to resources of the first. When a_{12} is 1, individuals of both species depress the growth rate of the population of species 1 equally. When a_{12} is small, individuals of species 1 and 2 have different ecological requirements and the effect of 2 on the population of 1 is small. For species 2, the expression analogous to that for species 1 (16-2) is

$$dN_2/dt = r_2 N_2(1 - N_2/K_2 - a_{21}N_1/K_2)$$ (16-3)

If two species are to coexist, their populations must both reach a stable size greater than zero. That is, both dN_1/dt and dN_2/dt must both be zero at some combination of positive values of N_1 and N_2. From equations 16-2 and 16-3, we see that $dN_1/dt = 0$ when

$$1 - N_1/K_1 - a_{12}N_2/K_1 = 0$$

or

$$\hat{N}_1 = K_1 - a_{12}N_2$$ (16-4)

Similarly, $dN_2/dt = 0$ when

$$\hat{N}_2 = K_2 - a_{21}N_1$$ (16-5)

The little hats over the N's indicate that they are equilibrium values. By substituting $K_2 - a_{21}\hat{N}_1$ for N_2 in equation 16-4, and doing a little algebra, we find that

$$\hat{N}_1 = (K_1 - a_{12}K_2)/(1 - a_{12}a_{21})$$ (16-6)

Similarly,

$$\hat{N}_2 = (K_2 - a_{21}K_1)/(1 - a_{12}a_{21})$$ (16-7)

These equations express the equilibrium values of $\hat{N}_1$ and $\hat{N}_2$ solely in terms of carrying capacity and coefficients of competition between two species. When analyzed in greater detail, equations 16-6 and 16-7 reveal that two species may coexist stably only when a_{12} is less than the ratio K_1/K_2 and a_{21} is less than K_2/K_1. In general, the product of the competition coefficients $a_{12}a_{21}$ must be less than 1. In biological terms, this means that the less similar the ecological relations of two species, (i.e., the lower the values of a), the more likely they will coexist. The converse of this truism is that where species coexist, they must have mechanisms by which they avoid intense competition.

Species Diversification

Species coexist because their resource requirements differ. Whenever ecologists have examined groups of similar species in the same habitat they have found small but significant differences in size or foraging behavior that enable the species to use slightly different resources and avoid intense competition. Imagine that we are watching the great cormorant and the shag cormorant (large, dark-colored, diving sea birds) feed in the same area and apparently in the same manner. They both swim on the surface of the water and dive for food. The cormorants seem to be competing for food, but the appearance is deceptive. Had we followed the cormorants beneath the water and observed their feeding habits directly, we should have found that one feeds primarily at the bottom and the other at intermediate depths. Because the species feed in different parts of the habitat, their diets do not overlap greatly and they do not compete intensely for food. This fact was not appreciated until studies of their stomach contents revealed that more than 80 per cent of the diet of the great cormorant consisted of sand eels and herring, which swim well above the bottom, whereas the shag cormorant ate bottom-living forms, particularly flatfish and shrimp.

Four species of bumblebees common in English fields avoid competitive exclusion through morphological and behavioral specializations to slightly different food resources. All four species gather nectar and pollen from flowers, but each has a different tongue length and visits flowers of correspondingly different size. The species with the longest tongue feeds only at a few kinds of plants, those with long tubular flowers. The other species have more broadly overlapping flower preferences but two of them feed in shrubby habitats, the other in open fields. The two shrub-habitat bees further avoid competition by appearing at slightly different times of the year.

These observations suggest that where similar species co-occur, small differences in their ecological relationships reduce the level of competition to below the level within each of the populations. Over the geographical extent of the species' range of overlap, conditions may change so as to favor one of the species over the other and result in the exclusion of the other. Where the value of $\hat{N}_1$, or $\hat{N}_2$ is reduced to zero as K's and a's vary geographically, we may come to the edge of a species range.

Ecological Replacement of Species

When closely related species are thrown into direct competition in the laboratory, which species turns out to be the superior competitor depends upon the conditions of the experiment. When two species of flour beetle, *Tribolium castaneum* and *T. confusum*, are grown together in wheat flour, *castaneum* excludes *confusum* under cool, dry conditions, but itself is excluded when colonies are kept warm and moist (Figure 16-12). When grown separately, both species reach their highest densities in the moister conditions. Thus a species may be a superior competitor under sub-optimum conditions; the outcome of competition is not determined by the performance of a species in the absence of competition, rather it is determined by the relative productivity of species grown together. The point is further emphasized by the observation that the species which was superior in competition at each combination of temperature and humidity was not necessarily the most productive species when grown alone under the same combination. Competition involved the special interaction between species not often evident when observing each species alone.

In the complex mosaic of conditions in the natural environment, we would expect species to be superior competitors in one place and to be inferior elsewhere. As we travel from one locality to the next, changing environments are followed by a successive replacement of species, each superior to its competitors at a particular point along our route. We have seen, for example, that the ability of the parasitic wasp *Aphytis melinus* to tolerate winter cold enabled it to displace the closely related *A. lingnanensis* in the interior valleys of southern California. Whereas *lingnanensis* was

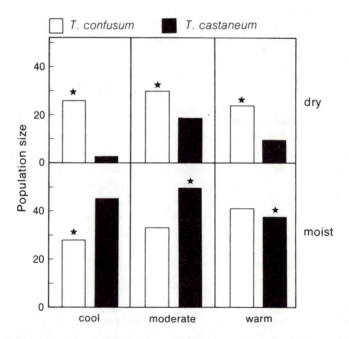

FIGURE 16-12 Competition between two species of flour beetles at different temperatures and relative humidity. Each section of the diagram shows the population density of the two species when grown separately (vertical bars); the superior competitor under each set of conditions is indicated by the star.

formerly widespread throughout the region, its range is presently restricted by a superior competitor to coastal regions.

If we were to climb to the summit of the White Mountains in southern California, we would pass through bands of vegetation types restricted to the narrow range of altitude where conditions give them superior competitive ability. Near timberline, a common species of fleabane (herbaceous perennial plants), *Erigeron clokeyi,* is abruptly replaced by the closely related *E. pygmeus.* Near the upper limit of the distribution of each species, low temperatures slow development so much that in many years flowering does not occur before the first frost. At the lower limits of their distribution, both species are adversely affected by the summer drought conditions characteristic of lower altitudes. Because *clokeyi* is more tolerant of drought and *pygmeus* is more tolerant of cold, the relative competitive superiority of *clokeyi* at lower elevations is reversed above an elevation of 11,000 to 12,000 feet. The exact altitude of their replacement depends in part upon the underlying bedrock. Sandstone weathers into a dark-colored soil that absorbs sunlight and retains warmth. Dolomite forms a light-colored, relatively cool soil. The sandstone causes temperature and moisture in a locality to resemble conditions at a lower elevation. Hence *clokeyi* replaces *pygmeus* at a higher elevation on sandstone than on dolomite.

Underlying parent bedrock has also influenced the outcome of competition between bristlecone pine and sagebrush in the White Mountains. Sagebrush is found primarily on south-facing and east-facing slopes where soils are derived from granite or sandstone; pine predominates on soils derived from dolomite (Figure 16-13). The competitive ability of the two species depends upon the relative moisture and phosphorus contents of the soil. Sandstone and granite weather to form soils with high levels of available phosphate, but poor water retention qualities. In addition, these soils are dark-colored and surface temperatures may be 5 C warmer than soils on dolomite. Heat, of course, increases water stress. Dolomite weathers into a soil with relatively good moisture retention, but low phosphate content.

Laboratory measurements of photosynthesis showed that sagebrush tolerates lower soil moisture than pine. Sagebrush seedlings reached peak photosynthetic rates at about five per cent soil moisture, which is three per cent lower than the point at which photosynthesis in pine seedlings begins to shut down.

By growing romaine lettuce in pots of each type of soil in the laboratory, dolomitic soils were shown to provide poor nutrition for plants. Of the total production on granite soil, lettuce produced only 36 per cent as much on sandstone soil and 18 per cent as much on dolomite soil. Production on dolomite was boosted by adding phosophorus to the soil, but not nitrogen or potassium. Dolomite soils have a low phosphorus content partly because the abundant calcium and magnesium carbonate in the rock produce a soil with a slightly basic reaction. As we have seen, phosphorus availability is greatest in slightly acid soils, such as the soils produced by granite and sandstone.

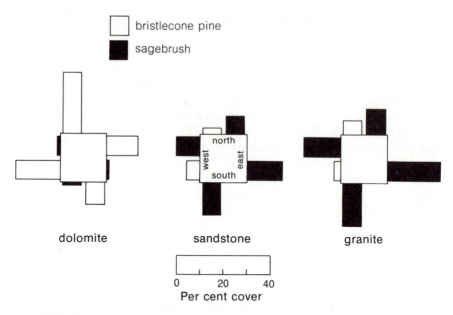

FIGURE 16-13 The percentage vegetation cover of bristlecone pine and sagebrush between 9,500 and 11,500 feet elevation in the White Mountains of southern California with respect to exposure and underlying bedrock.

The influence of the mineral content of soil on competition between pine and sage was determined by measuring the growth of seedlings of each species, grown separately and together. One-inch tall seedlings were planted in pots with soil taken from localities with dolomite, sandstone, and granite bedrock. The pots were placed out-of-doors and supplemented with distilled water to avoid drought stress. The seedlings were harvested, dried, and weighed after six months. The experiment demonstrated that the presence of a sagebrush seedling reduced the growth of pine seedlings by less than 30 per cent, and not at all on dolomite soils, whereas pine seedlings reduced the growth of sage by between 24 and 40 per cent, depending on the soil type (Figure 16-14).

The superiority of pine seedlings on poor soils can be related to a close association between the pine roots and fungi. Certain species of fungi form what are known as *mycorrhizal* associations in which the fungal mycelium extends from the cells of the pine roots into the soil. The mycorrhizal fungi acidify the soil, thereby increasing the availability of phosphorus, and decompose organic materials, thereby releasing nutrients to the pine. In return, the fungi obtain sugars from the pine to sustain their metabolism.

Mutual exclusion of two species of barnacles at different heights in the intertidal zone off the coast of Scotland provides an excellent example of how competitive ability varies with local conditions. Adult *Chthamalus* are

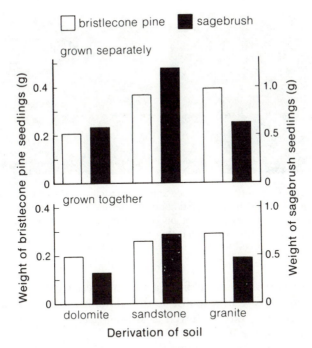

FIGURE 16-14 Relative growth of bristlecone pine and sagebrush seedlings planted separately and together in soil derived from different parent materials.

found higher in the intertidal zone than *Balanus*. The line of demarcation between the vertical distributions of adults of the two species is sharp, even though the vertical distributions of newly settled larvae overlap broadly. *Chthamalus* is not restricted to the zone above *Balanus* by physiological tolerance limits; if *Balanus* is removed from rock surfaces in the lower part of the intertidal zone, *Chthamalus* thrives there. The two species normally compete for space where the larvae grow up together (Figure 16-15). The heavier-shelled *Balanus* grows more rapidly than *Chthamalus*, and as individuals expand, the shells of *Balanus* literally pry the shells of *Chthamalus* off the rock. Hence rapid growth gives *Balanus* a distinct competitive edge in the lower parts of the intertidal zone. At higher levels, *Chthamalus* achieves competitive superiority because it resists desiccation better than *Balanus*.

Ecological Compression and Ecological Release

Exclusion is an extreme consequence of competition. Many of the species that coexist in any one place share a portion of their resources but do not overlap enough to cause exclusion. Nonetheless, each species depresses the populations of other species by an amount related to their ecological simi-

FIGURE 16-15 Competition for space among barnacles on the Maine coast. Above their optimum range in the intertidal zone, the barnacles are sparse and there are bare patches for the young to settle on (left). Lower in the intertidal (right) the barnacles are so densely crowded that there is no room for population increase; the young barnacles are forced to settle on older individuals.

larity. If a species is removed from a habitat, populations of ecologically similar species increase in response to the additional resources made available. This response is often referred to as *ecological release*, several examples of which are presented below.

An experimental demonstration of the influence of interspecific competition for light and water on plant growth comes from the tropical forest of Surinam, where forest ecologists set out to determine whether the growth of commercially valuable trees could be improved by removing species of little economic importance. The foresters poisoned 79 per cent of undesirable trees with girths greater than 30 centimeters in one area and greater than 15 centimeters in another, leaving the desirable species untouched. The increase in girth of the desirable trees was then measured over a year's time in experimental plots and in control plots which had not been selectively thinned (Figure 16-16). Removing trees greater than 30 centimeters in girth increased the penetration of light to the forest floor by six times. Removing smaller trees (15 to 30 cm girth) further increased light penetration by about one-third. Additional light stimulated the growth of the trees left on the experimental plot; improvement was greatest among small individuals, which are normally most shaded by competing species. Although removal of small trees did not increase light levels as much as removal of large trees, it produced a striking response in growth rate, particularly among the remaining large trees. The improved growth could not have been caused by increased light because many of the trees that responded were much taller than the trees which were poisoned. Therefore, the added growth stimulus probably resulted from reduced competition for either water or mineral nutrients in the soil.

The depressing effect of intraspecific competition on growth of trees has been demonstrated in many forest thinning experiments in temperate-zone

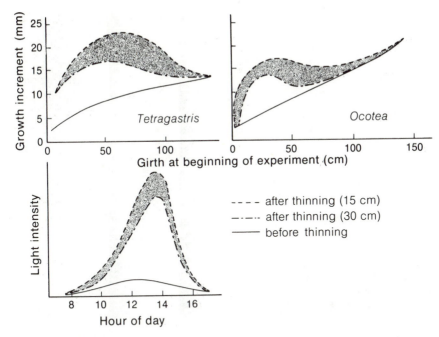

FIGURE 16-16 Effect on the increase in girth of two species of tropical trees. *Ocotea* and *Tetragastris*, achieved by removing competing trees greater than 15 centimeters or 30 centimeters in girth. Increase in light intensity which resulted from thinning is shown at left. Shaded area represents the difference between plots with 15-centimeter girth and 30-centimeter girth trees removed.

forests. The additional growth of young longleaf pine trees in response to selective thinning of trees over 15 inches in diameter is shown in Figure 16-17. Each core of wood was obtained by boring into a tree trunk, from the bark to the center, with a long, tubular device called an increment borer. The core of wood removed by the borer tube gives a record of annual ring growth without cutting down the entire tree. These cores show a very rapid increase in growth rate during the 18 years between the time the forest was logged and the time the cores were taken.

A study of competition between shrubs and grasses in the dry foothills of the Sierra Nevada Mountains in California demonstrated that the type of vegetation that gained an initial foothold effectively excluded the other type. Rapid-growing, deep-rooted annual grasses reduced soil moisture below the point that shrub seedlings could survive. On the other hand, because of their tall growth form, shrubs shade out grasses and other herbaceous vegetation. Competition of grasses on shrub seedlings was demonstrated by removing grasses from experimental plots a little more than six feet on a side (0.001-acre area), and following the growth and survival of shrub seedlings. One plot was completely denuded of grass and other

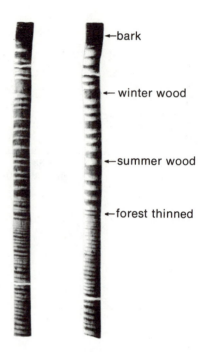

← bark

← winter wood

← summer wood

← forest thinned

FIGURE 16-17   Cores of two long-leaf pines obtained near Birmingham, Alabama, showing the effect of removing large trees on subsequent growth.

herbaceous vegetation and in another the grass was clipped to a height of one-half inch each week. A third plot was left undisturbed as a control. Sets of three such plots were established at ten localities within the study area.

At the beginning of the study in early spring, numerous seedlings of wedgeleaf ceanothus, chaparral whitethorn, and manzanita sprouted in all the plots. After the last rains of the season in April, the soil contained 13 to 14 per cent moisture at all depths. As the summer dry season progressed, soil moisture levels in control plots decreased quickly, reaching the wilting point in the top foot of soil by mid-May (Figure 16-18). Because grass roots penetrate to a depth of four to five feet, grasses can obtain moisture throughout the summer. The roots of shrub seedlings do not grow fast enough to reach below the moisture-depleted upper layers of soil and die rapidly after soil moisture drops below the wilting point. In clipped and denuded plots, the soil retained its moisture long enough for the seedlings to become firmly established. In mid-July, by which time most of the grasses had matured, set seed, and begun to die back, soil moisture profiles on clipped and control plots had reached their greatest difference (Figure 16-19). To the depth of penetration of grass roots, soil moisture was reduced below the wilting point, showing the influence of transpiration on soil moisture. Shrub roots had penetrated no farther than a foot in control plots and most of the seedlings had died. In denuded plots, soil moisture was sufficient for plant growth in all but the top foot of soil, and bush seedling roots had penetrated more than three feet into the soil.

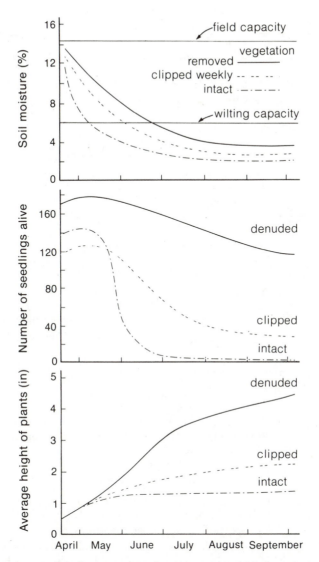

FIGURE 16-18 Seasonal progression of moisture content in the uppermost foot of the soil profile, survivorship of brush seedlings, and average height of surviving brush seedlings in plots where herbaceous vegetation was removed, clipped weekly, and left intact.

Once annual grasses become established in the dry foothills at low elevations, they reseed the area each year and exclude bush seedlings. At higher elevations, moisture levels in the soil are high enough to support bush seedling growth all summer, regardless of whether grasses and other herbs are present, and shrubs and small trees are the dominant vegetation.

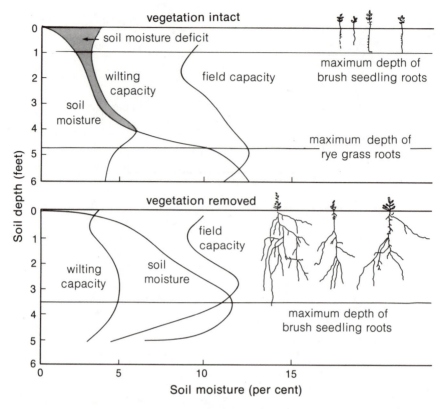

FIGURE 16-19 Soil moisture in relation to depth on control plots and plots denuded of herbaceous vegetation in early July. Depletion of soil moisture in control plots extends to the depth of grass roots, showing the influence of plant transpiration on soil moisture. Silhouettes of brush seedlings show the extent of their root systems.

At lower elevations, bushes sometimes become established in moist gullies or in the shade of rocks, and young bush seedlings grow at the edge of mature plants. Shrubby vegetation can invade grasslands in this way. If an area is denuded by fire, whichever type of vegetation reseeds the area in the greatest numbers will likely persist. Range management biologists have taken advantage of this fact to restore rangeland for cattle by controlled burning of bushland followed by heavy reseeding with annual grasses.

Ecological release following removal of competitors and, conversely, ecological compression following introduction of competitors maintain ecosystem function regardless of the specific composition of the community, or number of species in the community. Nowhere is this more clearly demonstrated than on oceanic islands. Distance from mainland sources of colonization has kept the number of species in most island ecosystems far below comparable mainland ecosystems. In response to the low diversity of competitors, island populations expand through habitats from which they would be excluded on the mainland and they attain greater abundance in each habitat. For example, censuses of land-bird species in small patches of wet tropical habitats of comparable area and variety revealed 135 species in Panama, 108 in Trinidad (a large island near the coast of Venezuela), 56 in Jamaica, 33 in St. Lucia and 20 in St. Kitts (small islands in the Lesser Antilles). Ecological release of bird populations on the small islands compensated for the reduction in number of species: the total populations of all species of birds did not vary greatly between island and mainland areas (Table 16-1).

The size of song sparrow territories on small islands with different numbers of competing bird species demonstrates the principle of ecological release for a single species. Where many species of bird are present, the types of food available to song sparrows are a relatively small fraction of the total, the rest having been eaten by competitors. Sparrows must defend large territories to include enough food resources to breed successfully; as territory size increases, population density decreases. Where few competitors occur, the sparrows utilize a wider variety of food types; smaller territories can satisfy food requirements, and population density thus increases (Figure 16-20).

How Competition Occurs

We have seen many examples in this chapter of competition through the mutual use of limited resources. Such *indirect competition* need never bring competitors face to face. They do battle by seige and starvation rather than by direct attack.

TABLE 16-1 Relative abundance and habitat distribution of birds in five tropical localities.*

Locality	Number of species observed	Habitats per species	Relative abundance per species per habitat (density)	Relative abundance per species	Relative abundance of all species
Panama	135	2.01	2.95	5.93	800
Trinidad	108	2.35	3.31	7.78	840
Jamaica	56	3.43	4.97	17.05	955
St. Lucia	33	4.15	5.77	23.95	790
St. Kitts	20	5.35	5.88	31.45	629

* Based on 10 counting periods in each of 9 habitats in each locality. The relative abundance of each species in each habitat is the number of counting periods in which the species was seen (maximum 10); this times number of habitats gives relative abundance per species; this times number of species gives relative abundance of all species together.

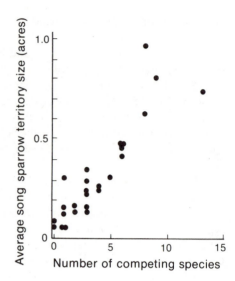

FIGURE 16-20 Territory size of song sparrows in relation to numbers of species of small land birds on small islands off the coast of Washington.

But one finds many cases in which individuals disputing some resource do attack each other directly. *Direct competition* usually occurs over space. Defense of territories by birds and other animals is a conspicuous example of competition by interference, although most conflicts are restricted to individuals of the same species. Because it would be impossible to defend individual prey within an area, competition is transferred from defense of the resource itself—food or water—to defense of an area containing it.

Space is the direct object of competition between barnacles; some species physically remove individuals of other species by prying them loose. One could construe the shading of competitors by plants as a form of direct competition, and possibly as severe a blow as a plant could deal. But not to be outdone by the behavior of which animals are capable, some species of plants inhibit the growth of other species by putting toxic chemicals into the soil. The decaying leaves of walnut trees release toxic substances which inhibit seedling growth in many other species of trees. Such toxic restraints are frequently referred to as *allelopathy*. Perhaps the best-known case of chemical interference involves several species of sage of the genus *Salvia* in southern California. Clumps of *Salvia* are usually surrounded by bare areas separating the sage from neighboring grassy areas (Figure 16-21). Sage roots extend to the edge of the bare strip but not into the bare area beyond. Thus it seems unlikely that a toxic substance was exuded into the soil by the roots. The leaves of these species produce volatile terpenes (a class of organic compounds that includes camphor) which apparently affect nearby plants directly through the atmosphere. Heavy rainfall washes the toxic compounds out of the atmosphere, reducing the prohibitory effect of sage on other species. This may explain the general absence of direct chemical competition in wet climates.

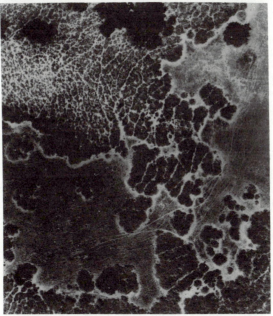

FIGURE 16-21 Top: Bare patch at edge of a sage clump includes a two-meter wide strip with no plants (A-B) and a wider area of inhibited grassland (B-C) lacking wild oat and bromegrass, which are found with other species to the right of (C) in unaffected grassland. Bottom: Aerial view of sage and California sagebrush invading annual grassland in the Santa Inez Valley of California.

A final example of competition, which is difficult to classify as either direct or indirect, involves the mutually depressing effects of closely related internal parasites. Parasitologists have long known that if a human is infected with a mild case of schistosomiasis caused by blood worms that normally infect cattle, he is unlikely to be severely affected if parasitized by the more virulent species of schistosome worm that normally attacks humans. Similar examples of cross-immunity to closely related parasitic species are common. It has been suggested that cross-immunity represents a form of interspecific competition between parasites. A parasite stands to benefit if it can stimulate its host's immune system to reject its competitors.

Indirect competition through exploitation of resources differs from direct competition because its effects are expressed slowly through differential survival and reproduction. Direct interference can immediately exclude a competing individual or population from a resource. Although interference reduces competition immediately, it also exacts a high price in terms of time, effort, and, potentially, survival. In evolving a strategy of competition, these costs must be weighed against the intensity of competition and the practicality of defending resources or resource substitutes. Interference competition occurs most frequently between individuals of the same species, hence between individuals that are ecologically most similar.

Coexistence and Resource Partitioning

Up to this point, competition has been discussed in terms of exclusion and success, elimination and persistence, superiority and inferiority. Although these terms describe what has *happened* in communities, coexistence describes what is found. For decades, ecologists have pondered the conditions necessary for species to coexist. Mathematical analyses of competition between species show that if a species limits its own numbers more than it limits the population of a second species, and vice versa, the two species can coexist. These conditions are fulfilled when each species uses somewhat different resources than the other. Consider the simple case of two species subsisting on three equally abundant resources. Species 1 consumes only resources A and B. Species 2 consumes only B and C. Because individuals of species 1 compete with others of their species for both A and B, but compete with individuals of species 2 only for resource B, intraspecific competition is more intense than interspecific competition and the two species will coexist.

This kind of coexistence can be seen in the replacement series diagram developed by botanists to evaluate the effects of competition. If both species are ordinarily inhibited more by intraspecific competition than by interspecific competition, the production of each species should lie above the line of expected production when intraspecific and interspecific competition are equivalent. Because individuals of both species are more productive when mixed than when separate, the total production of mixed species populations exceeds that of single species populations maintained at the same density (Figure 16-22).

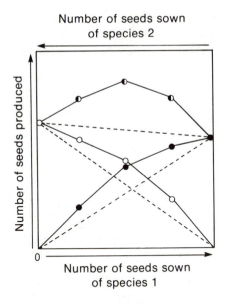

Number of seeds sown
of species 2

Number of seeds produced

0

Number of seeds sown
of species 1

FIGURE 16-22 Replacement series diagram of competition between two species with different ecological requirements. Mixed species plantings are more productive than separate plantings of either species maintained at the same density. Filled circles, species 1; open circles, species 2; half-filled circles, both species together.

Coexistence depends on avoiding competition. Any number of species can coexist so long as each exploits a portion of the resources in the habitat more efficiently than all the other species. To coexist with other species in a community, each species must excel in some way; otherwise, it will be excluded by superior competitors.

Species avoid ecological overlap by partitioning the available resources according to their size and form, their chemical composition, their place of occurrence, and their seasonal availability. Tolerance of extreme physical conditions allows some species to feed where others would succumb. Species that would compete intensely if together usually do not occur in the same locality or habitat. Resource partitioning within habitats often involves spatial separation of species according to their size and behavior. For example, each of five species of warblers that breed in spruce forests in Maine feed in different parts of the trees and use somewhat different foraging techniques as they search for insects among the branches and foliage (Figure 16-23). Four species of lizards of the genus *Anolis* coexist on the island of Bimini in the Bahamas, where they forage in different parts of the habitat. *Anolis sagrei*, a large, brown lizard, is the only species that commonly ventures to the ground to feed. The American chameleon, *A. carolinensis*, hunts for its prey among leaves. The remaining two species feed along branches, but the larger *A. distichus* is found more commonly than *A. angusticeps* on branches of large diameter.

Food specialization is often based on prey size. Large hawks usually eat larger prey than small hawks (Figure 16-24). Among species that partition

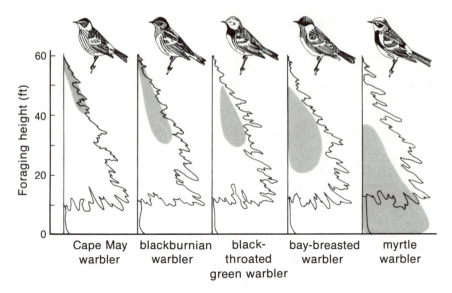

FIGURE 16-23 Location and method of foraging used by five species of warblers (genus *Dendroica*) in spruce forests in Maine.

resources solely by prey size, the body length of one predator is rarely less than 1.3 times the length of the next smallest predator. This difference apparently represents the smallest difference in resulting prey size that allows coexistence. The absence of size differences between some similar species of predators living in the same habitat does not necessarily imply intense competition; other avenues of resource partitioning may be used.

The morphology and behavior of predatory snails along the Gulf Coast of Florida emphasize the variety of ways ecological overlap can be avoided. The small murex snail, for example, cannot pry or chip open the shell of a large clam or oyster. Instead, it uses the filelike teeth (radula) at the tip of its proboscis to drill a neat hole in the shell of its prey, through which it scrapes away the flesh. Because *Murex* has a poorly developed foot, it can neither pursue other predatory snails nor dig clams out of the sand or mud. Its diet is therefore restricted to such prey as the rock cockle, which rests exposed on the sea floor. The large conches of the genus *Busycon* use the heavy edges of their shells to chip at the edges of large clam shells or to wedge them open. *Busycon contrarium* opens the thick-shelled *Chione* by grasping the prey and aligning the shell margin at a 45-degree to 90-degree angle to the lip of its own shell. Using its shell as a hammer, the conch then chips away enough of the clam's shell to insert its proboscis or to wedge the clam open. Conches pry open thin-shelled clams and species that do not close completely simply by using the edge of their shell as a wedge and pressing hard on the margin between the two halves of the clam's shell. *Busycon spiratum* has a thinner shell than the larger species,

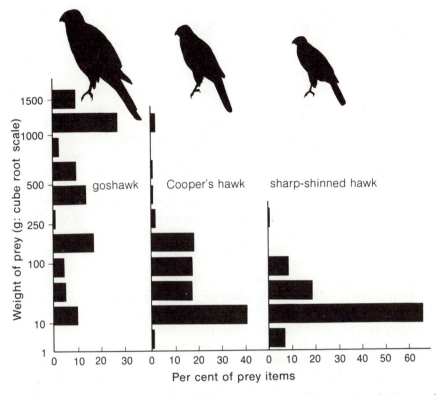

FIGURE 16-24 Size range of prey eaten by three widely distributed species of hawks. All three species belong to the accipiter group, which is adapted to pursue small birds and mammals.

B. contrarium. It usually preys upon clams that have thin shells or do not close completely. The horse conch and tulip shell attack other predatory snails smaller than themselves. The smaller *Fasciolaria hunteria* has a particularly long proboscis which is adapted to attack tube-building worms in their burrows beneath the mud. These differences in size and feeding technique enable the snails to avoid intense competition and permit their coexistence.

Competition and Coexistence Among Species of Plankton

Coexistence in natural situations is usually based upon ecological segregation, which results in reduced competition. Diversity is clearly related to the variety of resources available—in general, the complexity of the environment. With this rule of thumb firmly entrenched, ecologists have wondered how so many species of diatoms and green algae can live together in the open waters of lakes and oceans. When closely related species are maintained together in simplified laboratory environments they are frequently made to compete intensely for the limited variety of resources available to them. The environments of lakes and oceans, likewise, seem too uniform to support the many species found there. G. E. Hutchinson, limnologist and ecologist at Yale University, pointed to this phenomenon, the coexistence of many species in a seemingly uniform environment, as the "paradox of the plankton."

Hutchinson's paradox is by no means resolved, although ecologists now recognize that the aquatic environment varies greatly with season and locality, and, therefore, is not so homogeneous as previously supposed. The distributions of phytoplankton also are patchy in time and space. Hence local diversity at a particular season is much less than the total diversity in a lake, for example, and species replace each other during the course of the growing season. Even though we may encounter two species in the same liter of water, one may be replacing the other because it competes more effectively for resources under the prevailing conditions, and we cannot conclude that the two coexist stably.

In spite of its apparent uniformity, the water of streams, lakes, and oceans is a mosaic of continually changing conditions that shift the competitive balance between species. As we shall see later, predators, by preying upon one species more intensely than upon others, may shift the balance of competition and, under some circumstances, permit species with similar resource requirements to coexist.

Experiments on the nutrient requirements of two species of diatoms have illustrated how coexistence among phytoplankton might be mediated through differential resource use. David Tilman, of the University of Minnesota, studied the response of populations of *Asterionella* and *Cyclotella* to different concentrations of phosphate and silicate in the culture medium. Phosphorus, of course, is required by all plants to form nucleic acids and parts of cell membranes. Diatoms have an especial requirement for silicon to build their intricate glassy shells.

When he grew the species alone, Tilman found that the growth of each was limited by silicon when its concentration was relatively lower than phosphorus, and by phosphorus when the ratio of silicon to phosphorus (Si/P) was much higher. For *Asterionella*, the critical ratio of Si/P was 97 and for *Cyclotella* it was 5.6. Furthermore, at low levels of phosphate (Si/P greater than 97), *Asterionella* was able to assimilate phosphate more readily than *Cyclotella*, and Tilman therefore predicted that it would be the better competitor under those conditions. At low levels of silicate (Si/P less than 5.6), the two species' performances were reversed.

When *Asterionella* and *Cyclotella* were grown in culture together, the first excluded the second at low levels of phosphate and the second excluded the first at low levels of silicate, as predicted. When the Si/P ratio was intermediate—between 5.6 and 97—*Asterionella* was limited by silicon and *Cyclotella* by phosphorus, and neither species outcompeted the other. Tilman's experiment demonstrated that two species, which perceive the same environment differently because of their different requirements, can coexist. In theory, as many species could coexist as there were potentially limiting nutrients, provided that the species were differently adapted so as to require the nutrients in different proportions.

Evolutionary Divergence

Resources are not partitioned among species haphazardly. Competitive exclusion eliminates cases of extreme ecological similarity. Suppose we were to pick species at random from several localities and introduce them to an island whose climate resembled their place of origin. By chance, some of the species would have nearly identical resource requirements and compete intensely. The inferior competitors among these species would soon disappear from the community, leaving only those species that are ecologically distinct. But we would also notice that some of the species would begin to diverge in appearance due to evolutionary changes in each population. When similar species coexist, interspecific competition promotes the evolution of ecological divergence to reduce resource overlap.

Consider two hypothetical species which use many of the same resources. Suppose that the size range of prey eaten by each species overlaps that eaten by the other, shown in the upper diagram of Figure 16-25.

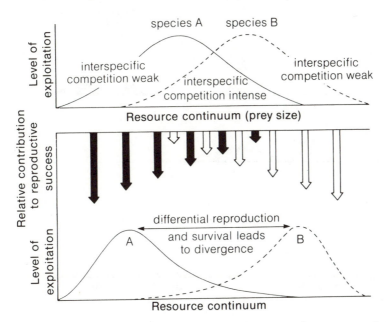

FIGURE 16-25 Diagram showing evolutionary divergence between populations caused by the influence of interspecific competition on the survival and reproduction of individuals that are adapted to eat prey of different sizes.

Where the species overlap most, they compete most intensely. Individuals of species A that eat prey smaller than the average have a greater evolutionary fitness than the population as a whole because their productivity is influenced less by competition. The same is true of individuals of species B that consume prey larger than average. Differential reproduction and survival caused by interspecific competition lead to the evolution of ecological divergence of close competitors. If three species feed off the same range of prey size, the middle-sized predator will become adapted to eat prey intermediate in size between the prey consumed by the largest and smallest species.

In most cases of resource partitioning we cannot determine whether the degree of ecological overlap between two species has been adjusted by evolutionary divergence. Evolution usually proceeds too slowly to be observed directly. We may, however, infer evolutionary divergence if a species displays appearance or behavior that differs in the presence of a competitor from appearance or behavior in the absence of a competitor. This phenom-

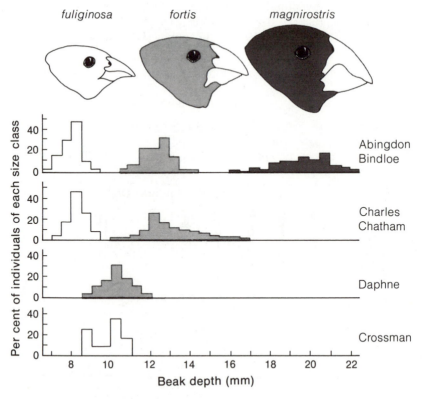

FIGURE 16-26   Proportion of individuals with beaks of different sizes in populations of ground finches (*Geospiza*) on several of the Galapagos Islands.

enon, referred to as *character displacement*, provides convincing evidence that competition can promote evolutionary divergence. When the ranges of two species do not overlap completely, interspecific competition influences evolution in the zone of overlap but not elsewhere. This principle is demonstrated in Figure 16-26 by illustrating the beak size of ground finches in the Galapagos Islands. The islands of Abingdon and Bindloe have three species of *Geospiza*, which partition seed resources according to size. With its large beak, *Geospiza magnirostris* is adapted to husk large seeds which smaller finches could not break open. With its small beak, *G. fuliginosa* can husk small seeds more efficiently than larger species. *G. fortis* feeds on seeds of an intermediate size range. The largest species does not occur on Charles Island nor on Chatham Island. Individuals of *Geospiza fortis* on these islands tend to have slightly heavier beaks, on the average, than on Abingdon or Bindloe. On Daphe Island, where *fortis* occurs in the absence of *fuliginosa*, its beak is intermediate in size between the two species on Charles and Chatham Islands. On Crossman Island, *fuliginosa* occurs in the absence of *fortis*, and *its* beak is intermediate in size there. Because habitats on the islands are similar in most respects, the simplest explanation for these patterns is that where the species occur together, competition has caused character divergence.

The evolutionary and ecological results of interspecific competition seen in this chapter organize the species on each trophic level according to the manner in which they exploit resources. Competitive exclusion and evolutionary divergence ensure, by reducing interspecific competition, that resources are utilized efficiently and that the overall flux of energy and nutrients in the ecosystem is maintained at the highest level possible with the array of populations present in the community.

17 | Predation

PREDATION IS THE PRIME MOVER of energy through the community. It is also a basic factor in population ecology and evolution. Prey populations largely determine the growth rate of predator populations because they provide the food necessary for growth and reproduction. Predators tend to reduce the growth rate of populations of their prey. To the extent that predators select prey with respect to genotypic variation, they cause evolutionary changes in the prey population. These changes are manifested in part by elaborate antipredator adaptations. When predators selectively prey on individuals of different ages, they alter the age structure of the prey population and change its dynamics.

Interest in predator-prey interactions was stimulated in the early part of this century by discovery of regular cycles in the abundance of some arctic mammals and birds. These cycles were thought to result from predator-prey interactions, whose properties were predicted by mathematical models devised during the 1920s. The relationship between predator and prey populations provides a useful system for the study of population dynamics and regulation, but the relationship also is important to community and ecosystem dynamics. The influence of predators on the size of prey populations affects competition between prey species, thus making predators an integral force in shaping the structure of the community.

The Impact of Predation

It is almost impossible to fully characterize the dynamic interactions between a single population of predators and a single population of prey in a natural habitat. The problems involved in such a task are varied and overwhelming. Measuring predation in natural populations, or even determining that the disappearance of an individual from the population is, in fact, due to predation, poses difficult technical problems. There are many causes of death, including starvation, disease, and exposure to the elements—not to mention emigration, a form of "local death." Predators are rarely caught in the act. Few predators restrict their feeding to one kind of prey, and few prey species are eaten by only a single species of predator; so additional difficulties are encountered in sorting out the interaction between two populations from the complex relationships that make up the community.

Predators do not capture prey at random. In prey species that display strong territorial behavior, predators tend to pick off surplus individuals that are forced into marginal habitats by intraspecific competition. One might expect extreme selectivity by predators that spend much time searching for and pursuing each prey; predators ought to choose the easiest catch, the individual with a relatively small chance of escape. For this reason, the inexperienced young and the decrepit old are most frequently captured. The risk of predation to healthy, reproductive-age individuals is much less. Other prey are more uniformly vulnerable. Indeed, many prey species have adopted a strategy of maximizing their production of offspring at the risk of increasing their vulnerability to predators. After all, what choice does an aphid have? If it is to suck the juices from the veins of a sycamore leaf, it must sit on a flat surface exposed to every passerby (Figure 17-1). The tiny

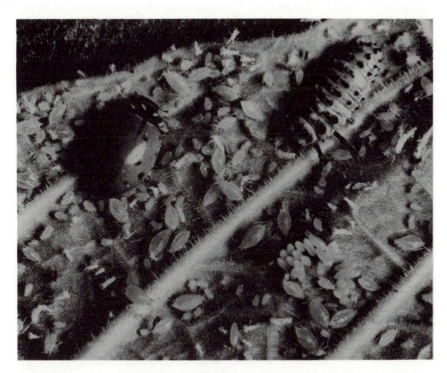

FIGURE 17-1 Adult and larvae lady beetles (family *Coccinellidae*) feeding on aphids in a laboratory culture. The flightless aphids are easy prey for the predatory beetles. Note the abundant hairs on the veins of the leaf. These help to deter the aphids from penetrating the plant and sucking its juices.

algae of the phytoplankton have nowhere to hide. Their survival depends purely on chance.

Whether predators can limit the size of prey populations below their carrying capacity is a question of basic importance in population and community ecology. If predators reduced the size of prey populations so much that the prey did not use all the resources potentially available to them, additional species might coexist on those resources.

Mites Versus Mites

Instances in which predators have been shown to depress prey populations below the carrying capacity of the environment are widely scattered. Several such studies are recounted here in detail because of the fundamental consequence of predation for community ecology. Most of these examples pertain to disturbed or artifically maintained situations in which prey populations are unusually high in the absence of predators.

The cyclamen mite is a pest of strawberry crops in California. Populations of the mites are usually kept under control by a species of predatory mite of the genus *Typhlodromus*. Cyclamen mites typically invade a strawberry crop shortly after it is planted, but their populations do not reach damaging levels until the second year. Predatory mites usually invade fields during the second year and rapidly subdue the cyclamen mite populations, which rarely reach damaging levels a second time.

Greenhouse experiments have demonstrated the role of predation in keeping the cyclamen mites in check. One group of strawberry plants was stocked with both predator and prey mites; a second group was kept predator-free by regular applications of parathion, an insecticide that kills the predatory species but does not affect the cyclamen mite. Throughout the study, populations of cyclamen mites remained low in plots shared with *Typhlodromus*, but their infestation attained damaging proportions on predator-free plants (Figure 17-2). In field plantings of strawberries, the cyclamen mites also reached damaging levels where predators were eliminated by parathion, but they were effectively controlled in untreated plots (a good example of an insecticide having the wrong effect). When cyclamen mite populations began to increase in an untreated planting, the predator populations quickly responded to reduce the outbreak. On the average, cyclamen mites were about twenty-five times more abundant in the absence of predators than in their presence.

The effectiveness of *Typhlodromus* as a predator owes to several factors in addition to its voracious appetite. Its capacity for population increase is of the same order as that of its prey. Both species reproduce parthenogenetically; female cyclamen mites lay three eggs per day over the four or five days of their reproductive life span; female *Typhlodromus* lay two or three eggs per day for eight to ten days. But even its high reproductive rate does not tell the whole success story of *Typhlodromus*. Seasonal syn-

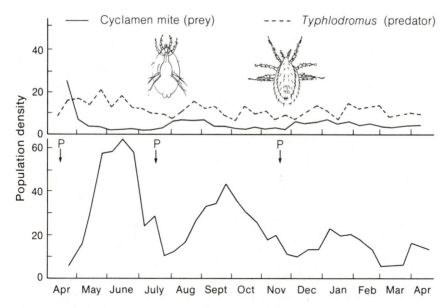

FIGURE 17-2 Infestation of strawberry plots by cyclamen mites (*Tarsonemus pallidus*) in the presence of the predatory mite *Typhlodromus* (above), and in its absence (below). Prey populations are expressed as numbers of mites per leaf; predator levels are the number of leaflets in 36 which had one or more *Typhlodromus*. Parathion treatments are indicated by p's.

chrony of reproductive activities with the growth of prey populations, ability to survive at low prey densities, and strong dispersal powers all contribute to its efficiency. During the winter, when cyclamen mite populations are reduced to a few individuals hidden in the crevices and folds of leaves in the crown of the strawberry plants, the predatory mites subsist on the honeydew produced by aphids and white flies, and they do not reproduce except when they are feeding on other mites. Whenever predators are suspected of controlling prey populations, one usually finds a high reproductive capacity compared with that of the prey, combined with strong dispersal powers and the ability to switch to alternate food resources when primary prey are unavailable.

Inadequate dispersal is perhaps the only factor that keeps the cactus moth from completely exterminating its principal food source, the prickly pear cactus. When prickly pear was introduced to Australia, it spread rapidly through the island continent, covering thousands of acres of valuable pasture and range land. After several unsuccessful attempts to eradicate the prickly pear, the cactus moth was introduced from South America. The caterpillar of the cactus moth feeds on the growing shoots of the prickly pear and quickly destroys the plant—literally by nipping it in the bud. The cactus moth exerted such effective control that, within a few years, the prickly pear became a pest of the past (Figure 17-3). The cactus moth has not eradicated the prickly pear because the cactus manages to disperse to predator-free areas, thereby keeping one jump ahead of the moth and maintaining a low-level equilibrium in a continually shifting mosaic of isolated patches. Indeed, one would probably not guess that the cactus moth keeps the prickly pear at its present low population levels; the moths are actually scarce in the remaining stands of cactus in Australia today. The same moth is probably responsible for controlling prickly pear in some areas of Central and South America, but its decisive role might have gone unnoticed if the appropriate experiment had not been performed in Australia.

The mite-mite and moth-cactus interactions prove that a predator can keep prey populations far below the capacity of the environment to support them. Yet in both situations, the prey originally occurred at unnaturally high levels in managed environments and the predators were introduced by man. These studies do not, therefore, help us to determine whether natural populations are ever controlled by predators.

Experiments on the effect of sea urchins on populations of algae have demonstrated predator control in some natural marine ecosystems. The simplest experiments consist of removing sea urchins, which feed on attached algae, and following the subsequent growth of their algae prey. When urchins are kept out of tidepools and subtidal rock surfaces, the biomass of algae quickly increases, indicating that predation reduces algal populations below the level that the environment can support. Different kinds of algae also appear after predator removal. Large brown algae flourish and begin to replace both coralline algae (whose hard shell-like coverings deter grazers) and small green algae (whose short life cycles and high reproductive rates enable algal population growth to keep ahead of grazing pressure by sea urchins). In subtidal plots kept free of predators, brown kelps formed thick forests below the ocean's surface and shaded out most small species.

Predator and Prey Cycles

Populations of predators and prey often vary in what appear to be closely linked cycles. The periodic fluctuations of the snowshoe hare, followed closely by fluctuations of lynx, one of the hare's major predators, is a classic example. Because the cycles persist for long periods, they represent a stable interaction between predator and prey.

The relationship between predator and prey populations is perhaps most readily seen in laboratory experiments in which prey are provided a constant, abundant source of food. When azuki bean weevils were maintained in cultures with predatory braconid wasps, the populations of predator and prey fluctuated out of phase with each other in regular cycles of population change (Figure 17-4). The weevil-braconid wasp interaction is often called a host-parasite system because the prey are consumed alive. Actually, bra-

FIGURE 17-3   Photographs of a pasture in Queensland, Australia, two months before and three years after introduction of the cactus moth to control prickly pear cactus.

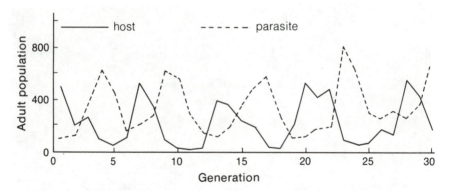

FIGURE 17-4 Population fluctuations of the azuki bean weevil preyed upon by the braconid wasp *Heterospilus*.

conid and other parasitic wasps are properly called *parasitoids*. In the life cycle of the wasp, eggs are laid on beetle larvae (Figure 17-5), which the wasp larvae proceed to consume after hatching. Abundance of prey, therefore, influences the number of adult wasps in the *following* generation, after the wasp larvae have metamorphosed into adults. This built-in time lag enhances the population fluctuation.

In one experiment, populations of braconids and weevils were maintained together in laboratory cultures for thirty generations (about one and a half years), although some of the prey or predator populations became extinct before the end of the experiment. The courses of the weevil and wasp populations are shown in Figure 17-4. When the predator and prey populations are both low, the prey increase rapidly. As the prey become abundant, the population of wasps begins to increase and, because *Heterospilus* is a very efficient parasite, the weevil population rapidly declines as the predators increase and begin to overeat their prey. Eventually, the prey are nearly exterminated and then the predator population, lacking adequate food, decreases rapidly to return the cycle close to its starting point. The wasp is never so efficient that all weevil larvae are attacked, hence a small but persistent reserve of weevils always remains to initiate a new cycle of prey population growth when the predators become scarce.

With an extremely efficient predator, prey populations are often eaten to extinction, and their predators soon follow. This type of predator-prey interaction can become stable only if some of the prey can find refuges in which they can escape predators. G. F. Gause demonstrated this principle with his studies on protozoa. He employed *Paramecium* as prey and another ciliated protozoan, *Didinium*, as predator. In one experiment, predator and prey individuals were introduced to a nutritive medium in a plain test tube. In that simple environment, the predators readily found all the prey. When they had consumed the prey population, the predators died from starvation.

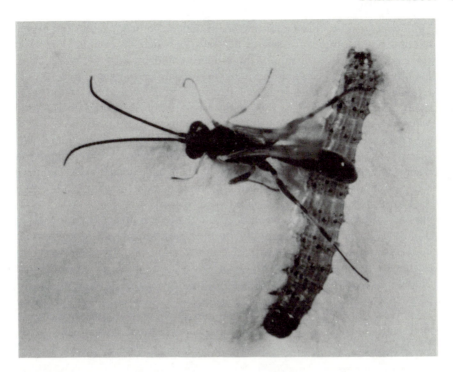

FIGURE 17-5 A braconid wasp laying an egg in a cotton boll worm. When the egg hatches, the wasp larva will consume the boll worm. Various species of these tiny wasps attack many kinds of insect larvae and are frequently major factors in pest population control.

In a second experiment, Gause added some structure to the environment by placing glass wool at the bottom of the test tube, within which the *Paramecium* could escape predation. In this case, the *Didinium* population starved after consuming all readily available prey, but the *Paramecium* population was restored by individuals concealed from predators in the glass wool.

Gause finally achieved recurring oscillations in the predator and prey populations by periodically adding small numbers of predators—restocking the pond, so to speak. The repeated addition of individuals to the culture corresponds, in natural predator-prey interactions, to repopulation by colonists from other areas of a locality in which extinction of either predator or prey has occurred. This is reminiscent of the interaction between the cactus moth and prickly pear cactus, in which the cactus escapes complete annihilation by dispersing to predator-free areas.

C. B. Huffaker, a University of California biologist who pioneered the biological control of crop pests, attempted to produce just such a mosaic environment in the laboratory. The six-spotted mite was prey; another mite,

Typhlodromus, was predator; oranges provided the prey's food. The experimental populations were set up on trays in which the number, exposed surface area, and dispersion of the oranges could be varied (Figure 17-6). Each tray had forty possible positions for oranges, arranged in four rows of ten each; where oranges were not placed, rubber balls of about the same size were substituted. The exposed surface area of the oranges was varied by covering the oranges with different amounts of paper; and the edges of the paper were sealed with wax to keep the mites from crawling underneath. In most experiments, Huffaker first established the prey population with twenty females per tray, then introduced two female predators eleven days later. Both species reproduce parthenogenetically.

When six-spotted mites were introduced to the trays alone, their populations leveled off at between 5,500 and 8,000 mites per orange area (Figure 17-7, left-hand diagram). The predators introduced to the system increased rapidly and soon wiped out the prey population. Their own extinction followed shortly (Figure 17-7, right-hand diagram). Although predators always eliminated the six-spotted mites, the position of the exposed areas of oranges influenced the course of extinction. When the orange areas were in adjacent positions, minimizing dispersal distance between food sources, the prey reached maximum populations of only 113 to 650

FIGURE 17-6 (Above) One of Huffaker's experimental trays with four oranges, half exposed, distributed at random among the forty positions in the tray. Other positions are occupied by rubber balls. (Right) An orange wrapped with paper and edges sealed with wax. The exposed area is divided into numbered sections to facilitate counting the mites.

individuals and were driven to extinction within 23 to 32 days after the prey were introduced to the trays. When the same amount of exposed orange area was randomly dispersed throughout the forty-position tray, the prey reached maximum populations of 2,000 to 4,000 individuals and persisted for 36 days. These experiments demonstrated that the survival of the prey population can be prolonged by providing remote areas of suitable habitat. The slow dispersal of predators to these areas delays the extinction of their prey.

Huffaker reasoned that if predator dispersal could be slowed further, the two species might coexist. To accomplish this, he increased the complexity of the environment and introduced barriers to dispersal. The number of possible food positions was increased to 120 and the equivalent area of six oranges was dispersed over all 120 positions. A mazelike pattern of Vaseline barriers was placed among the food positions to slow the dispersal of the predators. *Typhlodromus* must walk to get where it is going, but the six-spotted mite spins a parachutelike silk line that it can use to float on wind currents. To take advantage of this behavior, Huffaker placed vertical wooden pegs throughout the trays. The mites used the top of the pegs as jumping-off points in their wanderings. This arrangement finally produced a series of three predator-prey cycles over eight months (Figure 17-8). The

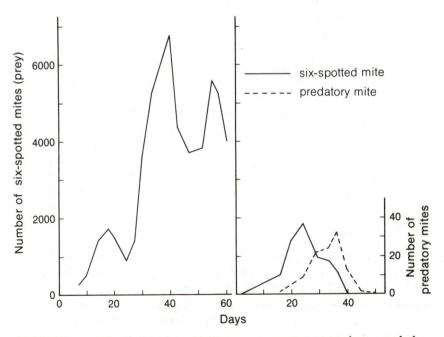

FIGURE 17-7 Number of six-spotted mites per orange area when raised alone (left) and in the presence of the predatory mite, *Typhlodromus* (right). The food was arranged in twenty small orange areas alternating with twenty foodless positions.

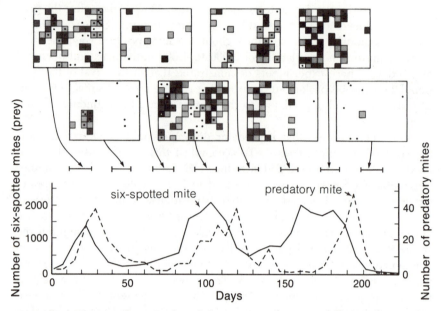

FIGURE 17-8 Population cycles of the six-spotted mite and the predatory mite *Typhlodromus* in a laboratory situation. The boxes above show the relative density and positions of the mites in the trays. Shading indicates the relative density of six-spotted mites; circles indicate presence of predatory mites.

distribution of the predators and prey throughout the trays continually shifted as the prey, exterminated in a feeding area, recolonized the next, one jump ahead of the predators.

In spite of the tenuousness of the predator-prey cycle achieved, we see that a spatial mosaic of suitable habitats allows predator-prey interactions to achieve stability. But, as we saw in Gause's experiment with protozoa, predator and prey also can coexist locally if some prey can take refuge in hiding places. And if the environment is so complex that predators cannot easily find scarce prey, stability will again be achieved.

A Predator-Prey Model

A simple mathematical formulation of the predator-prey interaction expresses the rate of growth of predator and prey populations by simple differential equations. We shall designate the number of predator individuals by P and the number of prey by H. Think of predators (P) and herbivores (H) to keep these straight. The rate of growth of the prey population, dH/dt, has two components. One represents the unrestricted reproductive rate of the prey population in the absence of predators, (rH), where r is the birth rate of a prey parent. The other represents the removal of prey from the population by predators, which, in our model, varies in direct proportion to the product of the prey and the predator populations (*HP*), and therefore is proportional to the probability of a chance encounter between predator and prey. Thus the rate of increase of the prey population is

$$\frac{dH}{dt} = rH - pHP \qquad (17\text{-}1)$$

where *p* is a coefficient representing the efficiency of predation.

For the predator population, birth rate is the number of prey captured times a coefficient (*a*) expressing the efficiency with which food is converted to population growth. Death rate is a constant (*d*) times the number of predator individuals. And the growth rate of the predator populations is

$$\frac{dP}{dt} = apHP - dP \qquad (17\text{-}2)$$

The model includes several simplifying assumptions: The prey do not interact with their own food supply; the relationship between predator and prey is linear; the death rate of individual predators is independent of population density.

Despite its limitations, this model proved successful in one respect: It predicted that predator and prey populations would oscillate, as we shall see below. When both predator and prey populations are in equilibrium (*dH/dt* = 0 and *dP/dt* = 0), *rH* = *pHP* and *apHP* = *dP*. These equations can be solved for

$$\hat{P} = \frac{r}{p} \quad \text{and} \quad \hat{H} = \frac{d}{ap} \qquad (17\text{-}3)$$

where $\hat{P}$ and $\hat{H}$ are the equilibrium population sizes of the predator and prey. Notice that $\hat{P}$ and $\hat{H}$ are both constant values.

When plotted on a graph of predator population size versus prey population size, the equilibrium population values $\hat{P}$ and $\hat{H}$ (the predator and prey *isoclines*) partition the graph into four regions (Figure 17-9). In the region below the prey isocline (*dH/dt* = 0), prey populations increase because there are few predators to eat them. In the region above the prey isocline, prey populations decrease because of overwhelming predator pressure. For the predators, their population increases in the region to the right of the predator isocline (*dP/dt* = 0) where prey are abundant. To the left, predator populations decrease owing to lack of food. The change in both populations simultaneously is shown by an arrow in each of the four sections of the graph. In the lower right, for example, both predator and prey increase and the population trajectory moves up and to the right. The vectors in the four regions taken together define a counterclockwise cycling

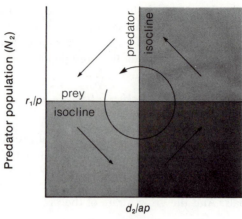

FIGURE 17-9   Representation of the predator-prey model on a population graph. The trajectories of the populations show that the predator and prey will continually oscillate out of phase with each other (full circle).

of the predator and prey population, that is, the populations cycle out of phase, with the prey population increasing and decreasing just ahead of the predator population.

The equilibrium isocline for the predator ($dP/dt = 0$) defines the minimum level of prey density ($H = d/ap$) that can sustain the growth of the predator population. That of the prey ($dH/dt = 0$) defines the greatest number of predators ($P = r/p$) that the prey population can sustain. If the reproductive rate of the prey (r) increased, the hunting efficiency of the predators (p) decreased, or both, the prey isocline (r/p) would increase— that is to say, the prey population would be able to bear the burden of a larger predator population. Similarly, if the death rate of the predators (d) increased and either the predation efficiency (p) or reproductive efficiency of the predators (a) decreased, the predator isocline (d/ap) would move to the right, indicating that more prey would be required to support the predator population. Increased predator hunting efficiency (p) would simultaneously reduce both isoclines; fewer potential prey would be needed to provide a given capture rate, and the prey population would be less able to support the more efficient predators.

As we have seen, time lags in the response of populations to their environments or to other populations can promote fluctuations in natural populations. We might ask how the introduction of a time lag into the predator-prey model would affect the trajectory of the populations on the predator-prey graph. In the oscillating system shown in Figure 17-10, the trajectory of the populations at point t moves directly toward the right. The number of prey is increasing, and the growth rate of the predator population

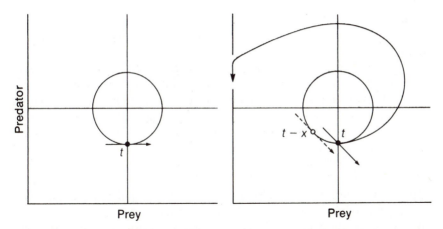

FIGURE 17-10 The effect of a time lag in the response of the predator population to prey density. At left, the response at time t is immediate, and continual oscillations are maintained. At right, the response has a time lag (x), which causes the amplitude of the oscillations to increase, eventually leading to the extinction of the predator, prey, or both.

is momentarily zero as the latter population reaches the low point of its cycle and is about to initiate a growth phase. If the response of the predator population to the abundance of its prey has a time lag, x, perhaps owing to a long gestation period before the birth of its young, the predator population will respond at point t as if it were at some earlier point $(t - x)$, and so will continue to decrease. The new trajectory produced by the time lag will spiral outward, eventually leading to the extinction of one or both populations.

Simple predator-prey models clearly do not adequately represent predator-prey relationships in natural communities. Time lags are almost certainly a feature of most interactions between predators and their prey, yet predator-prey oscillations appear to be highly stable.

Rules of Predator-Prey Stability

Experimental studies in conjunction with theoretical mathematical analyses of population dynamics have revealed four factors that enhance the stability of predator-prey interactions: (1) predatory inefficiency (or prey escape), (2) external ecological restrictions on either population, (3) alternative food sources for the predator, and (4) reduced time lags in predator population response.

A predator-prey system will achieve, or oscillate around, one of two possible equilibrium points. The first is set largely by the carrying capacity of the environment for the prey in the absence of predation. In this case, the predator exerts a minor influence on the prey population, which is ultimately limited by food or some other resource. The second equilibrium point is set at a much lower prey population level by the ability of prey to find refuges or hiding places. In this case, the predator reduces prey populations far below the carrying capacity determined by food, to a level determined by habitat complexity. The equilibrium point achieved by a particular pair of predator and prey species is decided by the hunting efficiency of the predator relative to the growth potential of the prey population. Efficient predators drive the prey population to its lower equilibrium point. Inefficient predators remove the vulnerable surplus of a prey population to a level near the carrying capacity of the environment. Predator inefficiency increases the stability of a predator-prey system around its upper equilibrium point by placing the burden of population regulation on density-dependent responses of the prey to its food resources. Near the lower equilibrium point, inefficiency increases because prey become fewer and farther apart and a larger proportion of them have access to good hiding places. In the case of the six-spotted mite, predator inefficiency eventually kept the prey from becoming extinct.

The upper equilibrium point of a predator-prey interaction is set by environmental limits to the prey population. External control of predator populations, owing to limited nest sites, water, or alternative food, reduces the impact of predators, and increases the control of external factors, on prey populations. Cold winter temperatures severely reduce populations of *Aphytis lingnanensis*, a wasp predator of citrus scale, in the interior valleys of southern California. Although the wasp controls citrus scale populations in the mild climate near the Pacific coast, it is ineffective in the interior.

The ability of predators to use alternative food sources when their prey are scarce or inactive greatly hastens their response to increases in prey populations and thereby helps to keep the prey population at the lower of its two equilibrium points. (We have seen this effect in studies on cyclamen mites, where alternative foods always kept predator populations at levels high enough to depress prey population growth.) Furthermore, use of alternative foods reduces the likelihood of a predator eating its principal prey to extinction.

Decreasing the lag time in the response of a predator population to prey population growth dampens population oscillations and increases the overall stability of the predator-prey interaction at either equilibrium point. Because predator response greatly influences the stability of predator-prey systems, we shall examine further two avenues of response to changes in prey population size.

The Functional Response

The relationship of an individual predator's rate of food consumption to prey density has been labelled the *functional response* by Canadian ecologist C. S. Holling. When tested in the presence of increasing levels of prey density in the laboratory, praying mantises increase their rate of food consumption in proportion to prey density, at first, but at high prey density, their feeding rate eventually levels off (Figure 17-11, curve B). Two factors dictate that the functional response of the individual should reach a plateau. First, as the predator captures more prey, the time spent handling and eating the prey cuts into hunting time. Eventually the two reach a balance and prey capture rate levels off. Second, predators become satiated—continually stuffed—and cannot feed any faster than they can digest and assimilate their food.

Hunger clearly influences a predator's motivation to hunt. The distance over which a praying mantis will strike at a fly depends on the time since its last feeding, which, since one cannot ask a mantis how hungry it is, serves as a reasonable index of hunger. Holling also experimented with

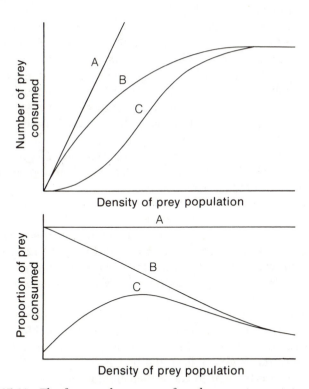

FIGURE 17-11 The functional response of predators to increasing prey density: (A) predator consumes a constant proportion of the prey population regardless of its density; (B) predation rate decreases as predator satiation sets an upper limit to food consumption; (C) predator response lags at low prey density owing to low hunting efficiency or absence of search image. The upper diagram portrays the functional response in terms of the *number* of prey consumed; the lower diagram, in terms of the *proportion* of the prey population consumed.

deer mice to determine how motivation affects the functional response of a vertebrate predator. Cocoons of the pine sawfly were buried at several densities in sand on the floor of large cages; dog biscuits were provided as an alternative food. Palatability of the cocoons influenced the functional response: deer mice dug up many more fresh cocoons than cocoons collected during the previous year and stored prior to the experiment. Deer mice apparently are not fond of stale cocoons. When fresh cocoons were buried deeper in the sand, the mice spent less time digging for them. The type of alternative food also influenced the functional response. With cocoons at a density of 15 per square foot, mice consumed 200 cocoons per day when dog biscuits were provided. When sunflower seeds were added to the menu, consumption of cocoons dropped to just over 100 per day. Deer mice evidently are not fond of dog biscuits, either.

Holling determined the functional responses of three small mammals to the density of pine sawfly cocoons in the relatively natural habitat of pine plantations in Ontario, Canada. The eggs of the sawfly, which are laid in live pine needles in the fall, hatch early in the spring. The larvae feed on the needles of the pine. In early June, full-grown larvae fall to the ground and crawl into the litter, where they spin cocoons. The adults do not emerge to lay eggs until September. For three months during the summer, the forest floor is scattered with varying numbers of cocoons from a few thousand to more than a million per acre, depending on the level of sawfly infestation. Small mammalian predators, particularly the common shrew, deer mouse, and short-tailed shrew, open the cocoons in characteristic ways enabling investigators to tally instances of predation separately for each species. Densities of the predators and prey can be assessed by trapping the rodents and sampling the forest litter for cocoons.

The functional response curve of each predator was unique. The short-tailed shrew, the least common species in the study area, increased its consumption of sawfly pupae in response to their density much more rapidly than either the common shrew or deer mouse (Figure 17-12). Because the short-tailed shrew increased its consumption of cocoons in response to small increases in their availability, it appeared to take full advantage of a highly variable food resource.

The functional response of many predators increases more slowly at low prey densities than at higher prey densities (see Figure 17-11, curve C). Two factors can cause this lag. First, hunting efficiency is decreased at low density because the few prey have the best hiding places. Second, vertebrate predators are thought to adopt their hunting behavior and prey recognition to the most worthwhile prey—usually an abundant, oft-encountered species. A preconception of what a given prey looks like and where it is found is called a *search image*. We use search images all the time. A lost object is easier found if we know its shape, size, and color. Predators presumably could base search images on prior experience. The more abundant the prey, the more often it would be found and the better prepared

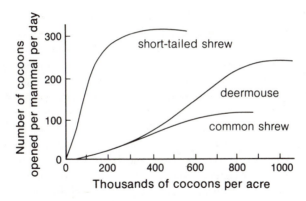

FIGURE 17-12 Functional response of three mammalian predators to the density of pine sawfly cocoons in the litter of pine plantations.

the predator to find it. Conversely, search-image formation works against a predator's being able to locate uncommon prey, hence the lag in the predator's functional response at low prey density.

Revealing observations on the factors that influence predator choice come from experiments of the Russian biologist V. F. Ivlev on the feeding behavior of fish on invertebrate prey. All the experiments were carried out in the laboratory. Ivlev used four species of predatory fish (carp, bream, roach, and tench) and four types of prey (chironomid fly larvae, amphipods, isopods, and molluscs), which could be provided at any selected density. He determined the selectivity of a predator for each prey type by calculating an *electivity index*

$$E = (r - p)/(r + p) \tag{17-4}$$

where p is the proportion of a prey species in the ration provided (all four prey species together sum to 100 per cent) and r is the proportion of that prey species in the diet of the predator. If, for example, 20 per cent of the prey items provided were amphipods, but 80 per cent of the roach's diet consisted of amphipods, the electivity of the roach for amphipods would be $E = (0.8 - 0.2)/(0.8 + 0.2) = 0.6/1.0 = 0.60$. Values of E can vary between -1.0, indicating complete avoidance of a particular prey, to $+1.0$, indicating exclusive preference. A value of 0.0 indicates no preference one way or the other.

Ivlev found that amphipods were generally preferred by the predators over other prey ($E = 0.10$ for carp, 0.21 bream, 0.24 roach, 0.02 tench). Comparable values for chironomid larvae, which burrowed in the silt on the bottom of the fish tanks and were less available to some of the predators, were 0.24, 0.05, -0.30, 0.27. The high electivities for chironomids of carp and tench reflect the benthic feeding habits of these predators.

Prey selection is influenced by a variety of factors. When Ivlev removed chironomid larvae as an alternative food source, electivities for amphipods increased to 0.24, 0.31, 0.31, and 0.18. Degree of satiation is also important. For example, hungry carp feed quite generally on all four prey types, but as the carp become satiated with food, molluscs, isopods, then amphipods are successively dropped from the diet; chironomid larvae are the only prey that a stuffed carp will show any interest in. Similarly, when prey are generally abundant, carp ignore molluscs and isopods to go after their preferred chironomid larvae. As the mobility of, and availability of hiding places to a certain prey type increase, electivity for that prey usually decreases. As one would expect, predators seek the easy catch.

In another series of experiments, using several different sets of predators and prey, Ivlev demonstrated that each predator has a preferred range of prey size over which its electivity for that prey is greatest (Figure 17-13). The electivity of predatory fish is usually greatest over some intermediate size range of prey; small prey are not worth eating and large prey are often too difficult to capture. Invertebrate predators, such as *Macrodytes*, often show increasing preference for prey as their size increases to some upper limit determined by the size of the predator's mouth.

Finally, Ivlev showed that if prey were not replenished and thus if their availability decreased as the predators ate them, each predator would consume its preferred prey first and then turn its attention to less desirable prey species. This phenomenon, called *switching*, has also been identified in predatory snails and several aquatic insects, but not in ladybird beetles. Where it occurs, switching acts to stabilize a predator's population and, perhaps, to reduce fluctuations in the community as a whole.

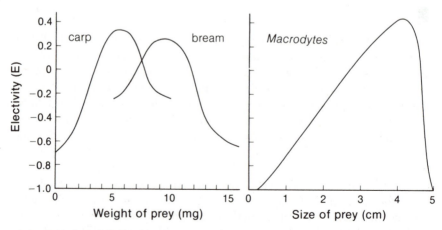

FIGURE 17-13 Electivity of predators as a function of prey size. Left, carp fed on chironomid larvae and bream fed on amphipods; right, larvae of *Macrodytes* fed on young roach.

The Numerical Response

Individual predators can increase prey consumption only to the point of satiation. Predator response to increasing prey density above that point can be achieved only through an increase in the number of predators, either by immigration or population growth, which together constitute the *numerical response*. In Holling's study of predation on sawflies, populations of the common shrew increased from about three individuals per acre at low prey density to twenty-four per acre at medium and high prey densities. The deer mouse population exhibited a less marked numerical response, and the short-tailed shrew, none. It may be only coincidental that the short-tailed shrew had the most rapid functional response to increasing prey density, but we would nonetheless expect the functional and numerical responses to be inversely related. The increased intraspecific competition resulting from the numerical response of a population reduces the availability of prey, upon which the functional response is based.

Numerical response by growth of a local population is relatively slow, particularly when the reproductive potential of the predator is much less than that of the prey, a condition that could lengthen the time lag in a predator-prey system. Immigration from surrounding areas is a major component of the numerical response of mobile predators, many of which opportunistically congregate where resources become abundant. The bay-breasted warbler specializes on periodic outbreaks of the spruce budworm, during which its population density may reach 120 pairs per 100 acres, compared with about ten pairs per 100 acres during nonoutbreak years. The bay-breasted warbler also responds to increased prey density by laying more eggs.

The numerical responses of birds to populations of the larch sawfly was studied in tamarack swamps in Manitoba. One of the swamps consisted of a nearly pure stand of tamarack with a well-developed shrub layer of alder, and the other area supported a mixed stand of tamarack and black spruce, with a shrub layer of dwarf birch. More than forty-three species of birds were known to eat sawfly larvae or adults, but the importance of these species as predators varied considerably. Bird censuses were taken over a period of nine years, during which the abundance of sawfly larvae varied from as few as 3,200 to more than 2 million larvae per acre. Although populations of all the species of birds varied directly with prey abundance, the numerical response was not nearly great enough to control the sawfly population. The pure tamarack stand had the greater infestation of larch sawflies, but although the density of breeding birds there was nearly twice as great as in the mixed tamarack-spruce habitat, the impact of avian predation on the sawfly population was only one-tenth as great (Table 17-1).

Three predatory birds, the pomarine jaeger, the snowy owl, and the short-eared owl, each respond in a different manner to varying densities of lemmings on the arctic tundra (Table 17-2). Lemming populations exhibit great fluctuations; high and low points in a population cycle may differ by a factor of one hundred. At Barrow, Alaska, during the summer of 1951, when lemmings were scarce, none of the predatory birds bred; short-eared owls did not even appear in the area. During the following summer, one of moderate lemming density, both the jaeger and snowy owl bred, but short-eared owls again were absent. In 1953, a peak year for lemmings, all three species of bird predators bred. Jaegers were four times more abundant in 1953 than in 1952. In contrast to the jaegers, the density of snowy owls did not increase. Instead, each pair of birds reared more young. Most snowy owls laid two to four eggs during the year of moderate lemming abundance, and up to a dozen during the peak year.

A Model of Predator-Prey Equilibrium

The diverse relations between predator and prey populations can be summarized in a diagram that compares the productivity of the prey population to the proportion of prey that are removed by predators, as both vary in relation to prey density (Figure 17-14). The two curves in the diagram represent the net addition of new prey to the population (in excess of deaths due to causes other than predation), either by reproduction or immigration (collectively called *recruitment*), and the removal of prey by predators. Both are expressed as a proportion of the prey population. The predation rate is a product of the functional and numerical responses taken together.

Recruitment and predation are analogous to birth and death. When recruitment exceeds predation, the prey population grows; when predation exceeds recruitment, prey numbers decline. Points at which the recruitment and predation curves cross are population equilibria for the prey.

The recruitment curve of the prey population declines with increasing prey density, owing to intraspecific competition for resources, and falls to zero when the prey are at the carrying capacity of the environment. In the

TABLE 17-1   Predation by birds on larch sawfly populations in two tamarack swamp forests.

	Pure tamarack	Mixed tamarack-spruce
Prey population per acre		
larvae	2,138,700	40,000
adults	205,400	4,700
Predator population per acre	23.5	12.6
Potential annual predation rate		
larvae (per cent)	10,665 (0.5)	2,344 (5.9)
adults (per cent)	10,601 (5.6)	3,049 (64.9)

TABLE 17-2 Response of predatory birds to different densities of the brown lemming near Barrow, Alaska.

	1951	1952	1953
Brown lemming (ind per acre)	1 to 5	15 to 20	70 to 80
Pomarine jaeger	Uncommon, no breeding	Breeding pairs 4 mi^{-2}	Breeding pairs 18 mi^{-2}
Snowy owl	Scarce, no breeding	Breeding pairs 0.2 to 0.5 mi^{-2} many nonbreeders	Breeding pairs 0.2 to 0.5 mi^{-2} few nonbreeders
Short-eared owl	Absent	One record	Breeding pairs 3 to 4 mi^{-2}

absence of predators, prey populations are regulated at this point by the balance between their biotic potential and resource limitations of the environment.

The shape of the predation curve is determined by the functional and numerical responses of the predators. Escape ability of the prey and the failure of predators to form search images limit the proportion of prey

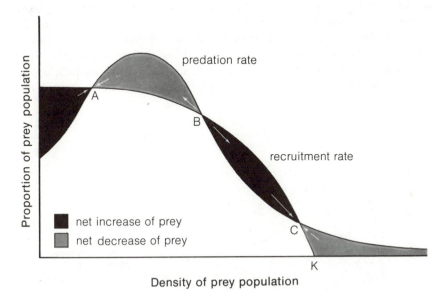

FIGURE 17-14 Predation and recruitment rates in a hypothetical predator-prey system. When predation exceeds recruitment, prey populations decrease, and vice versa (as shown by arrows). Points A and C are stable equilibria for the prey population; the lower point (A) represents population control by predators; the upper point (C) represents population control by food and other resources.

captured at low prey density. At moderate prey density, functional and numerical responses increase the effectiveness of the predators as a whole, and the proportion of prey removed increases. These responses are eventually satiated at high prey density, and although the number of prey consumed may continue to rise slowly, the proportion of individuals removed from the prey population decreases.

The recruitment and predation curves in Figure 17-14 were drawn to produce three equilibrium points for the prey population. The highest and lowest points represent stable equilibria around which populations are regulated, the middle equilibrium is unstable. The lower equilibrium point (A) corresponds to the situation in which predators regulate a prey population substantially below the carrying capacity of the environment. The upper equilibrium point (C) corresponds to the situation in which a prey population is regulated primarily by availability of food and other resources; predation exerts a minor depressing influence on population size.

Predators maintain a shaky hold on prey populations at point A. If a heavy frost or an introduced disease reduced the predator population long enough to allow the prey population to slip above point B, the prey would continue to increase to the higher stable equilibrium point (C), regardless of whether the predator population recovered. To the farmer, this means a crop pest, normally controlled at harmless levels by predators and parasites, suddenly becomes a menacing epidemic. After such an outbreak, predators could exert little control over the pest population until some quirk of the environment brought its numbers below point B, back within the realm of predator control. Outbreaks of tent caterpillars in the prairie provinces of Canada are generally preceded two to four years earlier by a year in which the winter is abnormally cold and the spring unusually warm. These conditions presumably upset the normal balance between tent cat-

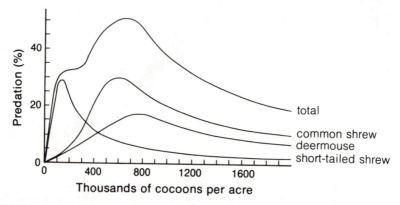

FIGURE 17-15 The combined functional and numerical responses of small mammals to density of sawfly cocoons showing the relative influence of several species of predator on prey populations.

erpillars and their predators and parasites. Infestations are subsequently brought under control by several cold winters that kill most of the tent caterpillar eggs.

The effectiveness of predators in maintaining prey populations at low densities depends on the relationship between predation and recruitment curves. The higher and broader the predation curve, the more effective predators are as control agents. Functional and numerical responses enhance predator effectiveness. Several species of predators attacking the same prey can control the prey population better than any one of them alone (Figure 17-15). Well-planned biological control programs take advantage of this principle in attempting to establish several species of predator and parasite, each with different predator tactics, to control pest populations.

Using the predation-recruitment rate diagram in Figure 17-14, we can examine the consequences of different levels of predation for prey population control (Figure 17-16). Inefficient predators cannot regulate prey populations at low density; they depress prey numbers slightly, but the prey population remains near the equilibrium level set by resources (upper-left diagram, point C). Increased predation efficiency at low prey density can result in predator control at point A (upper-right diagram). If functional and numerical responses are sufficient to maintain high densities of predators, or if prey are limited relative to predation by a low carrying capacity, predators may effectively control prey under all circumstances, and equilib-

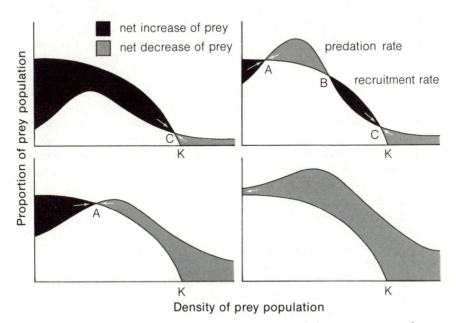

FIGURE 17-16 Predation and recruitment curves at different intensities of predation.

rium point C disappears (lower-left diagram). We could also envision predation being so intense at all levels of prey density that the prey are eaten to extinction (lower-right diagram, no equilibrium point). We would expect this situation only in simple laboratory systems, or when predators maintained themselves at high population levels by feeding off some alternative prey. Indeed, many ecologists have suggested that biological control of pests would be enhanced by providing parasites and predators of the pest with innocuous alternative prey.

Optimum Yield

Predators influence the dynamics of prey populations. There must be some optimum level of prey population density to support the greatest number of predators. Do predators prudently manage their prey populations to maximize their own productivity? And if so, how is this achieved by evolution?

A predator that eats its prey down to a low level takes the food out of its own mouth. Predators can capture prey more easily, and therefore be more productive, when prey are numerous. The ability of a prey population to support predators varies with its density. A small prey population can support correspondingly few predators because, while each prey individual's reproductive potential may be high, the total recruitment rate of a small population is low. Prey populations near their carrying capacity also are unproductive because although numerous, each individual's reproductive potential is severely limited as a result of intraspecific competition for resources. The total recruitment rate of every prey population reaches a maximum at some density below the carrying capacity. Because predators can remove a number of individual prey equivalent to the annual recruitment rate without reducing the size of the prey population, the prey population that yields the maximum recruitment also will support the greatest number of predators. This point is called the *optimum yield* or *maximum sustainable yield*. Ranchers and game management biologists are clearly concerned with maintaining populations of beef cattle, deer, and geese at their most productive levels to maximize man's ability to harvest these species without reducing their populations.

The achievement of optimum yield can be illustrated with populations of guppies maintained at different densities in aquaria. Recruitment (the number of immature fish produced in three weeks) reached a peak of 33 when there were 30 adult guppies per tank, and dropped to about 7 when adult populations exceeded 100 individuals per tank (Figure 17-17). In the absence of predators, the natural mortality of guppies would stabilize the population at about 120 adults, with the recruitment rate of 7 every three weeks just balancing mortality. The maximum sustainable yield would be achieved when predators removed between 40 and 50 per cent of the adult population every three weeks (about 2 per cent of the population per day).

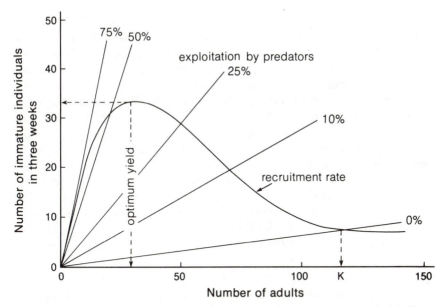

FIGURE 17-17 Recruitment curve and hypothetical exploitation rates for aquarium populations of guppies. In the absence of predation (0 per cent exploitation curve) the natural mortality of adult guppies would stabilize the population at about 120 individuals. The maximum exploitation rate possible is between 40 and 50 per cent per three weeks, at which point an adult population of about 30 and a yield of 33 would be maintained. A 75 per cent exploitation rate is more than the population can bear.

Maximum potential daily yields in laboratory populations have been calculated as 3 per cent for flour beetles, 13 per cent for unicellular algae, 23 per cent for water fleas (*Daphnia*), and 99 per cent for the sheep blowfly. Few estimates of maximum potential yield are available for natural populations of large animals. The population of ring-necked pheasants on Protection Island, Washington, could withstand removal at a rate of 1 to 3 per cent per day. Wolves kill about 25 per cent of the moose population of Isle Royale each year, 0.07 per cent per day, and perhaps 37 per cent of the white-tailed deer population of Algonquin Park in Ontario, 0.1 per cent per day. The population of ring seals on Baffin Island can withstand exploitation by Eskimos at a rate of 7 per cent per year, 0.02 per cent per day.

Could levels of exploitation observed in nature actually be maximum yields? Or would competition between individual predators cause overexploitation of their prey? Territorial animals, which exclude competitors from their feeding areas, could indeed space themselves with respect to their prey to achieve maximum yields. When the feeding areas of predators overlap, however, intraspecific competition dictates that each predator maximizes its immediate harvest at the expense of long-term yields. Man behaves no differently. Intelligently managed ranches, with fences to exclude competing livestock, can achieve maximum yields. Alas, in highly competitive situations—fishing in international waters, to name one—man has proved to be pathetically shortsighted and imprudent. Fishing and hunting practices that would attain long-term yields give way to practices that maximize today's harvest. For example, after World War II, the North Sea, between England and Norway, was fished so intensively that reducing fishing effort by 15 per cent and increasing the mesh size of nets to let more fish through actually would have *increased* the total catch by 10 to 20 per cent. Overexploitation of whale populations has similarly led to the near extinction of some species, and has virtually doomed the whaling industry.

The level of exploitation of a prey population is determined by the ability of the predator to capture prey compared to the ability of the prey to avoid being captured. Both skills are evolved characteristics of the population. Regardless of whether predators act prudently, they tend to achieve

TABLE 17-3   The relationship between populations of predators and their prey in several localities.

Locality	Predator	Principal prey	Density of predators (ind 100 mi^{-2})	Ratios of predator to prey populations	
				Numbers	Biomass
Jasper Nat'l Park	Wolf	Elk, mule deer	1	1:100	1:250
Wisconsin	Wolf	White-tailed deer	3	1:300	1:300
Isle Royale	Wolf	Moose	10	1:30	1:175
Algonquin Park	Wolf	White-tailed deer	10	1:150	1:150
Canadian arctic	Wolf	Caribou	1.7	1:84	1:186
Utah	Coyote	Jackrabbits	28	1:1000	1:100
Idaho primitive area	Mt. lion	Elk, mule deer	7.5	1:116	1:524
Ngorongoro Crater, Tanzania	Hyena	Ungulates	440	1:135	1:46
Nairobi Park, Tanzania	Felids	Ungulates	96	1:97	1:140
Alaska	Pomarine jaeger	Lemmings	—	1:1263	1:90

a characteristic equilibrium with prey populations. The relationship between wolves and various prey populations in several years demonstrates this equilibrium particularly well (Table 17-3). Population ratios and, particularly, biomass ratios (one pound of wolf for each 150–300 pounds of prey) are amazingly constant despite the fact that the species and density of the principal prey of the wolf vary considerably with locality. Exploitation rates also appear to be relatively constant—18 per cent of the moose on Isle Royale and 37 per cent of the deer in Algonquin Park. Still, not all predator-prey systems achieve the same equilibrium. The population ratio of mountain lions to deer in California is 1:500–600, which is equivalent to a biomass ratio of about 1:900, and the exploitation rate is only 6 per cent. A mountain lion-elk-mule deer system in Idaho had a biomass ratio of 1:524, and an exploitation rate of 5 per cent for the elk population and 2.7 per cent for the mule deer population. Evidently, wolves are more efficient predators than mountain lions, perhaps because of their social hunting habits. Where predators feed on more abundant populations of prey, as in savanna, grassland, and tundra habitats, predators not only are more numerous, but they achieve higher biomass ratios (1:50 to 1:150, Table 17-3). Such conditions seem to enhance both prey productivity and predator efficiency. Large cats—lions and cheetahs—remove 16 per cent of prey biomass in Nairobi Park, Tanzania, where their biomass ratio is 1:140.

Parasites and Host Populations

Parasites and predators use different strategies to exploit their prey or host populations. The death of a prey organism is the objective of predators, but the death of a host often causes the death of its parasite. Although predators evolve adaptations to maximize their ability to capture prey, selection acts strongly on parasites to adjust their food consumption to a level that the host can withstand. Parasite and host together evolve a balance. Adaptations of the parasite reduce its virulence, and adaptations of the host (immunity and resistance) reduce the parasite's hazard to health. The balance achieved between parasite and host is often upset when a parasite is accidentally transferred to a new host, frequently man, his livestock, or his crops, and sweeps through the population in a devastating epidemic.

Factors that limit populations of parasites extend far beyond the skin of their hosts. Most parasites undergo complicated life cycles involving changes in form and stages of dispersal that take the parasite into the hostile exterior environment. For an organism adapted to living within the tissues of a host, the environment through which its progeny travel to reach another host, often by way of intermediate species, must be forbidding.

Parasite populations fluctuate from year to year, as shown by a three-year study of the incidence of trypanosome parasites in populations of red squirrels and eastern chipmunks at Trout Lake, Manitoba. (Trypanosomes are parasitic protozoa that infect the bloodstream. In man, trypanosomes cause several fatal illnesses, of which sleeping sickness has received the most attention. The parasites are usually carried from host to host by such insects as tsetse flies.) Squirrels and chipmunks apparently are infected by different, host-specific strains of the trypanosome because the occurrence of trypanosomiasis in the two hosts fluctuated independently (4, 37, and 15 per cent in the squirrel population, and 42, 26, and 12 per cent in the chipmunk population, in 1961, 1962, and 1963). The incidence of the disease varied because adults that had previously been exposed to the disease became immune, and the number of juveniles in the population, all of which are susceptible, varied from year to year. In the eastern chipmunk, the incidence of the trypanosome in juveniles was more than four times that in adults (48 versus 11 per cent over three years), and the per cent of juveniles in the population during the summer months decreased from 68 per cent in 1961 to 29 per cent in 1963. This drop, combined with the reduction of disease incidence in adults from 19 to 4 per cent over the three years, accounted for the decrease of trypanosomiasis in the chipmunk population.

Infection by trypanosomiasis did not reduce the survival of squirrels and chipmunks. Endemic occurrences of disease organisms usually cause few detrimental effects in healthy organisms. Parasites can, however, intensify the harm caused by such stressful conditions as cold or lack of food, and thus can increase the incidence of death under these conditions. Infection by *Trypanosoma duttoni* reduces the tolerance of laboratory mice to stressful conditions. Groups of mice were given either full or half rations of food in either warm (19 to 22 C) or cold (3 to 8 C) environments. With full rations, none of the mice died over a nineteen-day period, regardless of parasitism or the temperature at which they were kept. The nonparasitized mice did, however, gain twice as much weight as the parasitized mice (14 versus 7 per cent increase). On half rations, all groups of mice had shorter average survival times in cold than in warm environments, as one would expect, but parasitism by *Trypanosoma* significantly reduced survival in both environments. Even in human populations, malnutrition often exerts its greatest effect on diseased or parasitized individuals. Undernourishment also tends to lead to an increase in the incidence of disease and parasitism.

Rapid development of immunity characterizes epidemic outbreaks of most diseases. Perhaps the most famous epidemics of all times have been the outbreaks of the Black Death in human populations. The bubonic plague organism, a bacterium, is endemic in many wild populations of rodents. The disease can be spread to man by rodent fleas, particularly the rat flea. Several times in history the natural balance between and among the bacteria, rodents, and fleas has been upset so badly that the plague spread to the human population, in which it caused epidemic disease. To produce a major plague epidemic, rat populations must be so great that hordes of rats, searching for food in houses, come into close contact with humans. Furthermore, rat fleas must heavily infest rats before they will abandon their preferred hosts for humans. These conditions have occurred infrequently, even in the crowded, garbage-ridden conditions in cities and towns of medieval Europe. But once the plague takes hold in a human population, its course runs swiftly and surely.

A typical epidemic of Black Death initially spreads rapidly, infects a large part of the population, and takes a high toll in human lives. But as susceptible individuals either die or become immune to the disease organism, the number of lethal cases and the mortality rate drop almost as rapidly as the disease first strikes. This is illustrated by data for a localized outbreak of the plague in India between 1953 and 1959.

Year	Contracted cases	Per cent lethal
1953	20,539	70.5
1954	6,670	84.5
1955	705	23.1
1956	331	20.5
1957	44	0
1958	26	0
1959	37	0

The great plague of the fourteenth century originated in 1346 during the siege of Caffa, a small military post on the Crimean Straits. From there, the epidemic spread to Italy and the south of France by 1347, and during the next year it had reached all of Europe. The plague did not disappear entirely until 1357. It reappeared in Europe three more times during the fourteenth century, in 1361, 1371, and 1382, but with lower incidence and mortality rate:

Year	Approximate percentage of population afflicted	Resultant deaths
1348	67	Almost all
1361	50	Almost all
1371	10	Many survived
1382	5	Almost all survived

The plague visited London three times during the seventeenth century, but unlike the epidemic waves that struck Europe during the fourteenth century, the effects were not attenuated during successive outbreaks:

Year	Population of London	Plague deaths	Deaths as a per cent of total population
1603	250,000	33,347	13
1625	320,000	41,313	13
1665	460,000	68,596	15

These outbreaks differed from those of the fourteenth century: Successive epidemics during the fourteenth century were separated by thirteen, ten and eleven years; during the seventeenth century, intervals were twenty-two and forty years. The longer interval increased the severity of successive epidemics in two ways. First, many more individuals lost their immunity over a twenty-year to forty-year period than over a decade. Second, the proportion of the population born since a plague epidemic, and therefore not immune, was much larger after intervals of twenty-two and forty years than after the shorter intervals during the fourteenth century. Because more humans were susceptible, the plague organism spread more rapidly through the population.

For many diseases, the frequency of outbreaks depends on the number of individuals born since the last outbreak and hence susceptible to the disease. Until the proportion of susceptible individuals reaches a certain critical level, which depends on the contagiousness and virulence of the disease, the disease organism cannot spread rapidly enough to cause an epidemic. Thus, although the smallpox virus brought to Mexico by the Spanish colonists in the sixteenth century remained at endemic levels in the native population, epidemics occurred only at eleven- to nineteen-year intervals, for example in 1520, 1531, 1545, 1564, and 1576.

The ecology of a disease vector (an organism that carries and transmits parasite organisms) frequently limits the spread of parasite populations. For example, when avian malaria and bird pox were introduced to the Hawaiian Islands in the last century, where neither disease had previously occurred in the avifauna, highly susceptible local populations of birds were quickly destroyed and several species became extinct. But malaria and bird pox organisms are carried from host to host by a species of mosquito that does not venture above an elevation of 600 meters, and so birds that lived at higher altitudes completely escaped the diseases. (The mosquito is also an introduced species in Hawaii.)

Herbivores and Plant Populations

Herbivorous animals that consume whole plants (usually seeds and seedlings) exert an influence on populations different from the influence of those that graze on vegetation. The first are predators. Grazers and browsers are akin to parasites. We can infer that herbivores play an important role in the lives of plants from the elaborate morphological and physiological defenses of plants against attack. Plant toxins like hypericin, digitalis, curare, strychnine, and nicotine are a fair match for herbivores. Thick bark, spines, thorns, and stinging hairs are also strong deterrents. Nevertheless, herbivores can greatly reduce plant populations.

Herbivorous insects are frequently used to control imported weeds. Klamath weed, a European species toxic to livestock, and the source of the drug hypericin, accidentally became established in northern California in the early 1900s. By 1944, the weed had spread over 2 million acres of range land in thirty counties. Biological control specialists borrowed an herbivorous beetle (*Chrysolina*) from an Australian control program. Ten years after the first beetles were released, the Klamath weed was all but obliterated as a range pest. Its abundance was estimated to have been reduced by more than 99 per cent.

The impact of herbivores in plant communities, measured by the total net primary production consumed, is least in forests, intermediate in grasslands, and greatest in aquatic environments. Herbivores consume between 2 and 10 per cent of the net production of forests. The rest enters detritus pathways in the community. Seed predators are much more efficient, consuming anywhere between 10 and 100 per cent of their food supply. Although they consume relatively little of the total biomass of the plant, seed predators attack a vital stage in the life cycle and can influence plant populations greatly.

Grazing herbivores, particularly large mammals, consume 30 to 60 per cent of grassland vegetation. The story of their influence on plant production is particularly well told by the results of exclosure experiments. A study in California employed wire fences to exclude voles (mouselike rodents) from small areas of grassland. Seed production and composition of the standing crop of plants were followed for two years after the experiment began and were compared with unfenced control plots. The results, summarized by the bar diagram in Figure 17-18, show that grazing by voles in the unfenced

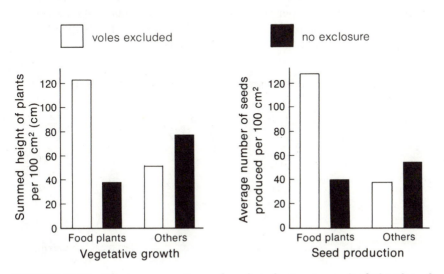

FIGURE 17-18 Species composition and seed production in grassland plots fenced to exclude voles and in unfenced control plots. The bar graphs present results of the experiment after two years. Food plants are mostly annual grasses; nonfood plants include perennial grasses and herbs.

plots reduced the abundance and seed production of food plants (mostly annual grasses) but did not affect perennial grasses and herbs absent from the vole's diet. Furthermore, competition from annual grasses in the fenced plots apparently depressed the growth of the nonfood species, suggesting a large role for the vole in determining the structure of the plant community.

Aquatic herbivores consume most of the net production of aquatic plants. The effectiveness of one type of herbivore, marine snails, as algal grazers is reflected in their abundance and annual energy flux. Representative figures for gross production (total energy assimilation) are 118 kcal m^{-2} yr^{-1} for the freshwater limpet *Ferrissia*, 290 kcal m^{-2} yr^{-1} for the periwinkle *Littorina*, and 750 kcal m^{-2} yr^{-1} for the turban shell *Tegula*. Comparable figures for terrestrial grazers range between 7 kcal m^{-2} yr^{-1} (field mouse) and 28 kcal m^{-2} yr^{-1} (grasshopper).

Although herbivores rarely consume more than 10 per cent of forest vegetation, occasional outbreaks of tent caterpillars, gypsy moths, and other insects can completely defoliate or otherwise eradicate entire forests. Long-term studies of growth and survival of trees after defoliation by tent caterpillars and other insects demonstrate that there may be a considerable lag between an infestation and the expression of its effects. A spruce budworm infestation on balsam fir caused varying defoliation mortality, and retardation of subsequent growth. In one area of light infestation, the defoliation exceeded 50 per cent during only three of nine years, reaching a maximum of 80 per cent in 1947 (Figure 17-19). No mortality was recorded in this area until 1951, but growth remained below one-half the normal rate for several years after the peak of the infestation. In an area of heavy infestation, defoliation exceeded 50 per cent for five years and reached 100 per cent in 1947, during the peak of the budworm outbreak. Growth was greatly suppressed and all trees in the area had died by 1951.

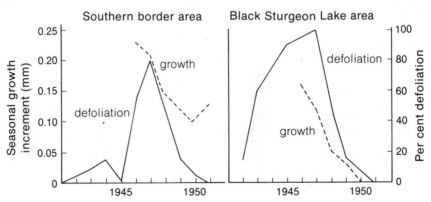

FIGURE 17-19 Defoliation of balsam fir trees in two areas during an infestation of the spruce budworm. In the area that was more heavily infested (right) all the trees died, whereas trees in the other area (left) had suffered no mortality by the end of the study period.

A Minnesota study of defoliation of quaking aspen by tent caterpillars showed that aspen usually survived defoliation and the increased intensity of insect and disease attack that inevitably followed. During the year of defoliation, however, growth was reduced by almost 90 per cent, and it was reduced by about 15 per cent during the following year. From studies of growth rings measured a decade later, foresters noted that among trees whose growth was normally suppressed by competition from dominant trees, mortality was independent of the history of defoliation and varied between 40 and 60 per cent following the tent caterpillar epidemic. Conversely, defoliation had a pronounced effect on survival of dominant trees. Only 2 per cent of the trees that were subjected to a single year of light defoliation died; at the other extreme, trees that had been badly defoliated three years in a row suffered almost 30 per cent mortality.

The interactions between predator and prey populations are so varied and complex as to defy summary. The examples we have looked at in this chapter lead us to the conclusion that predators, including herbivorous species, play an important role in the population processes of all species, even to the extent that efficient predators can regulate prey population density below the carrying capacity of the environment. Through their influence on prey populations, predators affect both the evolution of prey characteristics and the ability of prey to compete with other species utilizing the same resources. Evolutionary and ecological effects of predation are closely related. The efficiency of a predator as an exploiter of its prey is determined by the adaptations of both. In the next chapter, we shall examine how evolutionary responses of predator and prey populations, each to the other, influence the outcome of the predator-prey relationship.

18 | Evolutionary Responses

$$C$$ ARRYING CAPACITIES, competition coefficients, and predation efficiencies are products of the adaptations of individuals, and each is therefore subject to evolutionary change. Populations of predators, prey, and competitors are a part of the environment of every species. All such interacting populations select traits in each other that tend to alter their relationship.

When two populations interact, both can respond with evolutionary change. When their relationship is antagonistic, as it is between competitors and between predator and prey, the species can become locked into an evolutionary battle to increase their own fitnesses, each at the other's expense. Such a struggle can lead to an evolutionary equilibrium in which both antagonists continually evolve in response to each other, but the net outcome of their interaction does not change. This bubbling evolutionary pot is stirred by changes in the environment and the appearance of new species in the community, whether by natural range expansion or by human introduction.

We shall begin our discussion of evolutionary responses by describing a well-documented case of evolutionary change in a host-parasite system, that of the introduced European rabbit and myxoma virus in Australia.

Myxomatosis

As Western man immigrated to distant corners of earth, he brought a host of stowaways and invited guests who exceeded their welcome. Australia was not spared. In 1859, twelve pairs of the European rabbit were released on a ranch in Victoria. Within six years, the rabbit population had increased so rapidly that 20,000 were killed in a single hunting drive. By 1900, the several hundred million rabbits distributed throughout most of the continent became a critical problem for the Australians, whose economy has always been based largely on raising sheep. Being efficient grazers, rabbits destroyed range and pasture lands that otherwise would have been utilized for wool production. The Australian government tried poisons, predators, and other possible control programs—all without success. The answer to the rabbit problem seemed to be a myxoma virus, a relative of smallpox, that was discovered in populations of the related South American rabbit. Myxoma produced a small localized fibroma (a fibrous cancer of the skin). Its effect on South American rabbits was not severe, but a European rabbit infected by the virus died quickly of myxomatosis.

In 1950, the myxoma virus was introduced in one locality in Australia. A myxomatosis epidemic broke out in Victoria around Christmas time and spread rapidly. The virus was transmitted primarily by mosquitoes, which bite infected skin areas and carry the virus on their snouts. The first epidemic killed 99.8 per cent of the individuals in infected rabbit populations. During the following myxomatosis season (coinciding with the presence of mosquitoes), only 90 per cent of the remaining population was killed; and during the third outbreak, only 40 to 60 per cent of rabbits in the disease areas succumbed. At present, the myxoma virus has only a mild effect on the rabbits, which have increased in number to the point that they are again a nuisance and an economic problem.

Several factors contributed to the decline in the virulence of the myxoma virus in the rabbits. The few rabbits that survived the disease developed immunity and were unaffected by later outbreaks of the virus. More important, immunity was conferred to the offspring of immune females through the uterus. Over and above the immune response, evolutionary changes occurred that increased the resistance of the rabbit population to the myxoma virus and reduced the virulence of the virus population. Selection was strong on both the rabbit and virus. Genetically determined immunity to a disease reduced mortality, thereby increasing the fitness of the host. At the same time, virus strains with less virulence were favored because reduced virulence lengthened the survival time of the rabbits, and thus increased the mosquito-borne dispersal of the virus. A virus organism that kills its host quickly has small chance of being carried by mosquitoes to other hosts.

Samples of myxoma virus, collected in the field and tested on European rabbits with no previous exposure to the virus, demonstrated that the virulence of natural myxoma strains decreased over the years following its introduction. In 1950 and 1951, all strains of the virus collected in the field produced Grade I infections (more than 99 per cent mortality and a mean survival time of less than two weeks). By the 1958–59 season, however, no field strains of the myxoma virus had Grade I virulence; and by 1963–64, Grade II virulence (99 per cent mortality, 14-to-16-day survival time) was no longer evident. Similarly, resistance of wild rabbits to a particular grade of virus has increased steadily since 1950 (Figure 18-1). When the myxoma virus was first introduced to Australian rabbits, virtually all were mortally susceptible to myxomatosis, but by 1957 somewhat less than half the rabbits

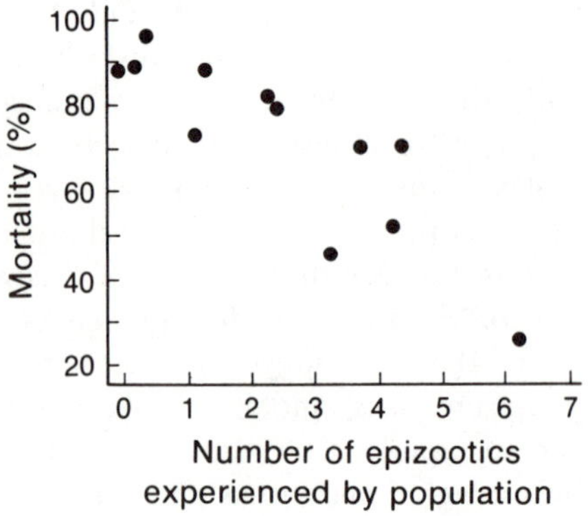

FIGURE 18-1 Decrease in susceptibility of wild rabbits to myxoma virus with Grade III virulence—90 per cent mortality in genetically unselected wild rabbits.

tested succumbed to virus with Grade III virulence. Studies indicated further that whether a rabbit died from myxomatosis, or not, depended largely on the inheritance of genetic immunity from its ancestors.

Evolution in Competing Populations

A change in environment, such as the introduction of a virus disease, can alter the fecundity and mortality of individuals in a population and, in doing so, change selective pressures on the population. Evolutionary responses will follow if the appropriate genetic variation is available. We might expect many of the genetic responses to an altered environment to have such subtle effects on the phenotype that we could not see them. But because the coexistence of two competitors and the stability of a predator-prey system can hinge on small changes in population processes, subtle genetic effects may exert a disproportionately large influence on the workings of the community. Species may drop out of, or they may be added to, a community on account of small differences in their adaptations.

The effects of subtle genetic differences can be seen in the different outcomes of competition experiments, according to the genetic strains of each species used. In competition experiments involving the flour beetles *Tribolium confusum* and *T. castaneum*, outcomes were uncertain within certain ranges of temperature and humidity. At 29 C and 70 per cent relative humidity, for example, *castaneum* won in about five-sixths of the experiments, and *confusum* in one-sixth.

These experiments by Thomas Park of the University of Chicago were begun with two pairs of each species. When Park's colleagues began experiments under identical conditions with ten pairs of each species, *T. castaneum* won twenty out of twenty times. These results suggested that in Park's experiments, genetic variation in the parent individuals might have altered the outcome of competition. Probably, the larger parental populations used in the later experiments more closely resembled the average genetic makeup of the species, thereby producing more consistent results. To test this hypothesis, several strains of each species were inbred to minimize their genetic variation, and then tested against each other in competition experiments. After the populations were inbred for more than twelve generations, certain strains of *confusum* consistently excluded certain strains of *castaneum*, the complete reverse of their normal competitive relationship. These experiments indicated that changes, even minor ones, in the genetic constitution of a population can greatly influence its competitive ability.

In experiments on competition between the fruit flies *Drosophila serrata* and *D. nebulosa*, the competitive ability of *D. serrata* appeared to increase during the course of the experiment. When the two species were established in population cages at 19 C, they quickly achieved a pattern of stable coexistence with 20 to 30 per cent *D. serrata* and 70 to 80 per cent *D. nebulosa*. In one experiment, but not in a replicate, the frequency of *D. serrata* began to increase after the twentieth week and attained about 80 per cent by the thirtieth week, a complete reversal of the initial equilibrium conditions. When individuals of both species were removed from the experimental populations after the thirtieth week and tested against stocks maintained in single-species cultures, the competitive ability of each species was found to have increased after exposure to the other in the competition experiment. In one of the replicates, the competitive ability of *D. serrata* evidently had evolved much more rapidly than that of *D. nebulosa*, and their equilibrium frequencies were greatly altered. The difference between these replicates also appeared in the competitive ability of *D. serrata* tested against the unselected stocks of *D. nebulosa*.

In similar experiments, David Pimentel of Cornell University showed that a poor competitor can evolve a competitive advantage (judged by relative population density) over a formerly superior adversary. Pimentel reasoned that when populations were rare, intraspecific competition would be greatly reduced, permitting the evolution of greater efficiency in interspecific competition. To test this model, he conducted laboratory experiments with flies to determine whether two species could coexist on one food resource by frequency-dependent evolutionary changes in their competitive ability.

The housefly and the blowfly which have similar ecological requirements and a comparable life cycle (about two weeks), were chosen for the experiments. Both species feed on dung and carrion in nature, and they are often found together on the same food resources. The flies were raised in small population cages at 27 C, with a mixture of agar and liver provided as food for the larvae, and sugar supplied for the adults. The outcomes of an initial series of four competition experiments between individuals from wild populations of the housefly and the blowfly were split two each. The mean extinction time for the blowfly, when the housefly won, was 92 days, and it was 86 days for the housefly when the blowfly won. The two species were

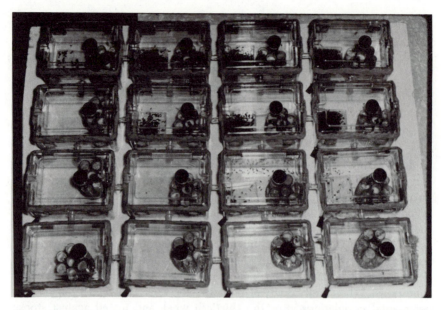

FIGURE 18-2 The sixteen-cell cage used by Pimentel to study competition between populations of flies. Note the vials with larval food in each cage and passageways connecting the cells. The dark objects concentrated in the upper-right-hand cells are fly pupae.

certainly close competitors, but the small cages used did not allow enough time for evolutionary change before one of the populations became extinct.

To prolong the competitive interaction, Pimentel and his colleagues started a population in a sixteen-cell cage, which consisted of single cages in four rows of four cages each with connections between them (Figure 18-2). Under these conditions, populations of houseflies and blowflies coexisted for almost seventy weeks, and showed a striking reversal of number between the two species (Figure 18-3). After thirty-eight weeks, when the blowfly population was still low and just a few weeks prior to its sudden increase, individuals of both species were removed from the sixteen-cell population cage and tested in competition with each other and with wild strains of the housefly and blowfly. Captured wild blowflies turned out to be inferior competitors to wild and experimental strains of the housefly. But blowflies that had been removed from the population cage at thirty-eight weeks consistently outcompeted both wild and experimental populations of the housefly (five cases of each). Thus, the experimental blowfly population apparently had evolved superior competitive ability while it was rare and on the verge of extermination.

The tenuous competitive edge of the blowfly over the housefly was revealed in the course of competition between populations removed from the sixteen-cell experiment. In most of these populations, the blowfly elim-

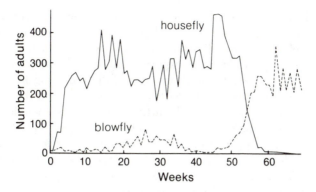

FIGURE 18-3   Changes in competing populations of houseflies and blowflies in a sixteen-cell cage.

inated the housefly in an average of 112 days, but in one of the five populations, the two species coexisted to a near draw for 519 days, during which time numerical dominance in the population cage shifted between the species several times. Apparently, the housefly had temporarily regained its competitive edge over the blowfly. Pimentel's experiment demonstrated emphatically that the outcome of competition can turn upon very small genetic differences. How such genetic adjustments influence the coexistence of species in nature is not known. Complex natural environments offer opportunities for species to avoid competition through divergence and ecological specialization.

Evolutionary Equilibrium of Predator and Prey

The outcome of competition, whether it leads to stable coexistence or to the elimination of one form, depends on a delicate balance between the genetic compositions of the competing populations. This is also true of interacting populations of predators and prey. Pimentel and his students explored the evolution of host-parasite relationships with the housefly and a wasp parasite, *Nasonia vitripennis* (Figure 18-4). In a control unit, *Nasonia*

FIGURE 18-4   Pupa of the housefly and the parasitic wasp *Nasonia*.

was allowed to parasitize a fly population that was kept at a constant level by replenishment from a stock population that had no contact with the wasp. None of the control flies that survived exposure to the wasp were returned to the population. In an experimental unit, the fly population was kept at the same constant level, but successfully emerging flies were retained so as to allow the population to evolve resistance to the wasp. The populations were maintained for about three years, long enough for evolutionary change to occur. In the experimental system, the reproductive rate of the wasps had dropped from 135 to 39 progeny per female, and their longevity had decreased from seven to four days. The average level of the parasite population also decreased (1,900 adult wasps, versus 3,700 in the control), and population size was more constant.

Experiments were then begun in thirty-cell cages in which the numbers of flies were allowed to vary freely. A control cage was started with flies and wasps that had no previous contact with each other, and an experimental cage was started with animals from the experimental population discussed above. In the control cage, wasps were efficient parasites, and the system underwent severe oscillations. In the experimental cage, however, the wasp population remained low and the flies attained a high and relatively constant population (Figure 18-5). This result strongly reinforced the conclusion drawn from the earlier experiments that the flies had evolved resistance to the wasp parasites.

As one might expect, parasites and diseases of economic crops have been scrutinized closely by agronomists. Control of pathogens like wheat rust is accomplished primarily by breeding strains of wheat with genes that confer resistance to the pathogen. The defense mechanisms provided by resistance genes have proven to be easily circumvented by rapid evolutionary change in the pathogen, probably a single gene replacement. Agricultural geneticists keep track of such changes in plant disease organisms by routinely exposing different genetic strains of the crop plant to a variety of races of the pathogen and recording the virulence. A survey in Canada of wheat rust—a fungus—had a surprising result: For a given race, virulence on strains of wheat with different resistant genes appeared and disappeared sporadically. Virulence apparently is altered by changes in single genes. For example, in 1969 race 15 B-1L of the wheat rust was virulent on strains of wheat with resistance genes (Sr) 8, 10, and 11, and it was avirulent on strains with genes 15 and 17. In 1970, a subrace of the rust was isolated that had become avirulent on Sr 11. In 1971, a new subrace appeared virulent on Sr 15. That same subrace lost its virulence on Sr 8 the following year: another lost its virulence on Sr 11. In 1973, a new subrace virulent on Sr 17 appeared. The significance of virulence changes for the rust is not clear. But the rust story does reveal that natural populations of predators and parasites continually generate genetic variants that challenge the resistance of their prey and hosts, which presumably respond in kind. Predator

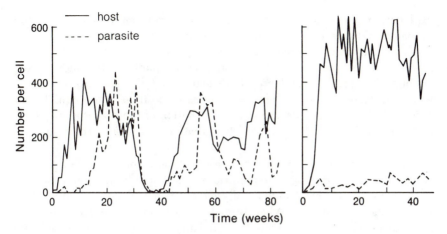

FIGURE 18-5   Populations of houseflies and a wasp parasite, *Nasonia vitripennis,* in thirty-cell laboratory cages. Control: flies had no previous experience with the wasp. Experimental: flies had been exposed to wasp parasitism for more than a thousand days.

and prey are clearly locked into an unending evolutionary struggle and a perpetual equilibrium.

The evolutionary responses of predator and prey populations can be depicted by a simple graphical model that relates the rates of evolution of the two to the efficiency of the predators. For the prey, the rate at which new adaptations to escape or avoid predators evolve should be expected to vary in direct proportion to the predation rate. If there were no predation, there would be no selection of adaptations for predator avoidance. As predation increases, selection increases, and so should evolutionary response, at least up to the limits set by genetic variation and balancing effects of correlated selection responses. The rate of evolution by the predator should vary in opposite fashion. If a particular prey species were not heavily exploited, predators that were adapted to utilize this resource would be selected. As the rate of exploitation of the prey population increases, intraspecific competition by the predators reduces the selective value of further increase in rate of predation on that prey. Very high rates of predation could conceivably select individuals that shifted their diets toward other prey species. Hence the rate of evolution by the predator to increase its efficiency on a particular prey could become negative.

In this simple model, the countervailing influences of predator and prey adaptation achieve a stable equilibrium where the two curves cross. If predator adaptations are relatively effective and the prey are exploited at a high rate, selection on the prey population will tend to improve the prey's escape mechanisms relatively faster than selection on the predator population will improve the predator's ability to exploit the prey. When the exploitation rate is low, the prey evolve relatively more slowly than the predators. This will lead to an equilibrium between the adaptations of predator and prey, producing a relatively constant rate of exploitation regardless of the specific predator and prey adaptations. In other words, although the predator and prey both evolve, neither gains on the other when the system is in equilibrium. The exact position of the equilibrium point will depend on potential rates of evolution of the predator and prey, and also on the availability of genetic variation and various constraints on evolutionary flexibility.

Pimentel's experiments on host-parasite interactions provide a demonstration of the predator-prey equilibrium. The housefly (host) and the parasitic wasp *Nasonia* have undoubtedly achieved an evolutionary equilibrium in their natural habitats. When brought into a simple laboratory habitat, *Nasonia* is able to exploit housefly populations at a greatly increased rate because little time is required to search out hosts. This is equivalent to shifting the exploitation rate of *Nasonia* on houseflies far above the equilibrium level in Figure 18-6. This shift increases selective pressure on the housefly to escape parasitism relative to selective pressure on the predator to further increase its exploitation rate. As a result, the ability of the housefly to escape parasitism increased and the level of exploitation by *Nasonia* decreased toward its natural equilibrium.

Reciprocal Evolution Between Populations

Plant and animal populations are caught in an eternal evolutionary struggle to outdo one another. Any new twist in a predator's hunting strategy is

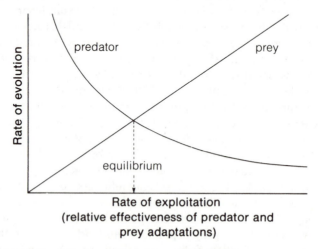

FIGURE 18-6 A graphical model of the evolutionary equilibrium between predator and prey adaptations which determine the level of exploitation (for explanation, see text).

countered by an evolutionary response in its prey. Natural insecticides in plants are rendered ineffective when herbivorous insects evolve biochemical mechanisms to detoxify the substance. Each evolutionary advantage is countered by novel adaptations in another population. The natural world is like Alice's looking-glass world where one has to run as fast as possible to stay in the same place. On our limited time scale, organisms seem static and conservative, yet their characteristics are being molded by selective forces in their physical and biological environments.

The protective coloration of many animals provides a striking demonstration of adaptations to the biological world. Each of the species portrayed in Figure 18-7 is uniquely adapted either to blend in with its background and thereby avoid detection by predators, or to announce to predators that it is bad-tasting or toxic and should be left alone.

Our fascination with the natural world is perhaps most strongly stimulated by the defensive adaptations of prey, particularly insects, against visually hunting predators. This is not so surprising, as we ourselves are visually oriented. But the perfection of cryptic appearances and attitudes to avoid detection by predators will always remain an impressive testament to the force and pervasiveness of natural selection.

Many prey organisms have evolved special resemblances to inedible objects. Among the most obvious of these are the small invertebrates that very closely resemble sticks, leaves, and flower parts—even bird droppings. The great pains that evolution has taken to conceal the head, antennae, and legs of organisms underscores the importance of these cues to predators for recognizing prey. In the stick-mimicking phasmids (stick insects) and leaf-mimicking katydids, legs are often concealed in the resting position either by being folded back upon themselves or the body, or by being protruded in a stiff, unleglike fashion. The dead-leaf-mimicking mantis *Acanthops* partially conceals its head under its folded front legs (Figure 18-7B). Symmetry is also a sure giveaway for animals, but it is difficult to conceal. The leaf-mimicking moth *Hyperchiria* produces the appearance of an asymmetrical midvein by folding one forewing over the other (Figure 18-7H). Moths will sometimes rest with a leg protruding to one side, but not to the other, or with the abdomen twisted to one side, both of which achieve the same effect.

Warning Coloration and Mimicry

Concealment is a strategy of palatable animals. When an individual is high on a predator's shopping list, it is best to be active only at night when motion is not an immediate giveaway; by day the individual hides in some protected cranny or fades away into the habitat, posing as bark, a leaf, a twig, or some equally commonplace object. Other organisms have rejected cryptic appearance for a bolder approach to predator defense. Instead, they produce noxious chemicals and advertise the fact with conspicuous color

FIGURE 18-7   Some examples of protective coloration and form in forest-dwelling animals in Panama. Three species of praying mantids (A, B, and C) avoid detection by blending with different backgrounds. The mantid in A is green and blends in with leafy vegetation; the mantid in B resembles a dead, curled up leaf; the mantid in C has a long, thin body that can easily be confused with a twig. The spider in D has a greatly enlarged, orange-colored abdomen whose appearance resembles flower parts that frequently drop into its web. The swallow-tailed butterfly in E is strikingly colored with a bold, iridescent green and black pattern as a warning to predators that it is unpleasant to taste. The frog in F escapes notice among the leaf litter of the forest floor. The caterpillar in G blends in perfectly with the twig on which it is resting when viewed from above; from the side, it is betrayed by its silhouette. The moth in H folds its wings unevenly over its back to break up its symmetry. The numerous spiny projections covered with toxic hairs on the body of the caterpillar in I tend to dissuade predators from eating it.

patterns. Predators learn quickly to avoid such conspicuous markings as the black and orange stripes of the monarch butterfly. One attempt to eat such a butterfly is a memorable enough experience to make a predator avoid monarchs for some time to come. It is interesting to note that many noxious forms adopt similar patterns of warning coloration: black and either red or yellow stripes characterize such diverse animals as yellow jackets and coral snakes.

Distasteful animals and plants that display warning coloration often serve as models for the evolution of mimicking color patterns by palatable forms.

This relationship is known as *Batesian mimicry*, named after its discoverer, the nineteenth century English naturalist Henry Bates. In his journeys to the Amazon region of South America, Bates found numerous cases of palatable insects that had forsaken the cryptic patterns of close relatives and had come to resemble brightly colored distasteful species. Bates rightly guessed that these insects were successfully tricking predators into avoiding them.

Mimicry may involve more extensive changes in morphology than altered color pattern. The buprestid beetle *Acmaeodera* mimics bees and wasps with yellow and black bands on its abdomen (Figure 18-8). Because the mimicry pattern must appear both in flight and at rest, the elytra (the hard forewings of beetles that are not used in flight, but only for protection) became fused, and remain fixed in position over the back even in flight. Most beetles spread their elytra in flight and do not resemble bees and wasps in the least. Some beetles have taken to mimicking flies, which are perfectly edible, but fly so fast that predators cannot catch them. Because of their speed, flies are effectively inedible and certainly not worth pursuing to some predators.

Experimental studies have demonstrated that mimicry does confer advantage to the mimic. Toads fed live bees thereafter avoid the palatable drone fly that mimics bees; if the toads were fed only dead bees from which the stings had been removed, they ate the drone fly mimics quite readily. Similar results were obtained with blue jays as predators; the distasteful monarch butterflies were the models and their viceroy butterfly mimics were the experimental subjects.

FIGURE 18-8 Mimicry of wasps by a buprestid beetle, *Acmaeodera pulchella* (left). The mimic bears yellow bands on a black abdomen, which is typical of wasps. Note that the elytra (forewings) of *Acmaeodera* are fused, whereas most beetles spread their elytra during flight (right).

Predators must learn to distinguish acceptable and unacceptable prey items. Naïve animals will often eat unpalatable prey—once anyway. Predators probably do not "remember" their experiences with prey indefinitely, but continually resample potential prey items. If the noxious taste or sting associated with a particular color pattern is not reinforced by eating an occasional model, a predator will continue to feed on the palatable mimics. It has been shown that a predator will remember an unpleasant experience with a prey item longer if the experience was strong. Thus, the greatest advantage is conferred to mimics by models that are both common and extremely distasteful.

Müllerian mimicry, named for the nineteenth century German zoologist Fritz Müller, involves the resemblance of two or more species of unpalatable organisms. Müllerian mimicry complexes often involve dozens of species, both closely and distantly related. The advantage of Müllerian mimicry to unpalatable species is that predators must learn to avoid fewer different types of noxious organisms. By resembling one another, diverse species can share the burden of the predator learning experience.

Structural and Chemical Defenses of Plants

Plant defenses against herbivores are based partly on the inherently low nutritional value of most plant tissues and partly on the toxic properties of so-called *secondary substances* produced and sequestered for defense. Structural defenses such as spines, hairs, and tough seed coats are also important.

The nutritional quality and digestibility of plant foods is critical to herbivores. Because young animals have a high protein requirement for growth, the reproductive success of grazing and browsing mammals depends upon the protein content of their food. Herbivores usually select plant food according to its nutrient content. Young leaves and flowers are frequently chosen because of their low cellulose content, and fruits and seeds are particularly nutritious.

Many plants are capable of making their proteins unavailable to herbivores. For example, the tannin deposited in oak leaves combines with leaf proteins in such a way that the proteins cannot be digested by caterpillars and other herbivores, thereby slowing their growth considerably. With the buildup of tannin in oak leaves during the summer, fewer and fewer leaves are attacked by herbivores. Tannins are usually deposited in vacuoles near the surface of the leaf where they will not interfere with the functioning of the plant's proteins. When herbivores eat the leaves, the vacuoles are broken and the tannins combine with the proteins. Leaf-mining beetle larvae have gotten around this problem by burrowing through and consuming the leaf's inner tissues, completely avoiding the tannin-filled vacuoles.

The seeds of legumes (the pea family) are frequently infested with larvae of bruchid weevils, a kind of beetle. The adult weevil lays its eggs on developing seed pods. The larvae then hatch and burrow into the seeds, which they consume as they grow. To counter this attack, legumes have evolved a variety of defensive adaptations. Each larva feeds on only one seed. To pupate successfully and metamorphose into an adult, the larva must attain a certain size, which is ultimately limited by the amount of food in the seed. As a result, many legumes have evolved extremely tiny seeds that do not contain enough food to support the growth of a single bruchid larva and thus they are not suitable as host plants. Most contain substances that reduce the digestibility of proteins by inhibiting the effectiveness of the proteolytic enzymes produced in the herbivore's digestive tract. While these inhibitors probably evolved as a general biochemical defense against insects, they are futile against bruchid weevils, which have metabolic pathways that either bypass or are insensitive to the presence of the inhibitors. Among legume species, soybeans are resistant to attack even from most bruchid species. If bruchid eggs are laid on soybeans, the first instar larvae die soon after burrowing beneath the seed coat; chemicals have been isolated from soybeans that completely inhibit the development of bruchid weevil larvae in experimental situations.

The production of toxic compounds forms a common defense by plants. Where predation is most intense, plants appear to have more varied and more concentrated toxins. This response only stimulates adaptations that enable some herbivores to detoxify some poisonous substances, and generally leads to biochemical specialization of herbivores to certain restricted groups of plants with similar toxins, or to single species. The tobacco hornworm (larval stage of the sphinx moth *Manduca sexta*) can tolerate nicotine concentrations in its food far in excess of levels that kill other insects. Nicotine disrupts normal functioning of the nervous system by preventing the transmission of impulses from nerve to nerve. The hornworm's solution to this problem is to keep nicotine from crossing the membrane of the nerve; in other species of moths, nicotine readily diffuses into nerve cells.

Resistance to nicotine enables *M. sexta* to feed on tobacco (*Nicotiana tabacum*), a member of the tomato family (Solanaceae), but some other species of *Nicotiana* have other alkaloid toxins that the tobacco hornworm cannot handle. Presumably, the mechanism that prevents nicotine from entering the nerve cells fails to recognize other related toxins, or these toxins have a different mode of action. Regardless of the physiological effect, when tobacco hornworms were grown on 44 species of *Nicotiana* in greenhouse experiments, growth was normal on 25 species, but was retarded or stopped completely on the others. In addition, fifteen of the species caused moderate to severe mortality. Specialization on tobacco has some favorable consequences for *M. sexta* over and above access to vast crops of food grown for it by man. In addition to being toxic to most herbivores, tobacco also dissuades many would-be parasites. Tiny wasps that might parasitize hornworm eggs become trapped in the gummy fluid produced by trichomes (fine hairs) on the tobacco plants. Eggs laid on other solanaceous plants that lack trichomes, like Jimson weed and tomato, are readily parasitized. In another plant-herbivore system, terpenoid resins produced by pines as a general herbivore defense are in turn used by sawflies as a chemical defense against parasitic wasps and flies.

The potential defenses of plants against herbivores extend beyond direct confrontation with structural and chemical weapons. Ecologist Daniel Janzen has suggested that many plants avoid intense herbivore pressure by being widely scattered in space and by breeding at irregular intervals. Populations of herbivores and seed predators cannot reach high levels when their food supply is erratic and difficult to locate. Many trees produce large crops of seeds at intervals of several years. During the interval between crops, populations of seed predators die out, or are maintained at low levels on alternate foods. During mast years, the seed crop floods the market, quickly exceeding the ability of herbivores to consume it. Most of the seeds, therefore, come through unscathed.

Adaptations of predators and their prey and of hosts and their parasites provide striking examples of the evolutionary responses that constantly alter the structure and function of the biological community. Traits that make good competitors are more subtle and have only been revealed indirectly in competition experiments, but they are nonetheless important to the evolution of interrelationships within the community.

Attributes of the community change in response to pervasive changes in the physical environment and to shifts in its own biological composition. Natural selection endows the community with a powerful homeostatic mechanism sensitive to changes in the environment over hundreds and thousands of generations. Evolutionary changes enable populations of short-lived insects and microorganisms to respond to the havoc wreaked by man on the earth's surface. Over longer periods, evolution has shaped the structure and function of present-day ecosystems.

Populations are sometimes put under stress by changes in the environment that are too rapid to be countered by evolutionary change or too great to be compensated by homeostatic responses of organisms. Under these extreme physical conditions or biological pressures, a population may decline in size and eventually become extinct. As we shall see in the next chapter, much has yet to be learned about extinction. But whatever the cause or causes, extinction is partly the result of a failed evolutionary response by the population.

19 | Extinction

IN 1810, American ornithologist Alexander Wilson observed an immense flock of passenger pigeons in the Ohio River Valley. For days, the column of birds, perhaps a mile wide, passed overhead in numbers to darken the sky. Wilson estimated that there were more than two billion birds. The last passenger pigeon died in the Cincinnati Zoological Garden just more than a century later, on September 1, 1914.

The demise of the passenger pigeon was caused by two unfortunate circumstances. First, the pigeons roosted and nested in huge assemblies, sometimes numbering several hundred million birds within a few hundred square miles. Second, roasted or stewed, the pigeons tasted very good. Pigeoners, as professional hunters were called, gathered in large groups to trap and slaughter them. Nesting trees were felled to collect the squabs. As farmers cleared the forests and railroads made vast areas within the pigeons' range accessible to eastern big-city markets, the persecution increased. By the mid-1800s increased killing disrupted breeding, and the pigeon population began to decline. By 1870, large breeding congregations were found only in the Great Lakes States, at the northern edge of the pigeon's former range. The last nest was found in 1894, and the bird was last seen in the wild in 1899.

With its extinction in 1914, the passenger pigeon joined a lengthening list of species that have vanished from the earth. Since naturalists began describing systematically the forms of plant and animal life two hundred years ago, 53 birds, 77 mammals, and numerous other animals and plants have disappeared, most of them at the hand of man. By exploiting species for food or hides or destroying them as pests, by subjecting them to depredation by domesticated animals and hangers-on like rats, and by destroying their habitats, we have managed to reduce the numbers or productivity of many populations below their self-sustaining level.

Perhaps many of these species, including the passenger pigeon, would have persisted had it not been for man's activities. But the fossil record reveals that virtually all lineages have become extinct without leaving descendants. The million or so living species of plants and animals are derived from a small fraction of those alive at any time in the distant past. It has been estimated that less than 1 per cent of all the species that have existed since the beginning of the Paleozoic Era (about 600 million years) are extant. If this were a reasonable estimate, it would mean that species have become extinct at a rate of one every six or so years on average. No doubt, therefore, that extinction is a commonplace event—in fact, extinction is the inevitable fate of virtually all lineages.

Extinction is important in the study of ecology for two reasons. First, the extinction of local populations of a species can play a role in the regulation of population size. As we have seen, many populations exist in mosaics of habitat patches. Population size is related to the balance between local extinction and immigration of colonists to patches left empty by extinct populations. Second, the local extinction of one species broadens the ecological opportunity open to similar species and may enable a species previously excluded by competition to establish itself in the community.

To the extent that extinction reflects the relative inability of a population to adapt to a changing environment, it also belies the relative evolutionary success of populations better able to adapt. Hence the replacement of species by others that are more successful may be a form of selection in which the entire population rather than the individual bears the favored trait.

The study of extinction ought to elucidate many aspects of ecology and evolution, but it has largely failed to do so. This failure can be traced directly to three facts: extinction is so infrequent and difficult to predict that direct observation is not feasible; extinction is merely the final event in a long sequence of subtle evolutionary and ecological processes leading to the demise of a population; and the fossil record is too incomplete to record the details of a dwindling lineage. Fossils open only a tiny obscured window on biological communities of the past.

Most of what we know about extinction comes from observations of the phenomenon on three different levels. First, we can observe the extinction of small local populations when local climatic conditions become severe or when an efficient predator is introduced to the community. Such extinctions are ecological rather than evolutionary events, and would not normally lead to extermination of an entire lineage. Second, we can observe or infer the local, and sometimes global extinction of a species resulting from competition from ecologically similar species. Such replacements usually would cause a change in the structure of the community and lead to ecological and evolutionary readjustments among the remaining species. Third, fossils record the extinctions of entire groups of organisms. Dinosaurs, ammonites, and trilobites are among numerous once-successful taxonomic groups that have disappeared, sometimes abruptly, from the face of the earth. Although the extinction of a major phylogeny must necessarily involve the extinction of all the numerous closely related species that comprise the taxon, the fossil record does not preserve so much detail and we are left guessing the ecological and evolutionary significance of the successive replacement of major phylogenetic groups.

Extinctions on Islands

When populations dwindle to small size, particularly when density-dependent mechanisms no longer operate owing to intense interspecific competition, they become extremely susceptible to extinction following random fluctuations in size. By virtue of the fact that populations on small islands are restricted geographically and are not frequently augmented by immigration, they are particularly susceptible to extinction.

Extinction occurs frequently enough on small islands that probability of extinction can be determined from historical records. For example, in the 51-year interval between 1916 and 1968, numerous cases of species disappearances from the Channel Islands off the coast of Southern California were revealed. At one extreme, 70 per cent of the avifauna (seven out of ten species) on Santa Barbara Island (one square mile in area) disappeared. At the other, 17 per cent (six out of 36 species) on Santa Cruz (96 square miles) were gone. On an annual basis, these figures amount to between 0.1 and 1.7 per cent of the avifauna, with extinction rate and island size inversely related. Comparable rates have been determined for two tropical islands: 0.2 per cent per year on Karkar, an island of 142 square miles located ten miles off the coast of New Guinea, and 0.23 per cent per year on Mona, ten square miles, located between Puerto Rico and Hispaniola in the Greater Antilles. Because some reported extinctions may be attributed to inadequate census procedures, extirpations, and habitat change, extinction rates may be overestimated by as much as an order of magnitude. It is apparent nonetheless that populations on small islands, more than populations on large islands, are susceptible to extinction, regardless of the cause.

When an entire species disappears, one can be sure that extinction has occurred. Of 53 species of birds that have become extinct in the past 300 years, only three (the crested sheldrake of eastern Asia, the passenger pigeon, and Carolina parakeet, both of eastern North America) were found on major continents. The remaining fifty species disappeared from islands the size of New Zealand and smaller. As we have seen, island populations are particularly vulnerable because of their small size and isolation. Many island species have adapted to habitats without predators and with few competitors and are unable to cope with newcomers introduced from comparable mainland habitats, or with man. Other island species that are restricted to forest habitats suffer when land is cleared for sugar cane and coconut palm plantations. Small isolated populations also are vulnerable to such perturbations in their environments as prolonged drought, hurricanes, and disease epidemics. And the local extinction of these endemic island populations means the extinction of an entire species. Prior to their extinc-

TABLE 19-1 Recorded extinctions of mammals on islands and continents.

	NUMBER OF		
	Large predators	Large herbivores	Small species
Continents	10	17	0
Islands	1	12	42

tion, many species of birds were regularly eaten by man, whose depredations led directly to the demise of ducks, pigeons, and rails (small henlike marsh birds). But small inconspicuous, forest-dwelling birds have also disappeared, indicating the general vulnerability of island populations.

In contrast to birds, mammals have become extinct almost as frequently on the major continents (27 species) as on outlying islands, including Australia (55 species). Most extinct continental species were large herbivores killed for food (ground sloths, sheep, deer, zebras) and large carnivores, killed because they competed with man for food or were thought to threaten man directly—bears, wolves, cats (Table 19-1). Losses of mammals less involved with man's well-being (small rodents, bats, shrews, small marsupials) are confined to islands.

Nonrandomness of Extinction

When tossing a coin, some trials come up heads, others come up tails, but the probability of each outcome is the same for each trial. Some species persist and others disappear, but evidence is beginning to indicate that the probability of extinction differs greatly among species. Some are close to their inevitable fate and will falter with the least ecological setback; others are resilient and productive, able to withstand perturbations in their environments. Such differences in probability of extinction can be inferred from patterns of geographical distribution and taxonomic differentiation of populations inhabiting groups of islands, like the West Indies and Polynesian Islands.

Immigrants to islands appear to be excellent competitors initially. Colonizing species are usually abundant and widespread on the mainland; these qualities make good immigrants. Most invaders of an island exhibit ecological release. Their populations increase greatly and spread into habitats not occupied by the parent population on the mainland. After immigrants become established, however, their competitive ability appears to wane; their distribution among habitats becomes restricted, and local population density decreases. These trends eventually lead to extinction.

We can judge the relative ages of populations on islands by their patterns of geographical distribution and by differences in their appearance from the appearance of mainland forms from which they are derived. Range maps of representative species of birds in the Lesser Antilles (Figure 19-1) demonstrate the progressive changes in distribution and differentiation of species with time, called the *taxon cycle* by Harvard biologist E. O. Wilson. On the basis of such distribution patterns, we can assign populations to one of four stages (Table 19-2): expanding (I), differentiating (II), fragmenting (III), and endemic (IV). Similar patterns have been described for ants on islands in the southwestern part of the Pacific Ocean, for birds in the Solomon Islands, and for some insects in the Solomon Islands. Among birds of the West Indies, species in late stages of the taxon cycle exhibit a loss of interisland movement and seasonal migration, reduced flocking behavior, and an increased tendency to occur in, or be restricted to, deep forest habitats, often montane forests. Similar habitat changes have been found in ants and beetles on islands in the Pacific.

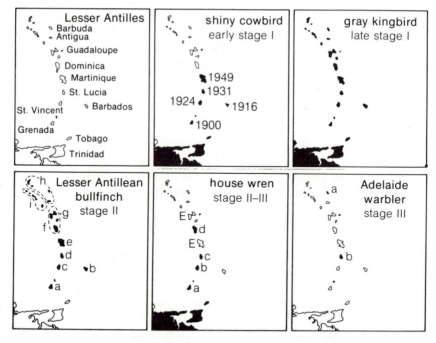

FIGURE 19-1 Distribution patterns and taxonomic differentiation of several birds in the Lesser Antilles, illustrating progressive stages of the taxon cycle. The shiny cowbird has expanded its range in the islands (dates of arrival are indicated), and the house wren has become extinct (E) on several islands during this century (a small population has been rediscovered recently on Guadaloupe). Lower case letters designate subspecies.

TABLE 19-2 Characteristics of distribution of species in the stages of the taxon cycle.

Stage of cycle	Distribution among islands	Differentiation between island populations
I	Expanding or widespread	Island populations similar to each other
II	Widespread over many neighboring islands	Widespread differentiation of populations on different islands
III	Range fragmented due to extinction	Widespread differentiation
IV	Endemic to one island	—

Populations become more vulnerable to extinction as the taxon cycle progresses (Table 19-3). More endemic species (stage IV) of West Indian birds have become extinct since 1850, or are currently in grave danger of extinction, than widespread species (stages I to III). Almost a quarter of the endemic species either are extinct or are sufficiently reduced that they are at present in danger of extinction.

The decrease in competitive ability of island populations with age may be caused by evolutionary responses of an island's biota to new species. Immigrants are thought to be relatively free of parasites, predators, and efficiently specialized competitors when they colonize an island, so that their populations increase rapidly and become widespread. Having reached this stage, new immigrants constitute a larger part of the environment of many other species, which then evolve to exploit, avoid exploitation by, or outcompete the newcomer. Apparently, a large number of species, when

TABLE 19-3 Rate of extinction of island populations of birds in the West Indies as a function of stage of the taxon cycle.

	TAXON-CYCLE STAGE			
	I	II	III	IV
Number of recently extinct or endangered populations	0	8	12	13
Total number of island populations*	428	289	229	57
Per cent extinct or endangered	0	2.8	5.2	22.8

* A widespread species has many island populations, an endemic species only one.

adapting to a single abundant new population, can evolve faster than the new species can adapt to meet their evolutionary challenge. Competitive ability of the immigrants is progressively reduced by counteradaptations of island residents until the once-abundant new species becomes rare. Species are eventually forced to extinction by subsequent arrivals from the mainland that are more efficient competitors. When a species becomes rare, other species no longer gain evolutionary advantage by adapting to it, and the evolutionary pressure upon the rare species is released, similar to the effect that Pimentel observed in his experiments on competition between houseflies and blowflies. If this occurred before a species' decline had proceeded too far, the species might again increase and begin a new cycle of expansion throughout the island. This apparently has occurred many times; species distributions provide ample evidence of secondary expansions within the West Indies (for example, Figure 19-1, lower left).

The Fossil Record of Proliferation and Extinction

Observations of contemporaneous extinctions suggest that species disappear when better competitors or, less commonly, more efficient predators reduce their numbers to the point that random fluctuations quickly lead to extinction. Biological factors are, without doubt, more important than stochastic factors. Extinction by competitive exclusion results in the replacement of a species by one or more ecologically similar species that are often closely related taxonomically. Hence, although species turnover occurs, the character of the community does not change substantially. The passage of evolutionary time—measured in millions of years—reveals, however, the appearance and proliferation of new groups of organisms, often with novel structures, and the demise of once dominant groups. If we could step back 20 million years in time to the middle of the Miocene Epoch, we would see few familiar species in the fields and woods. To be sure, we could recognize many rodents and carnivores and some birds, and we could identify several of the trees, but others would be completely alien to our experience. Transported back 100 million more years into the Cretaceous Period, we would find ourselves in a world of giant reptiles and probably would hardly notice the small primitive mammals and birds that would one day inherit the earth.

The fact that extinctions of major taxonomic groups have occurred has been known for more than a century. Charles Darwin summarized what was known of extinction in the mid-1800s in *On the Origin of Species*:

> On the theory of natural selection, the extinction of old forms and the production of new and improved forms are intimately connected together. The old notion of all the inhabitants of the earth having been swept away by catastrophes at successive periods is very generally given up, even by those geologists . . . whose general views would naturally lead them to this conclusion. On the contrary, we have every reason to believe, from the study of the tertiary formations, that species and groups

of species gradually disappear one after another, first from one spot, then from another, and finally from the world. In some few cases, however, as by the breaking of an isthmus and the consequent irruption of a multitude of new inhabitants into an adjoining sea, or by the final subsidence of an island, the process of extinction may have been rapid. Both single species and whole groups of species last for very unequal periods; some groups, as we have seen, have endured from the earliest known dawn of life to the present day; some have disappeared before the close of the palaeozoic period. No fixed law seems to determine the length of time during which any single species or any single genus endures. There is reason to believe that the extinction of a whole group of species is generally a slower process than their production. If their appearance and disappearance be represented, as before, by a vertical line of varying thickness the line is found to taper more gradually at its upper end, which marks the progress of extermination, than at its lower end, which marks the first appearance and the early increase in number of the species. In some cases, however, the extermination of whole groups, as of ammonites, towards the close of the secondary period, has been wonderfully sudden.

The historical record of extinction has been more recently reviewed by the paleontologist George Gaylord Simpson, who elaborated the basic observations of Darwin and others a century before. Reiterating the point that phylogenetic lines vary widely in their persistence and degree of evolutionary diversification, Simpson rejected the notion that higher taxa pass through inherently determined life stages similar to the life history of an organism. Extinction cannot in any way be construed from the fossil record as "racial senescence." He then developed the following typical profile of the history of a phylogenetic group.

1. Periods of evolutionary activity within a line are usually accompanied by rapid diversification, geographical expansion, and adaptive radiation. As a group dies out, many lines become extinct and the remainder usually represent highly specialized forms, not a uniformly thinned-out spectrum of the group at the point of its greatest diversification.
2. Periods of diversification reflect enlarged opportunities for the group. The reptiles diversified after they moved onto dry land; marsupials, after they crossed the water gap to Australia; mammals, in general, after the disappearance of large reptiles; birds, after they attained the ability to fly; and so on.
3. Major adaptive radiations were usually followed by long periods of little' evolutionary change, during which extinct species were replaced by close relatives, much as we can observe or infer from present faunas and floras.
4. The initial diversification of a group produced many false starts, innovative lines that were quickly replaced by even more successful offshoots. Thus, when diversification intensifies, so does extinction within the line.
5. Each large taxonomic group has a characteristic average duration of species, genera, and families within it. Thus the average time span of

extinct genera (that is, those for which the time span is known) is 80 million years for pelecypods (clams and their relatives) but only 8 million years for carnivore mammals.

The decline of one phylogenetic line is often accompanied by increase in another, sometimes derived from the first. The succession of major reptilian groups by mammals provides a well-documented example. The reptiles themselves arose during the Pennsylvanian Period, perhaps 300 million years ago, from a progressive line of amphibians. The most important advance in the new reptile line was the development of an egg that could withstand desiccation—the *amniote* egg of present-day reptiles and birds— which enabled reptiles to escape water completely, and finally to launch a full-scale invasion of the terrestrial environment. From primitive beginnings in the late Paleozoic, reptiles underwent a major diversification during the Triassic Period (Figure 19-2) and produced all the major groups, including modern lizards, snakes, and turtles, by the middle of the Jurassic, 150 million years ago. The Jurassic and Cretaceous were the Age of Reptiles.

One of the ironies of the fossil record of reptiles concerns the synapsid line, which underwent a major radiation and dominated all other groups during the Permian and Triassic periods. Of the synapsids, the therapsids

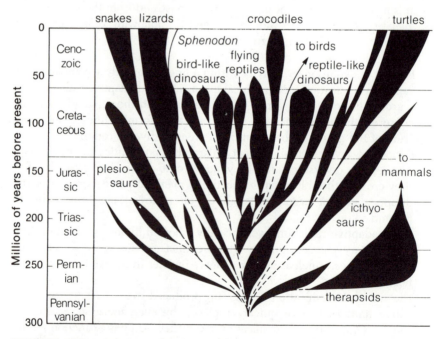

FIGURE 19-2 A schematic diagram of the phylogenetic tree of reptiles. The width of each line is roughly proportional to its diversity. The lines that led to birds and mammals were well-distinguished by the middle of the Jurassic Period.

were the most important, but with the rise of the dinosaurs during the late Triassic, the therapsids and, with them, the synapsid line, all but disappeared by the beginning of the Jurassic. Their only descendants eventually evolved to become what we recognize as mammals.

While the dinosaurs were having their heyday, an aberrant line of the dwindling synapsid group crossed the reptile-mammal boundary sometime during the late Triassic Period. This line then continued its precarious existence—occasionally producing abortive radiations—for nearly 100 million years, until the reptiles had finally waned, before embarking on a major period of diversification during the Paleocene Period.

The mammal body plan required considerable time to develop. Mammals differ from reptiles in being warmblooded, nursing their young, and abandoning the egg in favor of internal development of the embryo. Adopting a high body temperature and an active mode of life was accompanied by the development of insulating fur, a more efficient heart design, rearrangement of some parts of the skeleton, particularly the position of the appendages to elevate the body off the ground. Certain aspects of the skull and jaw became modified, and the teeth became fewer in number, more strongly rooted, and more diversified in form and function within the mouth. The manner in which the long bones of the appendages grow was also modified to allow the development of well-articulated joints in young, growing mammals. In reptiles, the bones grow at their ends by laying down cartilage that is subsequently ossified. The soft cartilage cannot be used to form an active, secure joint. In mammals, cartilage formation takes place in a region, called the *epiphysis*, under a bony cap at the end of the long bone which can assume the final shape of the joint at an early age. Mammals also developed a secondary palate at the roof of the mouth (the hard palate) that separates breathing and eating passages and allows mammals to continue breathing while they eat. This was not an important consideration for reptiles, owing to their low metabolic rates, but was a must for the more active mammals, which could suffocate in short time if breathing were cut off.

These attributes were perfected over a 100 million-year period during the Mesozoic Era, when reptiles reached their pinnacle of diversification. Then, at the end of the Cretaceous Period, most of the major lines of reptiles disappeared abruptly, followed shortly by the first major diversification of mammals. Earlier radiations of the mammal line during the Jurassic Period had been unsuccessful. None of the lines had attained any numerical importance, and only two survived the Cretaceous—the multituberculates, which became extinct in the early Eocene, and the line that gave rise to present-day marsupials and placentals.

Mass Extinctions

There can be little doubt that the dinosaurs disappeared within a few million years at most. It is also reasonably certain that their demise was not caused by mammals, which did not become abundant until well after the extinction of the dinosaurs. The end of the Mesozoic also brought the simultaneous extinction of other groups, the most notable of which was the ammonites, which resembled the present-day chambered nautilus, a type of predatory mollusc. Actually, the fossil history of the ammonites shows several phases of large-scale extinction followed by rapid diversification (Figure 19-3). These phases occurred at the end of the Permian Period (230 million years ago) and at the end of the Triassic, 50 million years later. In both cases, new phylogenetic lines do not appear in the fossil record until well after most of the old lines have died out, precluding competitive exclusion as a cause of mass extinction.

Extinctions of major phylogenetic groups reached highs at the ends of the Cambrian, Devonian, Permian, Triassic, and Cretaceous periods. It is no coincidence that periods of mass extinction correspond to the dividing lines of geologic periods, because geologists classified the ages of rock layers according to the fossils they contained. Any abrupt change in the animal families represented would form a natural division. It is also apparent that mass extinctions are well correlated with widespread unconformities in the geological record—lapses in the deposition of sediments. It has been suggested that mass extinctions accompanied a general rising of the continents, resulting in the draining of vast inland seas and destruction of immense areas of habitat. If we consider that most fossils were deposited in shallow seas and that small changes in sea level would cause large changes in the

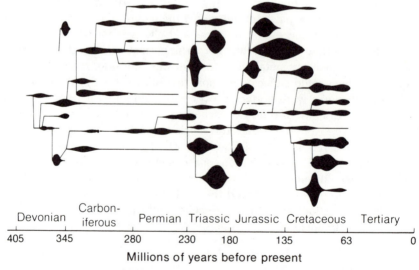

| Devonian | Carbon-iferous | Permian | Triassic | Jurassic | Cretaceous | Tertiary |

405   345   280   230   180   135   63   0

Millions of years before present

FIGURE 19-3 Diversification and phylogenetic relationships among superfamilies of ammonites. The group as a whole nearly went extinct at the end of the Permian and the end of the Triassic.

area of such habitats, mass extinctions become less mysterious. We may still, however, be hard put to explain the disappearance of a primarily terrestrial group like the dinosaurs. In the final analysis, we may never understand in detail the extinction of major taxonomic groups. Why were not some lines able to evolve rapidly enough to adapt to changing conditions? Was their extinction in any way hastened through competition with other groups or by the appearance of new, more efficient predators?

The Great American Interchange

The history of the mammal fauna of South America reveals a series of replacements by major groups brought about at first by the evolution and diversification of new forms within South America, and later by the immigration of North American species by way of Central America. During much of the last 60 million years, South America has been isolated from other continents; Central America was underwater or appeared as a string of islands until about 5 million years ago, when a solid land bridge finally joined the two continents and began a great interchange of the faunas of North and South America. The early mammalian history of South America tells us of the diversification of notoungulates and litopterns (primitive herbivores), marsupials, and many edentates (armadillos, sloths, and anteaters). Primates, primitive rodents, and raccoonlike forms apparently crossed the water barrier from North America and diversified well-before the modern land bridge formed.

Many changes occurred in the mammal fauna of South America during its extended isolation (Figure 19-4). Primitive ungulate herbivores, mostly notoungulates and litopterns, underwent a rapid diversification and reached their peak during the Eocene, 40 million to 50 million years ago. Some primitive ungulates and native marsupials evolved to fill the ecological roles of present-day rodents. Although this branch of the marsupial line did not persist, marsupials comprised most of the South American carnivorous species until the immigration of North American dogs, bears, cats, and weasels. Note that on Australia, the mammal fauna is dominated by marsupials. Few other types of mammals have ever existed there (other than bats), and Australia is still sufficiently isolated to prevent the immigration of potentially superior placental mammals.

The rodentlike ungulates of South America were gradually pushed aside by the diversification of native South American rodents, represented today by capybaras, agoutis, chinchillas, and their relatives. The success of these rodents was paralleled by a decline in the larger primitive ungulates, perhaps through competitive exclusion.

The Central American land bridge brought hordes of new immigrants from the north and caused the extinction of many South American ecological counterparts. The primitive ungulates were pushed aside by invading deer, tapirs, and camels. (Many of these are now extinct in North America. The

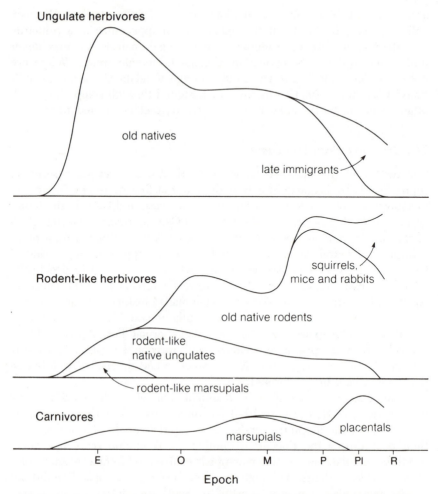

FIGURE 19-4 Diversification and reduction of several groups of mammals on the South American continent. Height of the graph in each case represents the relative number of genera. Time scale is only approximate: R = Recent, Pl = Pleistocene ($0 - 3 \times 10^6$ years before present), P = Pliocene ($3 - 13 \times 10^6$ years), M = Miocene ($13 - 25 \times 10^6$ years), O = Oligocene ($25 - 36 \times 10^6$ years), E = Eocene ($36 - 58 \times 10^6$ years). Recent exchange of species between North and South America began in the Pliocene.

camel family is represented in South America by llamas, vicuña, and alpaca.) Many of the early native rodents of South America vanished with the appearance of squirrels, mice, and rabbits, but several unique forms (guinea pigs, capybaras) survived the onslaught. The marsupial carnivores were replaced by placental counterparts (dogs, cats, bears).

The modern mammal fauna of South America is a nearly equal mixture of distinctly South American and North American forms, yet few southern mammals became established in North America. In fact, porcupines, the now extinct ground sloths, and the opossum are the only distinctly South American groups that have invaded the north temperate zone. Other southern forms, like armadillos, sloths, and some of the native rodents, have extended their ranges into Central America, but have not yet moved beyond the tropical zone. The failure of South American forms to colonize North America is all the more amazing in that South American mammals began to appear in North American fossil assemblages (mostly from Florida, Texas, and California) before North American forms appeared in fossil deposits in Patagonia. Yet the fact remains that most of the immigrants from South America either died out or their evolution was bogged down in North America, whereas many immigrants from North America underwent extensive adaptive radiation in South America during the last 2 million years.

It seems odd that the native fauna of South America, adapted to the particular environmental conditions on that continent, should have been overwhelmed by invaders from another land. It is true that South American forms had not evolved in competition with northern forms. By the end of the Miocene period, 13 million years ago, the native fauna had partitioned most of the possible ecological roles in the community, and the major radiations of new forms were largely over. In contrast, dispersal between Asia and North America occurred regularly by way of Alaska and the Bering land bridge. The mammalian fauna of North America has continually been exposed to immigrants from other areas, challenging its competitive ability. We may be tempted to apply terms like "evolutionary vigor" to the mammals of North America and "evolutionary stagnation" to the mammals of South America, but evolutionary biologists do not yet understand how the inherent characteristics of whole groups determine competitive ability. Why, for example, did *all* the litopterns die out? What common bonds of structure and function united these species in a fatal evolutionary pact?

The Conservatism of Evolution

Once the major radiation of a group has taken place, further change in body plan or way of life evidently becomes difficult. Most of the characteristics of present-day lungfish appeared during the first 50 million years of lungfish evolution. The line has changed relatively little in the last 250 million years. Genera of marine invertebrates and plants are recognized unchanged through hundreds of millions of years of the fossil record. These evolutionary lines have undoubtedly been buffeted by changing environments—shifts in temperature and moisture, the coming and going of predators and disease organisms—but they have retained their distinctive characteristics with a minimum of modification.

Evolutionary conservatism is well demonstrated by the floras of the southeastern regions of Asia and North America. During the late Tertiary Period, 70 million years ago, vast areas of the Northern Hemisphere were covered with broad-leaved forests. The remnants of these great forests—including primitive species like the rhododendron, tulip tree, magnolia, sweetgum, and hickory—are now localized in the southeastern United States and eastern Asia, where they have been isolated for more than 50 million years. Yet these plants still betray their common origin in the Tertiary flora, and although many have been given different species names, most of the genera remain unchanged. Moreover, genera with southern distributions in Asia tend to have southern distributions in North America; those with northern distributions in Asia tend to have northern distributions in North America; genera that are widespread in Asia also tend to be widespread in North America. Thus not only has form remained unchanged, but each genus has also retained a discernible portion of those physiological tolerances that determine geographical range—for more than 50 million years.

Extinction and Evolution

Evolutionary conservatism in the face of environmental change opens the door to extinction. Changes in climate, competitors, predators, and disease organisms all pose serious threats. Many changes come too quickly and too abruptly for evolution to respond. Certainly many of the results of human activities fall into this category. Natural changes in the environment take a more deliberate toll.

Climatic change by itself probably never drove a species to extinction. The physical environment sets the stage for competitive struggle between populations and higher taxonomic groups. The rate of extinction of species within a particular phylogenetic line, which is related to the probability of extinction of a particular population, may depend upon the rate at which the environment changes, on one hand, and the potential rate of evolutionary change relative to that of competitors and predators, on the other. Hypotheses concerning the causes of extinction usually spring from premises of environmental change or evolutionary stagnation. And hypotheses usually distinguish mass extinctions from the more usual replacement within a phylogenetic group, and rarely attempt to encompass both.

Mass extinctions remain something of an enigma, mostly because they occur in the absence of replacement by competitors and often involve several unrelated groups. Some hypotheses invoke extraterrestrial events, such as supernovae or solar flares showering killer doses of radiation on the earth's surface. But such events should bring more general havoc than the fossil record shows. Why should some, or most groups remain untouched, while others are disappearing? In addition, the fossil record indicates that mass extinctions were not sudden events but may have been drawn out over millions of years. The correspondence between mass extinctions and major worldwide geological changes suggests that mass extinctions were brought about by widespread change in climate or configuration of the continents and oceans that virtually eliminated certain types of habitats and the ways of life associated with them. The origination and proliferation of whales in the Eocene, followed by their near extinction in the Oligocene and subsequent reradiation in the Miocene have been attributed to the availability of concentrated food resources in coastal waters. Periods of adaptive radiations in marine mammals correspond to extensive deposition of rocks formed from the silicon skeletons of diatoms, a kind of alga that supports the marine food chain. These deposits suggest periods of high marine productivity, perhaps related by some unclear steps to wind patterns and upwelling currents, and increased availability of the marine crustacea and fish upon which present-day whales and porpoises feed. That dinosaurs, ammonites, or trilobites might have died out owing to the disappearance of a critical food source seems unlikely, but perhaps only because of our ignorance of the ecology of the earth tens and hundreds of millions of years ago.

For the study of modern ecology, extinctions involved in the replacement of species by competitors are of greater interest, for these processes play a role in shaping the structure of modern ecological communities. And the forces that lead to extinction ought to be directly observable in contemporary ecological interactions. It probably would suffice to say that we do not know why one species becomes extinct and another persists. Take two individuals, one from a population that is restricted in habitat and geographical range and contains, perhaps, only a few hundred individuals, and the other from an abundant, widespread species. One of these clearly is nearer the brink of extinction than the other, yet I would defy any biologist not familiar with the particular species, to choose with reason and certainty which individual was taken from the fading species.

Hypotheses about extinction usually are difficult to test adequately, and for each hypothesis proposed there are numerous counterexamples. Many have argued that the probability of extinction is directly related to the degree of fluctuation in the environment. But if one is willing to accept the premise that the environments of molluscs that burrow in the bottom sediments are more constant than those of molluscs that live on the surfaces of sediments, then the evidence is contrary. Rates of extinction are about the same in both groups. Another theory proposes that under constant environmental conditions, selection tends to make most genes homozygous and reduce a population's evolutionary potential. Such populations would be at greater risk of extinction should the environment suddenly change. Although there is ample evidence that species replacement rates are higher in constant environments than in fluctuating and unpredictable environments, there is no evidence that populations in constant environments are genetically impoverished.

Another hypothesis suggests that specialized organisms have greater probability of extinction than generalized organisms. Here again, we become bogged down trying to define "specialization." If one accepts that, for aquatic, free-living arthropods (trilobites, crabs, and their relatives), specialization can be measured by the diversity of limbs (claws, swimming appendages, walking legs, etc.) on the arthropod body, it turns out that specialized genera are as long-lived as generalized taxa. If we are skeptical of this index as a measure of specialization, then we are left without a conclusion.

Paleontologist Leigh Van Valen, of the University of Chicago, has followed a more general and abstract approach to the problem of extinction. He examined the survivorship curves of genera and families in dozens of taxonomic groups and discovered what appeared to be a constant probability of extinction for each age interval, just as a nonsenescing organism has a constant probability of death in each age interval. For Van Valen, this constant probability of extinction suggested that extinction is caused mainly by environmental change, not inherent evolutionary limitations, and that such changes are distributed among the environments of different organisms at random. But Van Valen's hypothesis has come under considerable criticism, both for his methods of constructing taxonomic survivorship curves and for his interpretation of their pattern, and neither has been resolved.

As you can see, biologists have found little agreement with respect to extinction, other than that it surely has occurred in the past and, by extension, that it must be an important on-going process in the modern world. We understand so little about extinction because its time scale is too long for our own experience and too short for detailed preservation in the fossil record. The changes in population parameters that lead to extinction undoubtedly are minute on the scale of processes that influence the short-term dynamics of populations with which we are more familiar.

At this point, we shall leave extinction behind and, with it, other topics in population biology, and turn our attention to the structure and dynamics of biological communities. But just as population characteristics express the adaptations of individuals, we should not forget that community characteristics express the interrelationships between populations.

20 | Community Ecology

IN PREVIOUS CHAPTERS, we have examined population processes, including the interactions between competitors, predators, and prey. The theory of population ecology and experimental tests of that theory are based on the interactions of a few species, frequently only two species. Yet in their natural habitats, most species interact with a wider variety of others, numbering, perhaps, in the tens or even hundreds. Many of these relationships are casual and individually exert little influence on the population. Others, such as competition with closely related species and avoidance of efficient predators and parasites, determine the carrying capacity, stability, and evolutionary course of the population. All these interactions together either permit or preclude the existence of a population at a particular locality. Summed over all species, these interactions determine the number and interrelationships of the inhabitants of a region. In this and the following three chapters, we shall study the community as a natural unit and trace the geographical, developmental, and evolutionary bases of its organization.

Definition of The Community

The term *community* has been given such a variety of meanings by ecologists that it borders on being meaningless. It usually is restricted to the description of a group of populations that occur together, but there ends any similarity among definitions. Throughout the development of ecology as a science, the term has often been tacked on to associations of plants and animals that are spatially delimited and that are dominated by one or more prominent species or by a physical characteristic. One speaks of an oak community, a sagebrush community, and a pond community, meaning all the plants and animals found in the particular place dominated by its namesake. Used in this way, "community" is unambiguous: it is spatially defined and includes all the populations within its boundaries.

Ecologists also define communities on the basis of interactions among associated populations. This is a functional rather than descriptive use of the term. But we may find communities difficult to delimit in this way because interactions among populations extend beyond arbitrary spatial boundaries. Some of the oxygen molecules that a squirrel inhales in New York might have been expired last month by a tree in the Amazon basin. Clearly, the Brazilian tree exerts a negligible influence on the squirrel, and we ought not to extend the boundaries of the community to which the squirrel belongs to the tropical rain forest. But biological interactions reach out in a complicated fashion, making it difficult to set limits—other than completely arbitrary ones—to a community.

Organisms and materials move slowly across the boundaries of most terrestrial associations of plants and animals compared to the turnover of organisms and flux of materials and energy within the association. "Community" and "association" are, therefore, nearly identical for many terrestrial habitats, so long as the boundaries of the habitats are not too narrowly drawn. Because most fresh-water and coastal aquatic communities receive inputs from the land, and because many organisms move freely between aquatic habitats, the boundaries of aquatic communities are more difficult to establish. Salamanders, which are aquatic as larvae and terrestrial as adults, link together the stream and forest communities; migratory birds carry a thread of interaction between many temperate and tropical communities.

Two extreme cases of associations, in which all the energy inputs come from outside, are the faunas of caves and of unlighted depths of the oceans. Strictly speaking, the inhabitants of these perpetually dark places depend upon primary producers in sunlit habitats for their source of energy, yet we may speak of competition among detritivore or carnivore members of the cave "community," if we wish. In fact, "community" may be usefully applied to any group of interacting populations—the herbivorous insect community, the oak leaf community, the kelp holdfast community—as long as community boundaries are clearly delimited.

The maintenance of community structure and functioning depends on a complex array of interactions, directly or indirectly tying all its members together in an intricate web. The influence of a population extends to ecologically distant parts of the community through its competitors, predators, and prey. Insectivorous birds do not eat trees, but they do prey on many of the insects that feed on foliage or pollinate flowers. By preying upon pollinators, birds indirectly affect the number of fruits produced, the amount of food available to animals that feed upon fruits and seedlings, the predators and parasites of those animals, and so on. The ecological and evolutionary impact of a population extends in all directions through its influence on predators, competitors, and prey, but this influence is dissipated as it passes through each successive link in the chain of interaction.

The Local Community as a Natural Unit

We may describe the functional relationships among an association of species just as the physiologist relates the various parts of the body. The obvious analogy between community and organism led many early ecologists, notably the influential plant ecologist F. E. Clements, to describe communities as discrete units with sharp boundaries. This view is reinforced by the conspicuousness of many dominant vegetation types. A forest of ponderosa pines, for example, appears distinct from the fir forests that occupy moister habitats and from the shrubby vegetation and grassland found on drier sites. The boundaries between these community types are often so sharp as to be crossed within a few yards along a gradient of climate conditions. Some community boundaries, such as that between deciduous forest and prairie in the midwestern United States, are respected by most species and spanned by relatively few.

An opposite view of community organization was held by H. A. Gleason, who suggested that the community, far from being a distinct unit like an organism, was merely a fortuitous association of organisms whose adaptations enabled them to live together under the particular physical and biological conditions found at a particular location.

Clements' and Gleason's concepts of community organization predict different patterns in the distribution of species over ecological and geographical gradients. On one hand, Clements believed that the species belonging to a community were closely associated with each other; the ecological limits of distribution of each species coincided with the distribution of the community as a whole. This type of community organization is commonly called a *closed community*. On the other hand, Gleason believed that each species was distributed independently of others that co-occurred in a particular association, an organization referred to as an *open community*. We would draw the boundaries of an open community arbitrarily with respect to the geographical and ecological distributions of its component species, which may extend their ranges independently into other associations of species.

The structure of closed and open communities is depicted schematically in Figure 20-1. In the upper diagram, the distributions of species in each community are closely associated along a gradient of environmental conditions—for example, from dry to moist. Closed communities represent natural ecological units with distinct boundaries. The edges of the community, called *ecotones*, are points of rapid replacement of species along the gradient. In the lower diagram, species are distributed at random with respect to each other. We may arbitrarily delimit an open community at some point, perhaps a dry forest community near the left-hand end of the moisture gradient, but some of the species included might be more characteristic of drier points along the gradient, while others reach their greatest productivity in wetter sites.

The separate concepts of open and closed communities both apply to

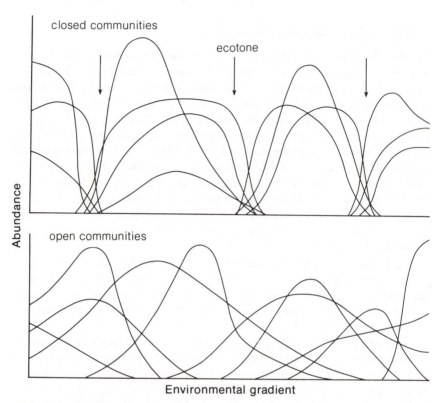

FIGURE 20-1 Hypothetical distributions of species organized into distinct associations (closed communities, above) or distributed at random along a gradient of environmental conditions (open communities, below). The species found at any point, or within any narrow region, along the lower gradient would be identified as an open community. Ecotones between communities in the upper figure are indicated by arrows.

associations of species in nature. We observe distinct ecotones between associations under two different kinds of circumstances: when there is an abrupt change in the physical environment—for example, at the transition between aquatic and terrestrial communities, between distinct soil types, or between north-facing and south-facing slopes of mountains—or when one species or life form so dominates the environment of the community that the edge of its range signals the distributional limits of many other species.

The transition between broad-leaved and coniferous forest is usually accompanied by an abrupt change in soil acidity. At the boundary between grassland and shrubland, or between grassland and forest, sharp changes in surface temperature, soil moisture, and light intensity result in many species replacements. Sharp grass-shrub boundaries occur because when one or the other vegetation type holds a slight competitive edge, it dominates the community. Grasses prevent shrub seedling growth by reducing the moisture content of the surface layers of the soil; shrubs prevent the growth of grass seedlings by shading them out. Fire evidently maintained a sharp boundary between prairie and forest in the midwestern United States. Perennial grasses resist fire damage that kills tree seedlings outright, but fires do not penetrate deeply into the moister forest habitats.

Sharp physical boundaries create sharp ecotones. Such boundaries occur at the interface between most terrestrial and aquatic (especially marine) communities (Figure 20-2) and where underlying geological formations cause the mineral content of soil to change abruptly. The ecotone between plant associations on serpentine-derived soils and nonserpentine soils in southwestern Oregon is shown in more detail by the diagrams of soil minerals and occurrence of plant species in Figure 20-3. Levels of nickel, chromium, iron, and magnesium increase abruptly across the boundary into serpentine soils; copper and calcium contents of the soil drop off. The edge of the serpentine soil marks the boundaries of many species that are either excluded from, or restricted to, serpentine outcrops. A few species are found only within the narrow zone of transition, and others, seemingly unresponsive to variation in soil minerals, extend across the ecotone.

Gradient Analysis

Sharp physical boundaries often create abrupt changes in vegetation. It is difficult for species to be fence sitters in such situations; they must adapt to the conditions on one side or the other. The few species specialized to live

FIGURE 20-2 A sharp community boundary (ecotone) associated with an abrupt change in the physical properties of adjacent habitats. Seaweeds extend only to the high tide mark. Between the high tide mark and the spruce forest, waves wash the soil from the rocks and salt spray kills pioneering land plants, leaving the area devoid of vegetation.

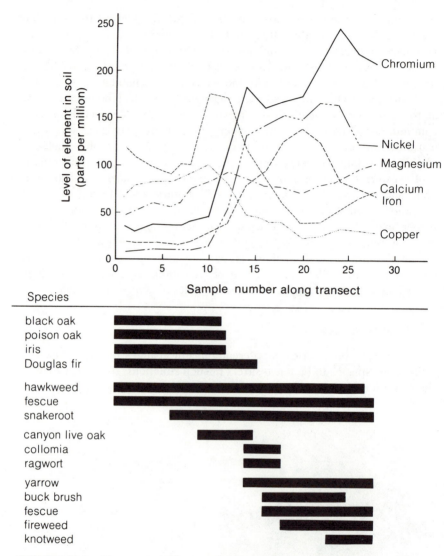

FIGURE 20-3   Changes in soil mineral content (above) and plant species (below) across the boundary between nonserpentine (left) and serpentine soils (right) in southwestern Oregon. The transect diagrammed here is somewhat atypical in that magnesium does not increase as abruptly as usual across the serpentine ecotone.

in the ecotone—often referred to as *edge species*—are necessarily restricted in numbers by the scarcity of their habitat.

Most large-scale ecological changes are more gradual than the abrupt changes between land and water, forest and field, or geological formations. The major biological communities of the earth occupy a *continuum* of gradually changing ecological conditions, varying along one dimension from cold to warm, along another from dry to moist, along a third from seasonal to moderate, and perhaps along many other minor gradients of ecological conditions. Changes along these *gradients* occur gradually over vast geographical distances; sharp physical boundaries do not intercept species ranges.

When we examine the occurrence of species along gradients of moisture or temperature, we find that the local plant community must be viewed as an open system. Groups of species generally are not restricted to particular associations, rather each species is distributed along environmental gradients almost without regard to the occurrence of others. The environments of the eastern United States form a continuum with a north-south temperature gradient and an east-west rainfall gradient. Species of trees found in any one region, for example those native to eastern Kentucky, have different geographical ranges, suggesting a variety of evolutionary backgrounds (Figure 20-4). Some species reach their northern limits in Kentucky, some their southern limits. Each has a unique evolutionary history, with a variable degree of association with other species in the local community.

FIGURE 20-4 Geographical distribution of twelve species of trees found in plant associations in eastern Kentucky.

A more detailed view of Kentucky forests would reveal that many of the tree species are segregated along local gradients of conditions. Some are found along the ridge tops, others along moist river bottoms, some on poorly developed rocky soils, others on rich organic soils. The species represented in each of these more narrowly defined associations would exhibit correspondingly closer ecological distributions, but the open community concept would still dominate our thinking about these associations.

Cornell University ecologist Robert Whittaker has examined plant distributions in several mountain ranges where moisture and temperature vary over short distances according to elevation, slope, and exposure. When Whittaker plotted the abundance of each species at sites at the same elevation distributed along a continuum of soil moisture, he found that the species occupied unique ranges with peaks of abundance scattered along the environmental gradient (Figure 20-5). The Oregon mountains have fewer species overall, but each species has a wider ecological distribution, on the average, than each Arizona species.

In the Great Smoky Mountains of Tennessee, dominant species of trees are widely distributed outside the plant associations that bear their names (Figure 20-6). For example, red oak is most abundant in relatively dry sites at high elevation, but its distribution extends into forests dominated by beech, white oak, chestnut, and even hemlock, an evergreen coniferous

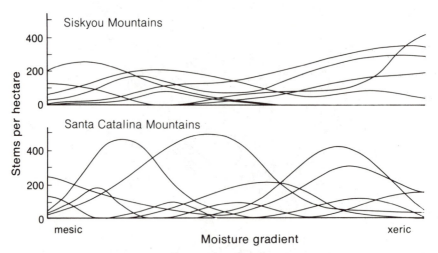

FIGURE 20-5 Distribution of species along moisture gradients at 460- to 470-meter elevation in the Siskyou Mountains of Oregon and at 1830- to 2140-meter elevation in the Santa Catalina Mountains of southern Arizona. Species in the more diverse Arizona flora occupy narrower ecological ranges; thus, in spite of the greater total number of species in the flora of the Santa Catalina Mountains, the Santa Catalina Mountains and Siskyou Mountains have similar number of species at each sampling locality.

species, and extends throughout the entire range of elevation in the Smoky Mountains. Beech prefers moister situations than red oak, and white oak reaches its greatest abundance in drier situations, but all three species occur together in many areas.

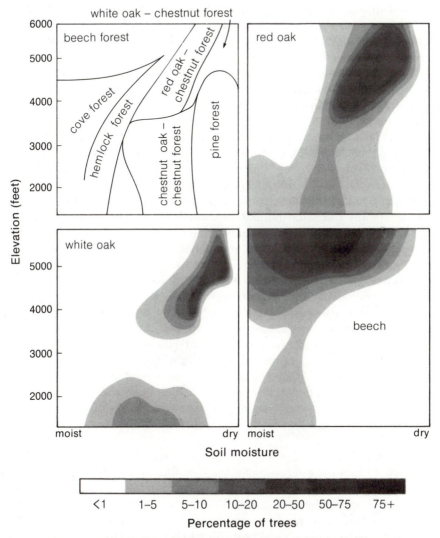

FIGURE 20-6 Distribution of red oak, white oak, and beech with respect to altitude and soil moisture in the Great Smoky Mountains of Tennessee. The approximate boundaries of the major forest associations are shown in the diagram at the upper left. Relative abundance, represented by degree of shading, corresponds to percentage of tree stems more than one centimeter in diameter in samples of approximately 1,000 stems.

The Community as a Unit of Adaptation

Community function is the sum of what individuals do and thus reflects the adaptations of individuals. We may ask, however, whether attributes of a community represent more than the evolved properties of individuals. Is the community itself a unit of adaptation having properties that can be interpreted only in terms of community function? Because adaptations are evolved through natural selection to improve the fitness of the individual carrying the selected trait, we would not expect to find adaptations enhancing community function that were inconsistent with this goal. What is good for the individual prevails, regardless of whether it is good for the community. But if evolutionary adjustments of prey to their predators, and of competitors to each other, tended to stabilize their relationships, these adaptations would improve the efficiency of ecosystem function and enhance community stability.

We may set down, at least tentatively, a basic ecological principle: Community efficiency and stability increase in direct proportion to the degree of evolutionary adjustment between associated populations. The action of this principle is shown quite clearly when foreign species are introduced to a community. In most cases they cannot successfully invade the community and so die out, but occasionally exotic species gain a foothold and rapidly come to dominate the community. Such species can upset the delicate balances achieved between members of the community and, in so doing, disrupt community function. Outbreaks of introduced pests like the European pine sawfly and gypsy moth can, in fact, almost destroy a community by defoliating its major primary producers.

The evolutionary adjustment of populations to one another depends upon their degree of association. If two species always occurred together, their interaction would exert an important influence on the evolution of each. If, however, the ecological and geographical ranges of two species were mostly nonoverlapping, each species would exert only a small portion of all the selective influences on the population of the other. A community cannot exhibit strong co-evolutionary adjustment among its members if the adaptations of its species are molded primarily by relationships in other communities.

The actual degree of association between species in a community lies somewhere between two extremes: those of obligate association, as we find among many pairs of mutually interdependent organisms, and independence, each species being distributed randomly with respect to others. Both can be identified in natural systems, but the concensus of ecologists is that communities are open associations without clear boundaries.

In spite of their general independence, species do adapt to the presence of other species in their environment. One would expect the degree of adaptation to be directly proportional to the degree of association between two species. British ecologist T. R. E. Southwood provided evidence for this point by examining the association between species of insects and trees in England. He found that the more abundant a species of tree in the recent history of England's forests, measured by the number of records in fossil pollen samples, the greater the number of species of insects that can be collected from it. Species that are poorly represented in the British flora—mountain ash, hornbeam, maple—are such minor components of the environment that few insects have evolved to exploit them compared with the numbers of insects that feed on the more abundant willow, oak, and birch. Furthermore, most of the recently introduced species of trees in England have very few insect species associated with them. But because host species are islands or patches of habitat, the number of species associated with them may result from a balance between colonization and extinction. Regardless of adaptations to particular host species, extinctions of herbivores and parasites should be more frequent on the least common hosts, which would therefore support the fewest species. Only further research will tell us the extent of co-evolution in these systems.

Mutualism

Not all interactions between species are antagonistic. You may recall that under certain environmental conditions the relationship between oropendolas and cowbirds is beneficial to both; the oropendola provides food for the cowbird young, and the cowbird young protect the oropendola nestlings from fly parasites. In a sense, the oropendola buys its anti-predator protection from the cowbird. Interactions that benefit both parties in this way are described as mutualistic. *Mutualism* is not as common in the natural world as predation and parasitism, but for many species mutualistic interactions are helpful in obtaining food or avoiding predation: the pollination of flowering plants by animals involves a mutualistic interaction; lichens are a mutualistic association of algae and fungi; ruminants and termites both have the ability to digest cellulose because of their mutualistic association with specialized microorganisms in their intestinal tracts; the ability of legumes to fix nitrogen from the air depends on mutualistic bacteria in their root nodes; and so on.

Most mutualistic relationships probably evolved by way of host-parasite, predator-prey, or plant-herbivore interactions. This origin is particularly evident in the association between cowbirds and oropendolas, in which the cowbird is parasitic on the oropendola in certain colonies. The presence of insects flying from flower to flower to eat the nutritious pollen may have provided the opportunity for plants to adapt their flower structure to enhance and control insect pollination. Plants have evolved showy flowers to attract pollinators; nectar, which is little more than a weak sugar solution, provides an inexpensive food incentive for insects (and also for many birds and bats) to visit the flowers; the structure of flowers has also been modified to ensure that pollinators will efficiently transfer pollen from one flower to another.

Plant-pollinator relationships are highly developed in the orchid family, with its variety of flower shapes, colors, and smells. The intricate tie between flower and pollinator is exemplified by the orchid *Stanhopea grandiflora* and the euglossine bee *Eulaema meriana*. The attraction of euglossine bees to the particular orchids they visit is unusual in that no nectar is produced by the flowers, and only male bees visit them. The flowers are extremely fragrant, and each species of *Stanhopea* orchid has a unique combination of fragrances that attracts only one species of bee pollinator.

When a male euglossine bee visits an orchid, it brushes parts of the flowers with specially modified forelegs, and then appears to transfer some substance to the tibia of the hind leg, which is enlarged and has a storage cavity. The function of the bee's behavior is not understood, but the orchids may provide the bee with a sort of perfume used for mate attraction. For *Eulaema meriana* to pollinate *Stanhopea grandiflora*, the bee enters the flower from the side and brushes at a saclike modification of the lip of the orchid (Figure 20-7). The surface of the lip is very smooth, and the bee

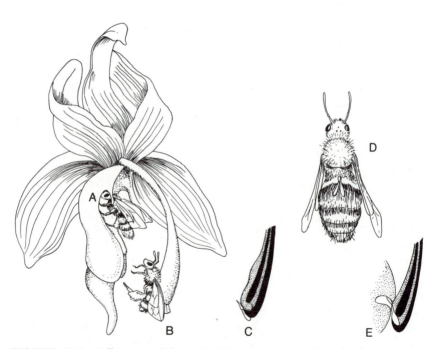

FIGURE 20-7   Pollination of the orchid *Stanhopea grandiflora* by *Eulaema meriana*. The bee enters from the side and brushes at the base of the orchid lip (A). If it slips (B) the bee may fall against the pollinarium, which is placed on the end of a column (C), and the pollinarium becomes stuck to the hind end of the thorax (D). If a bee with an attached pollinarium falls out of a flower, the pollinarium may catch in the stigma (E), which is so placed on the column that the flower cannot be self-fertilized.

often slips when it withdraws from the flower. (The orchid fragrances may also intoxicate the bees and cause them to lose their footing on the lip of the orchid.) If the bee slips, it may brush against the column of the orchid flower, where the pollinaria are precisely placed to stick to the hindmost part of the thorax of the bee. If a bee with an attached pollinarium slips and falls out of another flower, the pollinarium will catch on the stigma and pollinate the flower (Figure 20-7).

An even more unusual plant-pollinator relationship occurs between species of yucca plants and moths of the genus *Tegeticula* (Figure 20-8). Their curious interrelationship was first described a century ago, but has been considerably elaborated since. The moth enters the yucca flower and deposits one to five eggs on the ovary. Later, when the eggs hatch, the larvae burrow into the ovary, where they feed on the developing seeds. But after the moth has laid her eggs, she scrapes pollen off the anthers in the flower and rolls it into a small ball, which she grasps with specially modified mouthparts. She then flies to another plant, enters a flower, and proceeds to place the pollen ball onto the stigma of the flower before laying another batch of eggs.

The relationship between the moth and the yucca is *obligatory*. The moth can grow nowhere else, the yucca has no other pollinator. In return for pollinating its flowers, the yucca seemingly tolerates the moth larvae feeding on its seeds, but the extent of this loss of potential reproduction is small, rarely exceeding 30 per cent, and more nearly half that value, on average, in *Yucca whipplei*.

FIGURE 20-8 The mohave yucca and the yucca moth that pollinates it.

Daniel Janzen, of the University of Pennsylvania, has described a fascinating case of complete mutualistic interdependence between certain kinds of ants and swollen-thorn acacias in Central America. The acacia plant provides food and nesting sites for ants in return for protection that the ants provide from insect pests. The bull's-horn acacia has large hornlike thorns with a tough woody covering and a soft pithy interior (Figure 20-9). To start a colony in the acacia, a queen ant of the species *Pseudomyrmex ferruginea* bores a hole in the base of one of the enlarged thorns and clears out some of the soft material inside to make room for her brood. In addition to housing the ants, the acacias provide food for the ants in nectaries at the base of their leaves, and in the form of nodules, called Beltian bodies, at the tips of some leaves (Figure 20-9). As the colony grows, more and more of the thorns on the plant are filled and, in return, the ants protect the plant from insect pests. A colony may grow to more than a thousand workers within a year, and eventually may have tens of thousands of workers. At any one time, about a quarter of the ants are actively gathering food and defending the plant against herbivorous insects. The relationship between *Pseudomyrmex* and *Acacia* is obligatory: Neither the ant nor the acacia can survive without the other. Other ant-acacia associations are *facultative*. That is, the ant and the acacia can co-occur to mutual benefit, but they can both exist independently as well. Species of acacia that lack the protection of ants altogether frequently produce toxic compounds in their leaves.

The ant *Pseudomyrmex* grooms the acacia plant in return for food and protection. A similar function, cleaning parasites from the skin, is performed

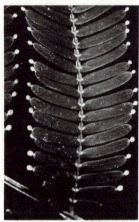

FIGURE 20-9   Modifications of swollen-thorn acacias for mutualistic interactions with ants. Many of the thorns are enlarged and have a soft pith that the ants excavate for nests (left). The acacias provide food for the ants from nectaries at the base of leaves (right) and in the form of nutritious nodules at the tips of modified leaves (center).

for many marine fish by other fish or by shrimp that benefit from the food value of the parasites they remove. Such relationships, often referred to as *cleaning symbiosis*, are most highly developed in the clear warm waters of the tropics, where many cleaners display their striking colors at particular locations, called cleaning stations, to which other fish come to be groomed. As you might expect, a few species of predatory fish mimic the cleaners; when other fish come and expose their gills to be groomed, they get a bite taken out of the gills instead.

The importance of cleaning and grooming mutualism in the overall functioning of natural communities is overshadowed by the many mutualistic relationships between organisms that fulfill specialized nutritional requirements. Among animals, both ruminant mammals (cows and sheep, for example) and termites have specialized microorganisms (bacteria and protozoa, respectively) in their digestive tracts that are biochemically equipped to break down the cellulose in plant food. Cellulose would normally be unusable to most organisms without the help of specialized symbionts.

Many plants, particularly the legumes, which include peas, peanuts, and alfalfa, have symbiotic nitrogen-fixing bacteria in nodules in their roots. These bacteria can utilize nitrogen directly from air spaces in the soil and make it available to their hosts. Most higher plants require nitrogen in the form of nitrates that diffuse into the nodules or of ammonia; where soils are deficient in these compounds, leguminous plants may thrive while other plants cannot. Alfalfa crops are often planted in rotation with other crops, such as corn, which deplete the soil of nitrogen, because the alfalfa can enrich the soils with nitrogen obtained from the atmosphere. The biochemical relationship between the nitrogen-fixing bacterium *Rhizobium* and its host plant is extremely complex. Nitrogenase, the enzyme needed to fix nitrogen, is found only in bacteria and blue-green algae, but the enzyme cannot function efficiently in the presence of dissolved oxygen. The nodules in the roots of legumes provide an anaerobic environment because although oxygen diffuses into the nodule along with nitrogen, it is utilized in respiration by root tissues before it reaches the center of the nodule where the bacteria are found. Nitrogen-fixing bacteria are also found in mutualistic association with green algae, which provide specialized, nonphotosynthetic cells for their guests. (The oxygen liberated by photosynthesis in normal algae cells would seriously reduce the rate of nitrogen fixation.)

Although free oxygen reduces nitrogen fixation, *Rhizobium* must have a source of oxygen for its own respiration. In the nodules of legumes, this is transported into the nodule bound to a special type of hemoglobin, called leghemoglobin. The mutualism between bacteria and plant has proceeded so far that *Rhizobium* carries the gene for the heme part of the molecule and the legume has the gene for the globin component. No one knows how this arrangement got started, but the early stages in the development of the mutualism may have been like the loose association that currently exists between another nitrogen-fixing bacterium, *Azotobacter*, and the roots of certain grass species.

Lichens are a close association of algae and fungi. They are frequently found on such surfaces as tree trunks and bare rock, substrates from which most plants cannot obtain the water or nutrients necessary for growth. In the lichen symbiosis, algae provide the photosynthetic mechanism for producing organic compounds, while the fungi secrete substances that dissolve nutrients out of bare rock surface, making them available for synthetic processes.

Soil fungi grow together with the roots of trees in what are called *mycorrhizal* associations. These fungi sometimes penetrate the cells of the roots, and they fulfill the important role of dissolving mineral nutrients that would otherwise be unavailable to the trees. In return, the fungi obtain organic nutrients from the roots.

Evolutionary History and Community Structure

The distribution of life forms over the surface of the earth is by no means uniform. Some regions lack groups that are abundantly represented elsewhere. Many irregularities in distribution patterns are linked to major climatic patterns: for example, snakes and lizards cannot tolerate the cold of arctic environments. Historical accidents of distribution, caused by geographical barriers to dispersal, have also played an important role in the distribution of major groups of animals and plants. Anomalies of distribution are most obvious on islands. Australia lacks most groups of mammals except for marsupials and bats, which are highly diversified there. Few species of any kind reach small remote islands, so communities are relatively simple compared with mainland associations.

Distributional problems are not limited to islands. The major continental land masses are sufficiently isolated to reduce the exchange of forms evolved in each area. The terrestrial environments of the Western and Eastern Hemispheres have been joined only sporadically by a land bridge between Alaska and Siberia during the last several million years. Many groups that have evolved and radiated to prominence in one hemisphere are absent from the other. Plants, insects, and other small invertebrates disperse easily across water barriers, and few of the major groups are missing from any continent. But terrestrial vertebrates, including birds, exhibit conspicuous differences between the faunas of the continents. The Iguanidae are the most conspicuous family of lizards in the Western Hemisphere, but they are replaced by agamid lizards in most of the Old World. Many ecological types of birds are represented by unrelated families in different areas. For example, the hummingbirds (*Trochilidae*) of North and South America, which feed primarily on nectar and tiny insects, have ecological counterparts in the sunbirds (*Nectariniidae*) of Africa and Asia. Sunbirds resemble hummingbirds in their food habits, small size, bright metallic-colored plumage, strong sexual dimorphism, and mating systems. Similarly, toucans (*Ramphastidae*), which are relatives of woodpeckers and are found in Central and

South America, are replaced ecologically in Africa by the hornbills (*Bucerotidae*), which are related to the kingfishers. Hornbills not only have greatly enlarged beaks like toucans, but both are gregarious and nest in holes. The adoption of similar feeding habits by diverse families of birds causes convergence of their morphology and behavior as well.

Convergence tends to obliterate the evidence of historical accidents in the composition of a local biota. Rain forests in Africa and South America are inhabited by plants and animals with different evolutionary origins but having remarkably similar adaptations. The rodents of North and South American deserts are more similar in morphological characteristics than one would expect from their different phylogenetic origins. Many similarities have been noted in the behavior and ecology of Australian and North American lizards, in spite of the fact that they belong to different families and have been separate for, perhaps, 100 million years. Mediterranean-climate plants in Chile and California have remarkably similar morphology and physiology.

Wherever one looks one finds convergence, and this reinforces our belief that community organization depends on local conditions of the environment more than it does on the evolutionary origins of the species that comprise the community. In many instances, species-for-species matchings have been made, suggesting that environments may closely specify the particular characteristics of species that inhabit them and that these specifications depend only on climate and other physical factors, not on phylogenetic background. This view is, however, much too simple. Detailed studies of convergence are as likely to turn up remarkable differences between the plants and animals in superficially similar environments. The ancient Monte Desert of South America is the only desert region of the world lacking bipedal, seed-eating, water-independent rodents like the kangaroo rats of North America and gerbils of Asia. Among frogs and toads, however, several South American forms have carried adaptation to desert environments a step further than their North American counterparts: They construct a foam nest in which their eggs are kept from drying up. Differences between the Australian agamid lizard *Amphibolurus inermis* and its North American iguanid analogue *Dipsosaurus dorsalis* include diet, optimum temperature for activity, burrowing behavior, and annual cycle, even though the species appear to be dead ringers for each other at first glance. Co-evolved relationships between species also may reveal the unique biogeographic position of each region. Unfortunately, little information has been gathered to compare such features of community organization, but one example may be drawn from the dispersal of seeds by ants. This interrelationship—a type of mutualism—is encouraged by the presence of edible appendages, called elaiosomes, on the seeds. Ants gather these, with the seeds attached, and carry them into underground nests, whereby they effectively plant the seeds. This seed trait is not common in most of the world and is usually restricted to trees in mesic environments. In Australia, however, the trait is exceedingly common among xerophytic shrubs, and it is associated with ecological and morphological features that are lacking in ant-dispersed plants elsewhere. The cause of these particular traits in Australian plants is an unresolved subject of speculation, but we nonetheless should be cautioned about hastily drawing ecological parallels between regions. Whether the peculiarities of seed dispersal among Australian plants reflect a different environment or the particular evolutionary history of Australian plants and animals cannot be determined at this point, but the traits clearly merit a closer look.

As we examine community structure in more detail, we shall assume that community attributes are more the product of environment than the product of evolutionary history. This is conventional ecological wisdom. Still, it is better to be somewhat skeptical of the generality of community attributes described in studies of limited geographical scope.

21 | Community Organization

BECAUSE IT IS IMPOSSIBLE to measure the interactions of each species with every other species in a community, ecologists have sought simpler indexes of community organization. The most widely studied of these is the number of species in the community. For the last two decades, ecologists have tried to understand why the number of species found in a particular habitat varies from place to place. Interest in this problem was stimulated by several essays of Yale ecologist G. E. Hutchinson in the 1950s and by his student Robert MacArthur. Naturalists had known for more than a century that more species live in a tropical habitat than in a temperate habitat with similar vegetation, but it was not until the late 1950s that the principles of ecology and population biology were applied to understanding variation in the diversity of communities.

One may ask whether a study of the number of species can teach us anything about the organization—the inner workings—of the community. Many have argued that the number of species varies in direct proportion to the productivity of the habitat, which is influenced by temperature, sunlight, rainfall, and nutrients. If this were true, variation in number of species would seem to be a trivial problem. But this explanation, and others like it, are unsatisfactory on several counts. First, there are too many exceptions. For example, salt marshes are among the most productive habitats, yet they support relatively few species. Second, any complete theory of species number must explain how the number of species in a particular habitat is regulated; how the species are formed; where they come from; how interactions between species set an upper limit to their number. A simple correlation is not a sufficient explanation. Third, species number is too important an attribute of community structure to dismiss easily. It is possible that many species can exploit different kinds of resources more efficiently than a few species can, because each species may become specialized to fill a different ecological role. It is also likely that the environmental heterogeneity created by many species provides further opportunity for the diversification of life forms. That is, diversity tends to breed more diversity. Variation in the number of species brings into focus many of these related problems and has played a major role in the development of ecological thought.

In this chapter we shall examine the major patterns of variation in the number of species in a community and discuss various explanations of these patterns. Then we shall consider briefly another attribute of community organization: variation in the relative abundance of species within a community.

The High Diversity of Tropical Communities

In virtually all groups of organisms, the number of species increases markedly towards the equator. For example, within a small region at 60° latitude one might find as many as ten species of ants; at 40° there may be between 50 and 100 species; and within 20° of the equator, between 100 and 200 species. Greenland is inhabited by 56 species of breeding birds, New York, 105, Guatemala, 469, and Colombia, 1,395. Diversity in marine environments follows a similar trend: arctic waters harbor 100 species of tunicates, but more than 400 species are known from temperate regions, and more than 600 from tropical seas.

Some groups vary from this general pattern. Among butterflies, the milkweed butterflies (monarchs and their relatives) are more restricted to tropical latitudes than the swallow-tailed butterflies are (Figure 21-1). The number of species of benthic marine invertebrates—forms that burrow in the sand and mud on the ocean floor—does not vary markedly. The prevalent trend sometimes is reversed by small, specialized taxonomic groups.

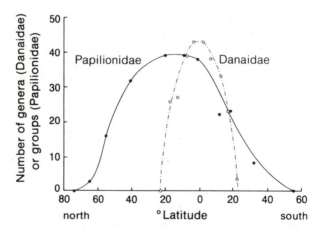

FIGURE 21-1   Number of genera of milkweed butterflies (Danaidae) and swallow-tail butterflies (Papilionidae) as a function of latitude.

Whereas birds generally increase in variety as one approaches the equator, sandpipers and plovers are more diverse in the arctic. These shorebirds are adapted to tundra and other open habitats, which do not occur commonly in tropical areas but are extensive in the arctic.

Within a given belt of latitude around the globe, the number of species varies widely among habitats according to productivity and degree of structural heterogeneity. For example, censuses of birds in small areas (usually ten to fifty acres) of relatively uniform habitat reveal about six species of breeding birds in grasslands, fourteen in shrublands, and twenty-four in floodplain deciduous forests (Table 21-1). But because not all communities fit this pattern, other factors must influence the number of species. Marshes are productive, but they have relatively few species of birds. Deserts, by contrast, are relatively unproductive but have many more species of birds than grasslands, which are more productive but structurally much simpler. The marsh is a uniform habitat, providing little opportunity for ecological specialization; deserts often are complex, supporting a variety of plant forms. Number of species appears to increase, therefore, with the productivity and structural heterogeneity of the vegetation.

Although tropical habitats are usually more productive than temperate and, especially, arctic habitats, the factors that determine the number of species are not necessarily related to production.

In certain groups of marine crustaceans, the number of species is inversely related to the organic productivity of the marine community. For example, species of calanid crustacea in surface waters of the Pacific Ocean are more numerous in tropical regions (more than eighty species) than in temperate regions (about thirty species) and the Bering Sea (about ten species). The productivity of these habitats, indicated by the number of individuals of crustaceans per cubic meter of water, *increases* toward the north from about 100 in tropical regions to almost 3,000 in temperate and arctic regions. The number of individuals per cubic meter of water may not

TABLE 21-1 Plant productivity and the number of species of birds in several representative temperate zone habitats.

Habitat	Approximate productivity $(g\ m^{-2}\ yr^{-1})$	Average number of bird species
Marsh	2,000	6
Grassland	500	6
Shrubland	600	14
Desert	70	14
Coniferous forest	800	17
Upland deciduous forest	1,000	21
Floodplain deciduous forest	2,000	24

be a valid measure of productivity, but this example nonetheless runs counter to the general correlation between productivity and species number. It is, however, consistent with the usual arctic-tropical trend in species diversity.

Variation in the number of species may be caused by historical accidents of geography that have isolated some areas from centers of species production or, conversely, that have placed areas in ideal situations for the generation of new species or for receiving immigrants. In North America, the number of species in most groups of animals and plants increases toward tropical latitudes, but the influence on species number of geographical heterogeneity and the isolation of peninsulas is apparent. Paleontologist G. G. Simpson tabulated the number of species of mammals in 150-mile-square blocks distributed over the entire area of North America to the Isthmus of Panama. The number of species per block increased from 15 in northern Canada to more than 150 in Central America, following the well-known latitudinal gradient in species diversity. Going from west to east across the middle of the United States, Simpson found the greatest number of species of mammals in the western mountains, where environmental heterogeneity provides effective isolating barriers to dispersal. The ecological heterogeneity of mountains creates opportunities for allopatric speciation and allows more species to coexist in a given area. Environments in the eastern United States are more uniform and, therefore, support fewer species of mammals (50 to 75 species per block) than are found in the west (90 to 120 species per block). The number of species of breeding land birds follows a similar pattern throughout North America, but reptile and amphibian faunas do not. Reptiles are more diverse in the eastern half of the United States than in the mountainous western regions; amphibians are strikingly underrepresented in the deserts of the Southwest.

The number of species of mammals, birds, and reptiles decreases strikingly towards the end of the Florida Peninsula. Most of Florida was covered by water during the last interglacial period, so Florida may yet be in an early stage of recolonization. But it is difficult to believe that birds, at least, could not have fully repopulated Florida in a few years. Uniformity of habitats also is not a likely candidate for the cause of the so-called peninsula effect, even though the lower half of the Florida Peninsula is flat and homogeneous. The number of species in many groups of plants and animals decreases along the peninsula of Baja California, which was not submerged during the Pleistocene Epoch and which is not homogeneous. Possibly, the peninsula effect is caused by isolation from centers of speciation in western North America and the tropics.

The reduced number of species in isolated areas is exaggerated on islands. Because water gaps are more effective barriers to the dispersal of terrestrial plants and animals than are most terrestrial habitats, islands usually have many fewer species than comparable mainland areas. The number of species on the tops of mountains, particularly in the tropics, behaves in much the same manner as on oceanic islands.

Qualitative Variation in the Composition of Communities

Variation in the number of species among areas often is the result of the selective addition or removal of particular kinds of species. We have already seen that the diversity of each taxonomic group has a characteristic geographical pattern, but what about the numbers of species that fill different ecological roles?

Part of the increase in the number of species of birds between temperate and tropical regions is related to an increase in frugivorous and nectarivorous species, and in insectivorous species that hunt by searching for their prey while quietly sitting on perches. These types of feeding behavior are uncommon among species in temperate regions. Among mammals, the increase in number of species between temperate and tropical areas results from the addition of bats to tropical communities. Terrestrial mammals are no more diverse at the equator than they are in the United States and other temperate regions at a similar latitude, although their variety does decrease as one goes farther to the north. Not only do tropical areas have more species than temperate and arctic areas, tropical communities appear to comprise more ecological roles. This finding provides a clue to one cause of variation in species diversity, as we shall see.

Specialization, Ecological Overlap, and Environmental Heterogeneity

Added species can be accommodated within a community in several ways. Consider the distribution of several species along a resource continuum—perhaps the size of prey items or the height at which feeding takes place within a forest (Figure 21-2A). If the variety of resources were increased, species could be added to the longer resource continuum without diminishing the ecological range of other species (Figure 21-2B). If the productivity of the habitat did not increase in direct proportion to the variety of resources, the added species would reduce the productivity of the others.

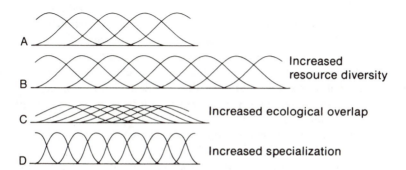

FIGURE 21-2 Schematic diagram showing how resource utilization along a continuum can be altered to accommodate more species.

Without an increase in the breadth of the resource spectrum, species can be added only by increasing the degree of resource overlap between species (Figure 21-2C) or by increasing the degree of specialization (Figure 21-2D). Once more, if the productivity of the habitat did not increase in direct proportion to the number of species, each species would become less productive when another was added to the community. Whether variation in species diversity is related to resource heterogeneity, ecological overlap, degree of specialization, or a combination of these factors, is a major question in the study of community organization.

When an ecologist considers the number of species in a small patch of uniform habitat, the degree of specialization with respect to resources within that habitat is an important consideration. When he considers the number of species in a large region, the degree of specialization with respect to habitat is an additional consideration. Suppose a region had five distinctive habitats each of which could support no more than ten species. If all species were habitat generalists—that is, if they could live in all the available habitats—the region could support only ten species, the number that occurred in each habitat. If, however, all species were habitat specialists and could maintain populations in only one habitat, the region would support 50 species: ten different species in each of five habitats.

This simple example illustrates two kinds of diversity: local (or *alpha diversity*) and regional (or *beta diversity*). Local diversity presumably depends upon the ability of a habitat to support a variety of species and upon the degree of competition between individuals in different populations. Specialization to exploit part of a habitat is a trait of the individual—a reflection of its adaptations to the environment and to the species with which it co-exists. Regional diversity depends on the replacement of populations by others in different habitats. Specialization to a particular habitat depends on the presence or absence of individuals of each species and, therefore, is a reflection of population processes as they are influenced by the physical environment, food, predators, and competitors. Because population size is more sensitive to environmental influences than is gene frequency, we might expect that where diversity increases, it does so because of habitat specialization rather than resource specialization.

Measurements of habitat specialization have shown that where many species coexist, each occurs in relatively few kinds of habitats. Islands usually have fewer species than comparable mainland areas, and island species often attain greater density than their mainland counterparts and expand into habitats that would normally be filled by other species on the mainland. These phenomena, called *density compensation* and *habitat expansion,* are collectively referred to as *ecological release.* On the island of Puerto Rico, Robert MacArthur and his co-workers found that many species of birds occupied most of the habitats on the island. In Panama, which has a similar variety of tropical habitats, species were more narrowly restricted, often to a single type of habitat.

If habitat specialization completely compensated for variation in species diversity between large regions, the number of species in small areas of uniform habitat—the local diversity—would be similar within regions of high and low diversity. The local diversity in samples of organisms obtained from tropical rivers and streams does, in fact, appear to be no greater than that in comparable temperate zone habitats, although the regional diversity of the freshwater fauna in tropical regions is greater than that of temperate regions. This example is not typical of terrestrial environments, in which local diversity usually varies in parallel to regional diversity to some degree.

In streams and rivers, the number of species in most taxonomic groups increases from the headwaters to the mouth of the river. The upper region of one stream in Ontario, at a point where the water temperature averaged 9 C, was inhabited by seven species of mayflies; in lower regions, where the water temperature exceeded 20 C, between twenty and thirty species of mayflies were found. Only three species had lower distributional limits along the stream, whereas twenty-six reached upper distributional limits (Table 21-2). In other words, local diversity increased downstream by the addition of species to the set that inhabited the higher reaches of the stream. Few species were replaced by others along the length of the stream. Similar patterns appear in fish communities. In the Rio Tamesi drainage of east-central Mexico, a headwater spring supported only one species of platyfish, a detritus feeder (Figure 21-3). Further downstream, three species occurred: the platyfish, plus a detritus-feeding molly that preferred slightly deeper water than the platyfish, and a carnivorous mosquito fish. Species that appeared in the community farther downstream included additional carnivores and other fish that fed primarily on filamentous algae and vascular plants. None of the species dropped out of the community downstream from any of the sampling localities. Diversity increased as the stream became larger and presented more kinds of habitats and a greater variety of food items.

TABLE 21-2 Number of upper and lower distributional limits of species of mayflies along a stream.

Collecting station	Water temperature (C)	Number of species	Number of species not found	
			Higher	Lower
1	9.0	7	0	0
2	16.3	15	8	1
3	19.5	16	2	0
4	21.5	22	8	1
5	20.5	21	2	1
6	24.0	29	6	0

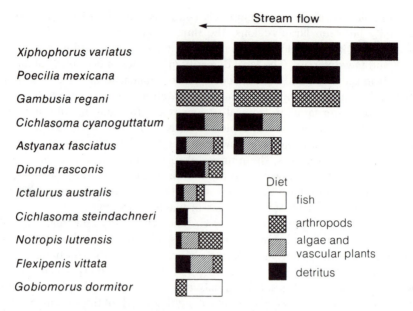

FIGURE 21-3   Food habits of fish species in four communities (vertical columns) from a headwater spring with one species (right) to downstream communities with up to eleven species (left). The communities sampled were in the Rio Tamesi drainage of east-central Mexico.

The examples discussed above show that variation in number of species in a region depends, in part, on the variety of habitats and the variety of resources within habitats and, in part, on the degree of specialization to particular habitats and resources. Still, we have not touched on the factors responsible for the production of species or for the regulation of their number. Regional diversity is determined by the balance between species production (speciation and immigration) and disappearance (extinction).

The Balance of Immigration and Extinction on Islands

Robert MacArthur and Edward Wilson developed a simple model to explain the variation in number of species among islands. For islands that are not large enough to permit speciation through geographical isolation of populations, increase in the number of species is brought about solely by immigration from other islands or from the mainland. Whereas we know practically nothing about rates of speciation within continents, we may reasonably assume that rate of immigration to an island by new species decreases as the distance from the island to the mainland increases. Hence the advantage of the island model.

Let us consider an island located at some distance from the coast of a mainland region whose flora and fauna comprise the species pool of potential colonists for the island. The rate of immigration of new species to the island decreases as the number of species on the island increases; that is, as more and more of the potential mainland colonists are found on the island, fewer of the new arrivals constitute new species (Figure 21-4). When all the mainland species occur on the island, the immigration rate is zero. The number of extinctions per unit of time increases with the number of species present on the island. Where the immigration and extinction curves cross, the number of species on the island is at an equilibrium ($\hat{S}$).

Immigration and extinction curves probably are not strictly proportional to the number of potential colonists and the number of species established on the island. Some species are undoubtedly better colonizers than others and they reach the island first. The rate of immigration to the island initially decreases more rapidly than it would if all mainland species had equal potential for dispersal and colonization. Hence, the immigration rate follows a curved line. Competition between species on islands probably accelerates extinction, so the extinction curve rises progressively more rapidly as species diversity increases (Figure 21-4).

If probability of extinction increased as absolute population size decreased, extinction curves for the inhabitants of small islands would be higher than for those of larger islands and small islands would have fewer species than large islands (Figure 21-5). If the rate of immigration to islands decreased with distance from mainland sources of colonists, the immigration curve would be lower for far islands than for near islands, and one would expect that the equilibrium number of species for distant islands would lie to the left of the equilibrium point for islands that are close to the mainland (Figure 21-5). These predictions have been verified for several groups of organisms on various islands throughout the world. For example, on the Sunda group of islands in the East Indies—including the Philippines, New Guinea, Borneo, and many smaller islands—the number of species of birds is closely related to the area of the island (Figure 21-6). The islands are so

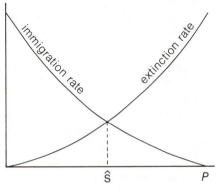

Number of species on island

FIGURE 21-4 Equilibrium model of the number of species on islands. The equilibrium number of species ($\hat{S}$) is determined by the intersection of the immigration and extinction curves. P represents the number of potential colonists from the mainland. The immigration rate initially drops rapidly as the best colonists become established on the island. The extinction rate increases more rapidly at high species number because of increased competition between species.

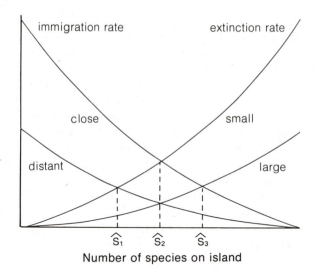

FIGURE 21-5   Relative number of species on small, distant islands ($\hat{S}_1$) and large, close islands ($\hat{S}_3$) predicted by the MacArthur-Wilson equilibrium model. The number of species on small, close islands and large, distant islands is intermediate.

close to each other and to mainland sources of colonists that no distance effect is evident. On islands to the east of New Guinea, however, which are located at a greater distance in the Pacific Ocean, the number of species of birds falls below a curve representing the number of species on islands of similar size close to the mainland (Figure 21-6). In principle, the equilibrium model should apply to islands of all kinds, including patches of habitat, widely-dispersed species of plants, and mountain tops.

If the number of species on an island behaved according to the equilibrium model, a change in the number of species would lead to a response tending to restore the equilibrium diversity. A natural test of this prediction was begun quite spectacularly in 1883 when the island of Krakatau, located between Sumatra and Java in the East Indies, blew up after a long period of repeated volcanic eruptions. At least half the island disappeared beneath the sea and its remaining area was covered with hot pumice and ash. The entire flora and fauna of the island was certainly obliterated. During the years that followed the explosion, plants and animals recolonized Krakatau at a surprisingly high rate: within 25 years, more than 100 species of plants and 13 species of land and freshwater birds were found there (see Table 21-3). During the ensuing 13 years, two species of birds disappeared and 16 were gained, bringing the total to 27. During the next 14-year period, the number of species on the island did not change, but five species disappeared and five new ones arrived, indicating that the number of species had reached equilibrium. The number of species of plants continued to increase. The number of species of birds on Krakatau after recolonization

TABLE 21-3 Recolonization of Krakatau by land and fresh water birds, and by plants, after it exploded in 1883.

	NUMBER OF SPECIES					
	1886	1897	1908	1920	1928	1934
Plants	26	64	115	184	214	272
Birds	—	—	13	27	—	27

was about what would have been predicted from the species area curve for an island eight square miles in area in the Sunda group, where Krakatau lies (see Figure 21-6).

Loss and replacement of species on islands is to be expected from MacArthur and Wilson's model; when the number of species is at equilibrium, immigration and extinction still occur and the species that inhabit an island are continually replaced. On Krakatau, five of the 27 species of birds present in 1920 had been replaced by other species 14 years later, indicating a turnover rate of 1.3 per cent per year [5 species / (27 species × 14 years)].

Equilibrium Theory of Mainland Diversity

MacArthur and Wilson's approach to the number of species of organisms on islands can be applied equally well to mainland biotas by adding speciation

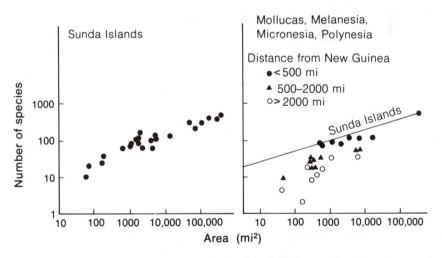

FIGURE 21-6 Species-area curves for land and freshwater birds on the Sunda Islands, together with the Philippines and New Guinea (left), and various islands of the Moluccas, Melanesia, Micronesia, and Polynesia (right). The latter islands show the effect of distance from the major source of colonization (New Guinea) on the size of the avifauna.

within the region to the immigration rate. The shape of the speciation curve for a mainland area differs from that of the immigration curve on an island. Because the island biota is derived from an established mainland source of colonists, immigration of new species is frequent when the diversity of species on the island is low. On large mainland areas, however, new species are generated from populations already present, so the rate of speciation varies in direct relation to the number of species in the area (Figure 21-7). As the number of species increases, and there are fewer opportunities for geographical expansion and subsequent isolation of subpopulations, the possibilities for speciation could decrease and result in a leveling off of the speciation rate. The shape of the extinction curve for mainland species should be similar to that for island species, but it will be lower owing to the larger area being considered.

Most of the theories that have been proposed to account for geographical variation in the number of species on mainland areas can be grouped into three categories, deriving from the equilibrium model: (1) the number of species is not in equilibrium and diversity reflects the time period in which communities have developed; (2) the combined immigration and speciation rate varies from area to area; and (3) the extinction rate varies from area to area. These hypotheses are not mutually exclusive, and all of them, or any combination of them, could describe the observed differences in number of species between areas.

The Time Hypothesis

Several ecologists have suggested that more species live in the tropics than in temperate and arctic regions because the tropical environment is older, allowing time for the evolution of a greater variety of plants and animals. This is a nonequilibrium hypothesis: diversity is thought to increase so long

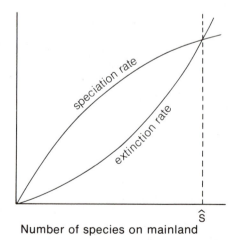

FIGURE 21-7 Equilibrium model of the number of species in a mainland region with a large area. When smaller areas are considered, immigration from neighboring mainland regions becomes important and the immigration-speciation curve more closely resembles that of an island with a high rate of immigration.

Number of species on mainland

as the environment remains undisturbed. The time hypothesis and the MacArthur-Wilson hypothesis are not incompatible, however. Supporters of the time hypothesis suggest that the diversity of today's biological communities is much less than the equilibrium level determined by the balance between speciation and extinction rates. Furthermore, because tropical environments are thought to be older and more stable than temperate and arctic habitats, the diversity of tropical communities is closer to the equilibrium.

The time hypothesis is not new; it was fully stated in 1878 by Alfred Russel Wallace, the co-author, with Darwin, of the theory of evolution:

> The equatorial zone, in short, exhibits to us the result of a comparatively continuous and unchecked development of organic forms; while in the temperate regions there have been a series of periodical checks and extinctions of a more or less disastrous nature, necessitating the commencement of the work of development in certain lines over and over again. In the one, evolution has had a fair chance; in the other, it has had countless difficulties thrown in its way. The equatorial regions are then, as regards their past and present life history, a more ancient world than that represented by the temperate zones, a world in which the laws which have governed the progressive development of life have operated with comparatively little check for countless ages, and have resulted in those wonderful eccentricities of structure, of function, and of instinct—that rich variety of colour, and that nicely balanced harmony of relations which delight and astonish us in the animal productions of all tropical countries.

Evidence bearing on the age of major climate belts is sketchy and varied. Because the tropical zone girdles the earth about its equator—the earth's widest point—the total area included within the tropics is much greater than that included within temperate and arctic regions. The earth's climate has undergone several cycles of warming and cooling, which have been discovered by records, in sediments and fossils, of their influence on vegetation and ocean temperature. As the climate of the earth warmed, as it last did during the Oligocene Epoch, the area of the tropics and subtropics expanded, reaching what is now the United States and southern Canada, and the temperate and arctic zones were squeezed into smaller areas closer to the poles. During the last 25 million years, the climate of the earth has become cooler and drier, and the tropics have contracted. Regardless of their area, the climate of the tropics is undoubtedly influenced less by major climate trends than that of temperate and arctic regions.

Northern and tropical regions both underwent drastic fluctuations during the Ice Ages of the past 2 million years. Temperate and arctic areas witnessed the expansion and retreat of glaciers, causing major habitat zones to be displaced geographically and, possibly, disappear. Periods of glacial expansion were coupled with high rainfall in the tropics. The Amazonian rain forest, which today covers vast regions of the Amazon River's drainage basin, was repeatedly restricted to small, isolated refuges during periods of drought. Restriction and fragmentation of the rain-forest habitat could have caused the extinction of many species; conversely, the isolation of populations in patches of rain forest could have facilitated the formation of new species. How these events have affected local and regional diversity is not known.

The only unequivocal test of the time hypothesis would be found in the number of species as it changed through time, but the fossil record is so fragmentary that this test could be applied to few taxa and would be restricted to certain types of habitats, particularly marine habitats. Clearly, worldwide diversity has increased greatly from the beginning of the Paleozoic Era to the present, particularly as plants and animals invaded new adaptive zones, and the end probably is not in sight. In terms of the equilibrium hypothesis, each new adaptive radiation represents a resetting of the speciation and extinction curves. To test the time hypothesis adequately on a scale that is relevant to the ecological structure of today's communities, we should measure the number of species in communities whose composition has not changed qualitatively. Such data are difficult to come by, but one study has reported that the diversity of temperate zone communities of foraminifera (marine, shelled protozoa) has not changed during the last 15 million years.

Speciation Rate

New species would be produced at a greater rate in the tropics than in temperate and arctic regions if geographical isolation were enhanced and if species-specific characteristics evolved more rapidly in isolated populations in the tropics. Tropical populations tend to be sedentary, partly because the tropical environment is relatively unseasonal and individuals, therefore, have little need to move from one locality to another to find favorable habitats. Temperate and arctic seasons change so drastically that many species migrate to other regions during the winter. It is widely believed that seasonal movements promote widespread gene flow and reduce local differentiation, but this notion has not been adequately tested. Whether plant populations are more sedentary in the tropics depends on how wide-ranging their dispersal mechanisms are.

Sedentariness reduces gene flow and speeds the genetic differentiation of subpopulations. Even for such highly mobile organisms as birds, a wide river or a small mountain range can effectively block dispersal in the tropics; the ranges of many species of birds stop abruptly at major river systems in the Amazon Basin; many families of deep-forest birds disperse poorly to nearby tropical islands. Because temperature varies so little seasonally within tropical habitats, mountains present a greater ecological barrier to dispersal in the tropics than in temperate regions; high mountain passes differ in climate from lowland habitats more in the tropics than in temperate zones. In the tropics, seasonal changes within habitats are small compared to the differences between habitats. This allows increased habitat specialization and thereby reduces the dispersal of populations across different habitats. In temperate and arctic regions, seasonal differences in climate within a habitat usually are greater than the average differences between habitats and environmental heterogeneity presents less of a barrier to dispersal.

The relationship between the biological diversity of a region and its geographical heterogeneity is obvious. Colombia has more species of birds, and probably most other types of organisms, than any other country; its geography is broken repeatedly by the Andes Mountains and by numerous valleys between the mountain ranges. Brazil and Zaire, which are topographically more uniform than Colombia yet of similar size and located within the same latitudes, have fewer species. A component of Colombia's high diversity may be the wide range of habitats created by its mountainous topography, yet it is also possible that the many opportunities for geographical isolation have fueled the fires of species production in Colombia. These factors are, however, difficult to distinguish with the only evidence being patterns of diversity.

If evolution proceeded more rapidly in the tropics than in temperate and arctic regions, the evolutionary divergence of populations might lead more rapidly to speciation. Rapid evolution is favored by many characteristics of tropical environments. A longer growing season permits many more generations each year, which creates more opportunity for selection, particularly on adaptations of courtship and mating that are responsible for the evolution of reproductive isolation between species. The late population geneticist Theodosius Dobzhansky suggested that, in temperate and arctic regions, physical factors in the environment select individuals for their degree of physiological adaptation to extreme conditions. In the tropics, physical factors are less important and biological agents assume a more important role in selection. Dobzhansky argued that the release of adapting systems from rigid physiological constraints allows evolution to respond more directly to the biological environment which, in turn, could enhance the evolution of divergence and reproductive isolating mechanisms through competition.

Competition and Extinction

Competition could be less among tropical species for several reasons: (1) greater resource availability, or use of fewer resources by each species; (2) greater heterogeneity of the environment, so competition can be reduced by specialization; and (3) limitation of population size by predation rather than by food resources.

If resources were more abundant in tropical regions than in temperate or arctic regions, tropical species could be more specialized without sacrificing productivity or population size; specialization would reduce interspecific competition and allow species to coexist more easily. The number of species in a community is consistently related to the organic productivity of the environment: tropical habitats are generally more productive than temperate and arctic habitats; mountains tend to be less productive and also have fewer species than lowland regions; deserts and grasslands are generally less productive, and also less diverse, than moister habitats. But if tropical and temperate habitats with similar productivity could be compared, the tropical habitat would likely have more species; differences in productivity cannot account for all variation in species number. Moreover, variation in productivity between regions is generally less than variation in organic diversity. Tropical forests are about twice as productive, on average, as temperate zone, broad-leaved forests, yet they are inhabited by perhaps five to ten times as many species of trees in small areas and have an equally greater diversity in other groups of plants and animals.

Heterogeneity in the environment allows specialization. No matter how productive a habitat may be, only one species could occur on each trophic level if the environment were completely uniform; competitive exclusion could not be avoided by specialization. Heterogeneity of habitats and geographical regions has not been quantified on a comparative basis in tropical and temperate regions, but tropical forests impress the casual observer with their variety of vegetational structure—including lianas, vines, and bromeliads (air plants)—each of which provides habitats for a variety of animals that could not otherwise live in the forest. For example, by trapping water at the base of their leaves, bromeliads create small aquatic environments inhabited by many types of organisms.

The variety of flowers and fruits in tropical communities is accompanied by an equally impressive variety of frugivores, seed predators, pollinating species, and the predators and parasites that prey on them. But although much of the animal diversity in tropical habitats can be attributed to the structural heterogeneity of the vegetation, one is forced to consider the more basic problem of the diversity of plants. The number of species of trees and other plants in tropical habitats greatly exceeds that of temperate habitats, but the diversity of soils, drainage, and exposure patterns for plants probably does not differ greatly. Does environmental heterogeneity limit the number of species that can coexist in an area; or does the degree of specialization within a community reflect competition within the community that exerts an influence on resource partitioning over and above the physical structure of the habitat? The relationship between diversity and heterogeneity will be difficult to disentangle because as species are added to a community, its structural heterogeneity increases.

Predation and Diversity

Predation could reduce competition among species if predators were so efficient that they could keep prey populations below the carrying capacity of the environment, as this would permit additional species to coexist on the resources used by the prey species. We have seen that efficient predators with high reproductive rates relative to their prey are able to limit prey populations, often to a spectacular degree. Many experiments have been conducted to assess the role predators play in regulating the number of prey species in a community. The most notable of these are a study on the diversity of intertidal marine faunas, and experiments on the control of diversity of annual plants by rabbits.

Robert Paine, a marine ecologist at the University of Washington, worked in the intertidal region of a rocky shore habitat along the Pacific coast of Washington that was dominated by several species of barnacles, gooseneck barnacles, mussels, limpets, and chitons (a kind of grazing mollusc); these were preyed upon by the starfish *Pisaster* (Figure 21-8). One

FIGURE 21-8 Congregation of starfish (*Pisaster*) at low tide on the coast of the Olympic Peninsula, Washington. The starfish, shown at lower left, is an important predator on mussels (lower right).

study area, eight meters in length and two meters in vertical extent, was kept free of starfish by physically removing them on a regular schedule. An adjacent control area was left undisturbed. Following the removal of the starfish, the number of prey species in the experimental plot decreased rapidly, from fifteen at the beginning of the experiment to eight at the end. Diversity declined in the experimental areas when populations of barnacles and mussels increased and crowded out many of the other species. Paine concluded that starfish were a major factor responsible for maintaining the diversity of the area. The crown-of-thorns starfish (*Acanthaster*) similarly enhances the diversity of coral reef communities near the Pacific coast of Central America by voraciously consuming one species of coral, *Pocillopora*, that otherwise crowds many species of coral out of the community (Figure 21-9). On a small island off the coast of England, rabbits had grazed the vegetation heavily, creating an evenly cut turf composed of many species of plants. When the rabbits were eliminated by myxomatosis virus, the character and species composition of the island's vegetation changed dramatically, and the number of plant species decreased.

A series of experiments on communities of protozoa and other small organisms living in pitcher plants, showed that predation by mosquito larvae reduced the number of prey species, in contrast to the results of Paine's predator removal experiment. But in the simple, test tube-like environment of the pitcher plant, mosquito larvae are such efficient predators that they drastically reduce populations of all prey species, driving some to extinction. It would appear, therefore, that the outcome of experiments on the role of predation in the maintenance of diversity among prey species will vary with the complexity of the habitat and the efficiency and abundance of the predators.

Daniel Janzen has suggested that the activities of herbivores could be responsible for the large number of species of trees in tropical forests

FIGURE   21-9 Crown-of-thorns starfish consuming a coral head in Panama.

compared to temperate forests. Janzen argued that herbivores specialize on the buds, seeds, and seedlings of abundant species to such an extent that they destroy young trees and eventually reduce the density of the species. This, in turn, allows other, less common species to grow in their place. Several lines of evidence support this hypothesis. For example, attempts to grow rubber trees in dense stands in their native habitats in the Amazon Basin have met with singular lack of success, but rubber tree plantations are successful in Malaya. In the Amazon Basin, epidemics of plant pests rapidly destroyed the rubber trees, but in the East Indies, where the natural diseases and pests of rubber trees do not occur, the trees can be grown successfully in large stands. Attempts to grow other commercially valuable trees in single-species stands in the tropics have frequently met the same disastrous end that befell large rubber plantations in South America. In contrast, cacao, the South American plant that is the source of chocolate, is infested with no more species of insect pests where it is planted in its native region than in plantations in Africa and Asia. The difficulties of monoculture, when they occur, are not restricted to the tropics, but for unknown reasons predators and parasites appear to be much more important in tropical environments.

Disturbance and Community Diversity

Predation is a special aspect of the much broader phenomenon of disturbance. Whether a mussel is killed by a starfish, a coral head broken off by pounding surf, or a tree toppled by a strong gust, individuals are removed from populations and their places are made available to others. Colonizers of bare patches of habitat or other vacancies in the biological community left open by death need not be of the same kind as their predecessors. To cite a familiar example, the first seedlings to sprout on bare ground after it has been cleared are not elms and maples, but crabgrass, horseweed, and other herbaceous inheritors of the land. Eventually these will be replaced by trees, which grow up to shade the herbs and shrubs out of existence. But before they disappear, their seeds may have been blown or carried to some other newly cleared patch, thereby maintaining the population. It is no different in the rocky intertidal, where the barnacle that will eventually be crowded out by more slowly growing mussels may mature and produce free-floating offspring that will settle onto some other bared area of rock.

Coexistence of trees and plants or barnacles and mussels depends on the occasional occurrence of bare patches where the less strong can become established quickly and reproduce before giving way to superior competitors. Disturbance is an equalizing force in biological communities in that it gives every species equal opportunity to become established from elsewhere in the species range. According to this view, the high diversity of such habitats as tropical forests and coral reefs is maintained in part by continual interruption of competitive exclusion and repopulation by a variety of colonists. Joseph Connell, an ecologist at the University of California at Santa Barbara, has argued that this effect is greatest at intermediate levels of disturbance. When there is little or no disturbance, competition eliminates all but a very few species which then dominate the community. At very high levels of disturbance, such as on exposed outer coasts or in hurricane belts, disruptions may be so frequent that only species that can colonize and grow quickly to maturity can persist. At intermediate levels, however, some parts of the habitat may have been recently disturbed while others may have remained untouched for long periods, just by chance. Such conditions would permit the greatest variety of species to coexist in an ever-changing patchwork. Although the role of disturbance in community organization is just beginning to be investigated, the idea has led many ecologists to question whether communities ever achieve stable equilibria in which the relative proportions of species are approximately fixed. Instead, what we observe at any point in time and space may be a transitional state whose future may continue to be shaped by interactions among its species or altered abruptly by a floating log or wind throw.

Are Communities at Equilibrium?

Traditionally, ecologists have adopted a simple view of nature in which populations achieve an equilibrium, or steady state, with respect to their environments. Factors in the environment, including competitors, exert constant pressure on the population. When one factor changes or a new one is added, the population responds quickly and achieves a new equilibrium. Therefore, we observe communities whose characteristics were molded by factors that should be readily apparent in the day-to-day functioning of the system.

Three kinds of evidence have led ecologists to begin to doubt the equilibrium viewpoint. The first is the common observation that all populations vary a little and most populations vary a lot. The tenure of most ecological studies—one to a few years—is too short to detect much of this change, but as the results of more and more long-term studies have become available the fact of change has become more apparent. One could argue, of course, that such variation merely represents fluctuations about average equilibrium values and that population change is just the response to variation in an unsteady physical world. Even if this were true, we would be forced to admit that such populations were more frequently returning to equilibrium than resting on top of it. But in many cases, communities have been observed not to return to a familiar equilibrium but to change directionally by the addition and subtraction of species, perhaps towards some distant steady state. A familiar example is the field reverting to forest, which might be considered the biological steady state in the absence of fires, hurricanes, ice storms, and agriculture. All communities are subjected to disturbances, if only the burrowing of worms through mud, which prevent the community existing within a small area from ever achieving a steady state. When we back off and see the biological world on a scale larger than the average disturbance covers, we may see an equilibrium of sorts with cycles of disturbance and regeneration occurring in a patchwork mosaic. Yet how large an area and on what time scale of occurrence are disturbances important to understanding the organization of communities today? The break up of Gwandanaland into the southern continents over one hundred million years ago was a global disturbance in the physical world. Is this event important to understanding the organization of today's biological world? In other words, how long are waves of disturbance propagated through communities before they dissipate and are submerged below the noise of other disturbances? This question brings us to the second kind of evidence.

Theoretical studies have only begun to address the problem of disturbance in communities, but one consistent result appears to be that recovery of equilibrium following disturbance can take much longer than the average disturbance cycle. Suppose for example that hurricanes sweep through a particular forest once every 500 years on the average. If the forest required 1,000 years to achieve a steady state of species composition and age-structure of populations, we would rarely observe the steady-state. Laboratory experiments with short-lived animals have shown that exclusion of one species by a competitor, even a strong competitor, can take tens of generations. Even among immobile plants and animals involved in direct competition for space, exclusion often requires a complete generation because only the juveniles are vulnerable. So when a weedy species of tree invades a disturbed habitat, its replacement by better competitors may take tens or hundreds of years, depending on its life span. And when the ecological requirements of the species competing for light, moisture, and nutrients are closely matched, and the seedlings of each can grow under the other, exclusion may require thousands of years. The slowness of such change compared to the frequency of many kinds of disturbance dictates that the species composition and age structure of a community might achieve a steady state only on a scale much larger than the average patch size created by disturbance.

Do such global geological events as continental drift, the formation of land bridges, and glaciation count as ecological disturbances? It depends upon how fast the recovery processes occur. The continental ice sheet receded from the northern United States about 10,000 years ago. Pollen grains in sediments of lakes left behind the glacier's retreating edge record the gradual repopulation of the northern states by species of plants whose ranges had been pushed southward by the expanding ice sheet. The sequence of forest development in particular areas—usually proceeding from spruce to pine and birch, to elm and oak—were determined partly by climate change. But the ability of some trees to disperse out of their ice age refuges altered the course of forest development in the southern Great

Lakes area. Pines, which were restricted to the Appalachian Mountains at the height of the glaciation simply did not disperse rapidly enough to Illinois to enter the sequence of species between spruce and broad-leaved hardwoods. In this case, recovery depended in part upon long-distance dispersal of slowly-migrating species. Adjustment of the biological world to the physical environment also involves evolutionary change. It is unlikely that there will ever be an evolutionary steady state because of continual long-term geological disturbance and generation of evolutionary novelty. Although such slow processes would not prevent the achievement of a local equilibrium, balancing more rapid ecological processes, evolutionary history and the circumstances of biogeography dictate that the local ecological steady state continually changes and that the steady states at any two localities probably differ. Evolutionary diversity makes the task of interpreting ecological processes more difficult.

The third kind of evidence about equilibrium in biological communities is negative and indirect, but nonetheless has persuaded many ecologists to adopt a nonequilibrium viewpoint. They argue that if populations achieve some equilibrium with respect to one another, one should observe well-ordered ecological arrangements between the species in their use of resources. Attempts to identify such patterns have frequently been futile and frustrating. Avian ecologist John Wiens of the University of New Mexico has pointed out that associations of grassland birds and the resources they feed upon vary so much during the season, from year to year, and from place to place, that populations rarely achieve their local carrying capacities and competition, therefore, is rarely intense. Wiens has found that during the breeding season, many species of grassland birds with highly overlapping food preferences coexist in the same habitats. There is little evidence that the abundances of populations are closely related to resources, that birds compete for limited resources (starvation of young is rarely observed, for example), or that patterns of resource use have been ordered by evolutionary divergence between species. How then is the structure of the communities regulated? Wiens has suggested that during occasional unusual years, conditions may force species into intense competition for brief periods that are sufficient to account for the patterns of co-occurrence observed. Alternatively, composition of the communities may be determined during winter, when the birds have migrated south from their breeding grounds and face very different ecological conditions than they experience during the summer.

The problem of identifying patterns in the organization of communities resulting from the effects of competition has been the concern of Daniel Simberloff and his colleagues at Florida State University. The difficulty is one of a statistical nature. Competition theory predicts that, in communities at equilibrium, species should partition resources so as to avoid intense competition. Indeed, too close competitors are excluded if they do not evolve rapidly to diverge in their ecological requirements. One expects, for example, that the body sizes of competitors be evenly spaced so that each exploits a unique size range of prey. Indeed, one often finds orderly progressions of predators from small to large, seemingly confirming the prediction. But Simberloff points out that if one selected species at random from different places, some would likely be smaller and some larger than the average. Such size differences would not have been the result of competition because the species were taken from different places and therefore had not interacted in the past, yet the artificial community assembled at random might have the superficial appearance of organization. The crux of Simberloff's argument is that to demonstrate organization according to a particular ecological process, one has to reject the hypothesis that the pattern observed might have occurred as the result of unrelated processes. The simplest statistical test is to compare the properties of natural communities to those of groups of species selected at random—that is, without regard to the presence of other species in the group. The random selection might represent the chance colonization of an island from the mainland or other process occurring on an historical or biogeographical scale much larger than that of local ecological interactions. In most of the cases to which Simberloff has applied this technique, he has been unable to distinguish patterns in natural communities from those of randomly-generated groups. Because this is negative evidence, it is less compelling than direct evidence supporting the organizing effects of competition in communities. It is possible that patterns created by interactions among species cannot be distinguished from patterns resulting from random processes. Simberloff's results nonetheless are forcing ecologists to refrain from concluding that the structure of biological communities is the equilibrium expression of internal ecological forces until more direct evidence is at hand.

Not only are the processes responsible for community organization unresolved, but whether or not communities even have a basic organization is not fully agreed upon. Competition and predation certainly occur and greatly affect the abundance—or perhaps, presence—of populations in the community. But it remains for ecologists to determine the degree to which these interactions dictate the number of species and their ecological roles.

22 | Community Development

OMMUNITIES are constantly changing. Organisms die and others are born to take their places. Energy and nutrients continually pass through the community. Yet the appearance and composition of most communities do not vary. Oaks replace oaks, robins replace robins, and so on, in continual self-perpetuation. When a community is disturbed— a forest cleared for agriculture, a prairie burned, a coral reef obliterated by a hurricane—the community is slowly rebuilt. Pioneering species adapted to the disturbed habitat are successively replaced by others until the community attains its former structure and composition. The sequence of changes on a disturbed site is called *succession* and the final association of species achieved is called a *climax*. These terms describe natural processes that caught the attention of early ecologists. By 1916, University of Minnesota ecologist Frederic Clements had outlined the basic features of succession, supporting his conclusions by detailed studies of change in plant communities in a wide variety of environments. In this chapter, we shall examine the course and causes of succession both from the traditional view of community development and in the light of recent studies on community succession.

Succession and Species Replacement

The creation of any new habitat—a plowed field, a sand dune at the edge of a lake, an elephant's dung, a temporary pond left by a heavy rain—invites a host of invading species to exploit the resources made available. The first colonizers are followed by others slower to take advantage of the new habitat but eventually more successful than the pioneer species. In this way, the character of the community changes with time. Successional species themselves change the environment. For example, plants shade the surface, contribute detritus to the soil, and alter soil moisture. These changes often inhibit the continued success of the species that cause them, and make the environment more suitable for other species which then exclude those responsible for the change.

The opportunity to observe succession is almost always at hand on abandoned fields of various ages (Figure 22-1). On the piedmont of North Carolina, bare fields are quickly covered by a variety of annual plants. Within a few years, most of the annuals are replaced by herbaceous perennials and shrubs. The shrubs are followed by pines, which eventually crowd out the earlier successional species, but pine forests are in turn invaded and then replaced by a variety of hardwood species that represent the end of the successional sequence. Change is rapid at first. Crabgrass quickly enters an abandoned field, hardly giving the plow furrows a chance to smooth over. Horseweed and ragweed dominate the field in the first summer after abandonment, aster in the second, and broomsedge in the third. The pace of succession falls off as slower-growing plants appear. The transition to pine forest requires 25 years. Another century must pass before the developing hardwood forest begins to resemble the natural climax vegetation of the area.

The transition from abandoned field to mature forest is only one of several successional sequences leading to the same climax. In the eastern part of the United States and Canada, forests are the end point of several different successional series, or *seres*, each having a different beginning. The sequence of species on newly formed sand dunes at the southern end of Lake Michigan differs from the sere that develops on abandoned fields a few miles away. The sand dunes are first invaded by marram grass and bluestem grass. Plants of these species established in soils at the edge of a dune send out rhizomes (runners) under the surface of the sand, from which new shoots sprout (Figure 22-2). These grasses stabilize the dune surface and add organic detritus to the sand. Numerous annuals follow the perennial grasses onto the dunes, further enriching and stabilizing them and gradually creating conditions suitable for the establishment of shrub species. Sand cherry, dune willow, bearberry, and juniper form shrub layers before pines become established. As in the abandoned fields in North Carolina, pines persist for only one or two generations, with little reseeding after initial establishment, giving way in the end to the beech-maple-oak-hemlock forest characteristic of the region.

Succession follows a similar course on Atlantic coastal dunes, where beach grass initially stabilizes the dune surface, followed by bayberry, beach plum, and other shrubs. Shrubs act like the snow fencing frequently used to prevent the blowout of dunes; they are called dunebuilders because they intercept blowing sand and cause it to pile up around their bases (Figure 22-2). Succession in estuaries leading to the establishment of terrestrial communities begins with salt-tolerating plants and progresses as sediments and detritus build the soil surface above the water line.

FIGURE 22-1 Stages of secondary succession in the oak-hornbeam forest in southern Poland. From upper left to lower right, the time since clearcutting progresses from 0 to 7, 15, 30, 95, and 150 years.

FIGURE 22-2 Initial stages of plant succession on sand dunes along the coast of Maryland. Top: beach grass on the frontal side of a dune. This grass is used widely to stabilize dune surfaces. Bottom: invasion of back dune areas by bayberry and beach plum.

Primary Succession

Ecologists classify seres into two groups, according to their origin. The establishment and development of plant communities in newly formed habitats previously without plants—sand dunes, lava flows, rock bared by erosion or exposed by a receding glacier—is called *primary succession.* The return of an area to its natural vegetation following a major disturbance is called *secondary succession.* We have already followed the course of primary succession on Lake Michigan sand dunes. The sequence of species colonizing habitats exposed by receding glaciers in the Glacier Bay region of southern Alaska is quite different (Figure 22-3). The surfaces of glacial deposits are stable but the thin clay soils are deficient of nutrients, particularly nitrogen, and pioneering plants are exposed to wind and cold stress. Here the sere involves mat-forming mosses and sedges, prostrate willows, shrubby willows, alder thicket, sitka spruce, and, finally, spruce-hemlock forest. Succession is rapid, reaching the alder thicket stage within ten to twenty years, and tall spruce forest within 100 years.

The development of vegetation on bare rock, sand, or other inorganic sediments is called *xerarch succession.* The low water retention by such habitats results in *xeric* (drought) conditions during the early stages of succession. At the opposite extreme, *hydrarch* succession begins in the open water of a shallow lake, bog, or marsh. Hydrarch succession can be initiated by any factor that reduces water depth and increases soil aeration, whether natural drainage, progressive drying up, or filling in by sediments.

Changes in the vegetation of bogs illustrate hydrarch succession. Bogs form in kettleholes or dammed streams in cool temperate and subarctic regions. Bog succession begins when aquatic plants become established at the edge of the pond (Figure 22-4). Some species of sedges (rushlike plants) form mats on the water surface extending out from the shoreline. Occasionally these mats grow completely over the pond before it is filled in by sediments, producing a more or less firm layer of vegetation over the water surface. The detritus produced by the sedge mat accumulates in layers of organic sediments on the pond bottom because the stagnant water of the pond contains little or no oxygen and thus does not support rapid decomposition of the organic matter by microbes. Eventually these sediments become peat, used by man as a soil conditioner and, sometimes, as a fuel for heating (Figure 22-5).

As the bog is filled in by sediments and detritus, sphagnum moss and bog shrubs, like Labrador tea and cranberry, become established along the edges, themselves adding to the development of a soil with progressively more terrestrial qualities. The shrubs are followed by a bog forest of black spruce and larch, which eventually is replaced by local climax species, including birch, maple, and fir, depending on the region.

In northern Michigan, the hydrarch succession that develops on bogs is only one of several seres of primary and secondary succession leading to a

FIGURE 22-3 A valley exposed by a receding glacier, visible at top center, in North Tongass National Forest, Alaska. The recently bared rock surfaces at the bottom of the valley just below the glacier have not yet been recolonized by shrubby thickets.

climax forest of spruce, fir, and birch (Figure 22-6). Following a fire, development of the climax follows a different course, passing through intermediate grass and aspen stages. If the soil is badly scorched and most of the humus is burned, a sere resembling that beginning on bare rock surface develops.

FIGURE 22-4 Stages of bog succession illustrated by a bog formed behind a beaver dam in Algonquin Park, Ontario. The open water in the center is stagnant, poor in minerals, and low in oxygen. Those conditions result in the accumulation of detritus from the vegetation at the edge and lead to a gradual filling-in of the bog, passing through stages dominated by shrubs and, later, black spruce.

FIGURE 22-5 A three-foot vertical section through a peat bed in a filled-in bog in Quebec, Canada. The layers represent the accumulation of organic detritus from plants that successively colonized the bog as it was filled in. The peat beds are probably several yards thick. Vegetation on the surface of the bog consists mostly of sphagnum, blueberry, and Labrador tea.

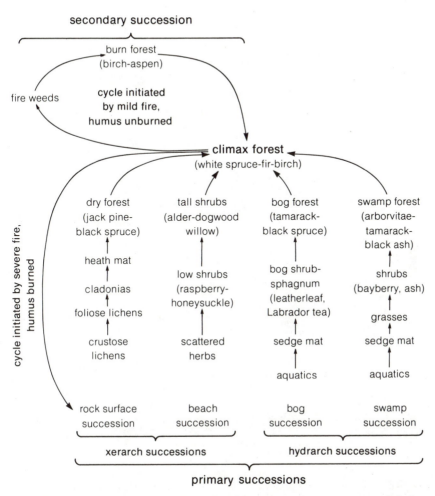

FIGURE 22-6 Trends of succession on Isle Royale, Lake Superior. Each habitat is characterized by a unique sere leading to the same climax. Secondary succession on burned sites follows different courses depending on the extent of the fire damage.

Ecoclines

The sequence of species in the sere parallels change in the physical environment as it is modified by the developing vegetation. Many of the stages in a time sequence through a sere may be found along geographical gradients in vegetation, often called *ecoclines* (Figure 22-7). For example, the xerarch succession from rock surface to forest in the eastern United States corresponds in structure, if not species composition, to the ecocline in vegetation from the nearly bare rock surfaces of the western deserts, through dry grasslands, prairie, shrubby oak woodland, to tall mixed hardwood forest that occurs along an increasing moisture gradient from west to east. The temporal sequence of primary succession in a particular place also follows

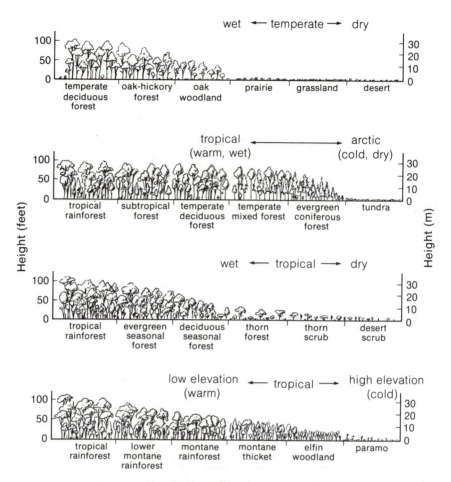

FIGURE 22-7 Schematic profiles of four ecoclines in vegetation type. Top, a wet-dry gradient from the Appalachian Mountains (left) to the southwestern United States (right); second, a warm-cold (and wet-dry) gradient from Panama to northern Canada; third, a wet-dry gradient within the tropics; bottom, an altitudinal gradient in the tropics from the Amazonian forest into the Andes.

an increasing gradient of moisture availability. The ecocline represents a series of stages of community development stopped at different points by lack of moisture. Each stage of the ecocline represents the local end point of succession. The species composition of a community along the ecocline differs from the corresponding stage of the complete sere in the eastern United States because species must be well adapted to the local conditions along the ecoclinal gradient, but, again, the general structure is the same.

Hydrarch succession on a bog is mirrored in the sequence of communities present at any one time from the open water at the center of the pond to the developing forest at its edge, where sedimentation and soil development are most advanced. Concentric bands of vegetation ring the bog, progressing from an inner circle of sedge mat outward through sphagnum and bog shrubs, larch, and finally black spruce.

The Climax

Succession traditionally is viewed as leading inexorably toward an ultimate expression of plant development, the *climax community*. In fact, studies of succession have demonstrated that the many seres found within a region, each developing under a particular set of local environmental circumstances, progress towards the same climax (see Figure 22-6). These observations led to the concept of the mature community as a natural unit, even as a closed system. Frederic Clements stated the concept in 1916: "The developmental study of vegetation necessarily rests upon the assumption that the unit or climax formation is an organic entity. As an organism the formation arises, grows, matures, and dies. Its response to the habitat is shown in processes or functions and in structures which are the record as well as the result of these functions. Furthermore, each climax formation is able to reproduce itself, repeating with essential fidelity the stages of its development. The life history of a formation is a complex but definite process, comparable in its chief features with the life history of an individual plant."

Clements recognized 14 climaxes in North America, including two types of grassland (prairie and tundra), three types of scrub (sagebrush, desert scrub, and chaparral), and nine types of forest, ranging from pine-juniper woodland to beech-oak forest. The nature of the local climax was thought to be determined solely by climate. Aberrations in community composition caused by soils, topography, fire, or animals (especially grazing) were thought to represent interrupted stages in the transition toward the local climax—immature communities.

In recent years, the concept of the climax as an organism or unit has been greatly modified, to the point of outright rejection by many ecologists, with the recognition of communities as open systems whose composition varies continuously over environmental gradients. Whereas in 1930 plant ecologists described *the* climax vegetation of much of Wisconsin as a sugar maple-basswood forest, by 1950 ecologists placed this forest type on an open continuum of climax communities extending both over broad, climatically defined regions and over local, topographically defined areas. To the south, beech increased in prominence, to the north, birch, spruce, and hemlock were added to the climax community; in drier regions bordering prairies to the west, oaks became prominent. Locally, quaking aspen, black oak, and shagbark hickory, long recognized as successional species on moist, well-drained soils, came to be accepted as climax species on drier upland sites.

Mature stands of forest in Wisconsin, representing the end points of local seres, have been ordered along a *continuum index* ranging from dry sites dominated by oak and aspen to moist sites dominated by sugar maple, ironwood, and basswood. The continuum index for Wisconsin forests was calculated from the species composition of each forest type, and its value varied between arbitrarily set extremes of 300 for pure bur oak forest to 3,000 for a pure stand of sugar maple. Although increasing values of the continuum index correspond to seral stages leading to the sugar maple climax, they may also represent local climax communities determined by topographic or soil conditions. Thus the so-called climax vegetation of southern Wisconsin is actually a continuum of forest (and, in some areas, prairie) types (Figure 22-8). Some botanists prefer to retain the term *climatic climax* for the furthest point of vegetational succession within a region, relegating all other endpoints of seres to terminated stages of succession or *subclimaxes*. This semantic difficulty might be resolved by the following logic: If interrupted stages of succession are so prevalent and persistent in a region that species have adapted to the particular environmental conditions in these subclimax communities, these species should be recognized as climax forms even though they enter transitional seral stages elsewhere. The concept of the climax is rooted in the self-perpetuation of an association under prevalent local conditions and in the adaptations of climax species that ensure their self-perpetuation.

The Causes of Succession

Two factors determine the positions of species in a sere: the rate at which species invade a newly formed or disturbed habitat, and changes in the environment over the course of succession. Some species disperse slowly, or grow slowly once they have become established, and therefore become dominant late in the sequence of associations in a sere. Rapidly growing plants that produce many small seeds, carried long distances by the wind or by animals, have an initial advantage over species that are slow to disperse, and they dominate early stages of the sere. Where fire is a regular feature of a habitat, many species have fire-resistant seeds or root crowns that germinate or sprout soon after a fire and quickly reestablish their populations.

Early colonists often change the environment in ways that favor the invasion of the community by species of superior competitors. Horseweed resists desiccation and rapidly colonizes abandoned farmland in the piedmont of North Carolina but, once established, horseweed plants modify the environment by shading the soil surface. Because horseweed seedlings require full sunlight, horseweed is quickly replaced by shade-tolerant species. As the community matures, the developing vegetation protects the surface layers of the soil from drying, permitting drought-intolerant species to get a foothold in the community. Progressive changes in the physical environment foster the replacement of species through seral stages.

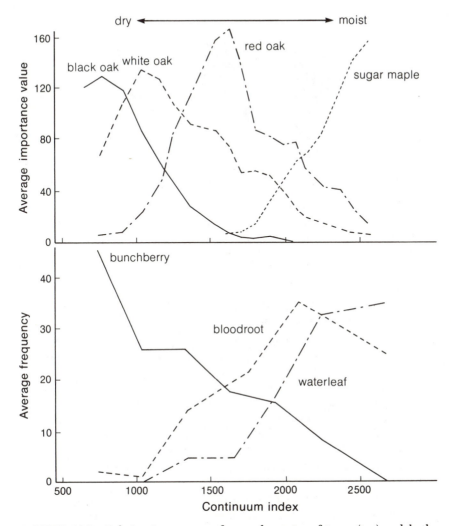

FIGURE 22-8 Relative importance of several species of trees (top) and herbs (bottom) in forest communities of southwestern Wisconsin arranged along a continuum index. Soil moisture, exchangeable calcium, and pH increase to the right on the continuum index.

Early stages of plant succession on old fields in the piedmont region of North Carolina demonstrate how succession is driven. The first three to four years of old-field succession are dominated by a small number of species that replace each other in rapid sequence: crabgrass, horseweed, ragweed, aster, and broomsedge. The life-history cycle of each species partly determines its place in succession (Figure 22-9). Crabgrass, a rapidly growing annual, is usually the most conspicuous plant in a cleared field during the

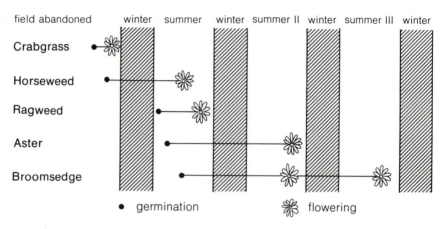

FIGURE 22-9 Schematic summary of the life histories of five early successional species of plants that colonize abandoned fields in North Carolina.

year in which the field is abandoned. Horseweed is a winter annual, whose seeds germinate in the fall. Through the winter, the plant exists as a small rosette of leaves, and it blooms by the following midsummer. Because horseweed has strong dispersal powers and develops rapidly, it usually dominates one-year-old fields. Ragweed is a summer annual; seeds germinate early in the spring and the plants flower by late summer. Ragweed dominates the first summer of succession in fields that are plowed under in the late fall, after horseweed normally germinates. Aster and broomsedge are biennials that germinate in the spring and early summer, exist through the winter as small plants, and bloom for the first time in their second autumn. Broomsedge persists and flowers during the following autumn as well.

Horseweed and ragweed both disperse their seeds efficiently and, as young plants, tolerate desiccation. These characteristics allow them to invade cleared fields rapidly and produce seed before populations of competitors become established. Decaying horseweed roots stunt the growth of horseweed seedlings; this self-inhibiting effect, whose function and origin are not understood, cuts short the life of horseweed in the sere. Such growth inhibitors presumably are the byproduct of other adaptations that increase the fitness of horseweed during the first year of succession. If horseweed plants had little chance of persisting during the second year, owing to invasion of the sere by superior competitors, self-inhibition would have little negative selective value. At any rate, the phenomenon is fairly common in early stages of succession.

Aster is a relatively successful colonist of recently cleared fields but, being a slow grower, it does not become dominant until the second year. The first aster plants to colonize a field thrive in the full sunlight, but, because aster seedlings are not shade-tolerant, they shade their progeny out of existence. Furthermore, asters do not compete effectively with broomsedge for soil moisture. In one experiment, circular areas, one meter in radius, were cleared around several broomsedge plants and aster seedlings planted at various distances. After two months, the dry weight of asters planted 13, 38, and 63 cm from the bases of the broomsedge plants averaged 0.06, 0.20, and 0.46 gram; available soil water at these distances was 1.7, 3.5, and 6.4 grams per 100 grams of soil. Broomsedge does not dominate early successional communities until the third or fourth year, in spite of its competitive edge over aster. Because their seeds do not disperse well, broomsedge plants do not increase rapidly in number until the first colonists of the field have themselves produced seeds.

The Endpoint of Succession

Succession continues until the addition of new species to the sere and the exclusion of established species no longer change the environment of the developing community. Conditions of light, temperature, moisture, and, for primary seres, soil nutrients, change quickly with the progression of different growth forms. The change from grasses to shrubs and trees on abandoned fields brings a corresponding modification of the physical environment. Conditions change more slowly, however, when the vegetation reaches the tallest growth form that the environment can support. The final biomass dimensions of the climax community are limited by climate independently of events during succession. Once forest vegetation is established, patterns of light intensity and soil moisture are not changed by the introduction of new species of trees, except in the smallest details. For example, beech and maple replace oak and hickory in northern hardwood forests because their seedlings are better competitors in the shade of the forest-floor environment, but beech and maple seedlings probably develop as well under their own parents as they do under the oak and hickory trees they replace. At this point, succession reaches a climax; the community has come into equilibrium with its physical environment.

To be sure, subtle changes in species composition usually follow the attainment of the climax growth form of a sere. For example, a site near Washington, D.C., left undisturbed for nearly 70 years developed a tall forest community dominated by oak and beech. The community had not then reached an equilibrium because the youngest individuals—the saplings in the forest understory, which eventually would replace the existing trees—included neither white nor black oak. In another century, the forest will be dominated by species with the most vigorous reproduction, namely red maple, sugar maple, and beech (Figure 22-10). The composition and age structure of a forest in northwestern Wisconsin having had minimal human disturbance over 200 years indicated a transitory state, perhaps towards the end of a sere, between oak dominance and a basswood-maple climax. At

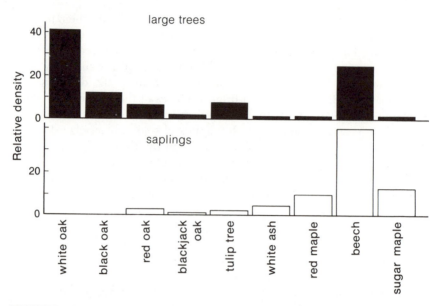

FIGURE 22-10   Composition of a forest undisturbed for 67 years near Washington, D.C. The relative predominance of beech and maple saplings in the forest understory foretells a gradual successional change in the community beyond the present oak-beech stage.

the time of that study, red oak was the commonest large tree in the forest, but basswood and, especially, maple were reproducing much more vigorously (Table 22-1). The ratios of seedlings and saplings (less than one inch diameter) to large trees (greater than ten inches diameter) were maple 186, basswood 155, red oak 18, and white oak 37. (White oak and bitternut are close to the northern edge of their ranges in northern Wisconsin and did not form a major component of the forest.)

The preponderance of red oak in the canopy of the Wisconsin forest and the evidence in the understory of successional changes yet to come indicate that the forest had been disturbed in some way, allowing seral species to enter and setting into motion the machinery of succession. The age structure of the tree populations suggested that fire destroyed much of the forest sometime between 1840 and 1850 (Figure 22-11). Most of the sugar maples were over 150 years old, indicating that they withstood fire damage. In fact, two-thirds of the sugar maple cores were so badly scarred by fire that their growth rings could not be counted accurately. Red oak and basswood both exhibited a period of rapid proliferation starting about 1850. Red oak gained its predominant position in the forest at that time and will not be excluded until present trees die and are replaced by basswood or maple seedlings.

The time required for succession to proceed from a cleared habitat to a climax community varies with the nature of the climax and the initial quality

TABLE 22-1 Number of trees of different species in a 2,500-square-meter forest area in northwestern Wisconsin. Individuals are separated into size classes.

	DIAMETER OF TRUNK (inches)			
Species	less than 1	1 to 3	4 to 9	greater than 10
Sugar maple	3,913	2	16	21
Basswood	931	22	21	6
Red oak	781	1	34	44
White oak	75	3	9	2
Bitternut	88	4	0	0
White pine	0	0	1	2
Ironwood	1,606	40	3*	0

* Maximum size class of ironwood, an understory species.

of the soil. Clearly, succession is slower to gain momentum when starting on bare rock than on a recently cleared field. A mature oak-hickory forest climax will develop within 150 years on cleared fields in North Carolina. Climax stages of western grasslands are reached in 20 to 40 years of secondary succession. On the basis of radiocarbon dating methods, complete primary succession to a beech-maple climax forest on Michigan sand dunes requires up to 1,000 years. In the humid tropics, forest communities regain most of their climax elements within 100 years after clearcutting, provided that the soil is not abused by farming or prolonged exposure to sun and rain. But the development of a truly mature tropical forest devoid of any remnants of successional species requires many centuries.

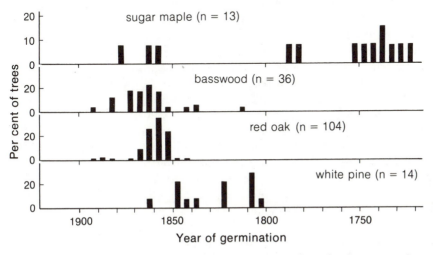

FIGURE 22-11 Age groups of sugar maple, basswood, and oak in a northern Wisconsin forest.

The Character of the Local Climax

Clement's idea that a region had only one true climax (the *monoclimax* theory) forced botanists to recognize a hierarchy of interrupted or modified seres by attaching names like subclimax, preclimax, and postclimax. This terminology naturally gave way before the *polyclimax* viewpoint, which recognized the validity of many different vegetation types as climaxes, depending on the habitat. More recently, the development of the continuum index and gradient analysis has fostered the broader *pattern-climax* theory of the late Cornell ecologist Robert Whittaker, which recognizes a regional pattern of open climax communities whose composition at any one locality depends on the particular environmental conditions at that point.

Many factors determine the climax community, among them soil nutrients, moisture, slope, and exposure. Fire is an important feature of many climax communities, favoring fire-resistant species and excluding others that otherwise would dominate. The vast southern pine forests in the gulf coast and southern Atlantic coast states are maintained by periodic fires. The pines are adapted to withstand scorching under conditions that destroy oaks and other broad-leaved species (Figure 22-12). Fire is even necessary to the life history of some species of pines that do not shed their seeds unless triggered by the heat of a fire passing through the understory below. After a fire, pine seedlings grow rapidly in the absence of competition from other understory species.

Any habitat that is occasionally dry enough to create a fire hazard but normally wet enough to produce and accumulate a thick layer of plant detritus is likely to be influenced by fire. Chaparral vegetation in seasonally dry habitats in California is a fire-maintained climax that is replaced by oak woodland in many areas when fire is prevented. The forest-prairie edge in the midwestern United States separates climatic climax and fire climax communities. Frequent burning eliminates seedlings of hardwood trees, but the perennial grasses sprout from their roots after a fire. The forest-prairie edge occasionally shifts back and forth across the countryside, depending on the intensity of recent drought and the extent of recent fires. After prolonged wet periods the forest edge advances out onto the prairie as tree seedlings grow up and begin to shade out the grasses. Prolonged drought followed by intense fire can destroy tall forest and allow rapidly spreading prairie grasses to gain a foothold. Once prairie vegetation is established, fires become more frequent owing to the rapid buildup of flammable litter. Reinvasion by forest species then becomes more difficult. By the same token, mature forests resist fire and rarely become damaged enough to allow the encroachment of prairie grasses. Hence, the forest-prairie boundary is stable.

Grazing pressure also can modify the climax. Grassland can be turned into shrubland by intense grazing. Herbivores kill or severely damage perennial grasses and allow shrubs and cacti unsuitable for forage to establish

FIGURE 22-12 A stand of longleaf pine in North Carolina shortly after a fire. Although the seedlings are badly burned (lower left), the growing shoot is protected by the dense, long needles (shown on an unburned individual, lower right) and often survives. In addition, the slow-growing seedlings have extensive roots, which store nutrients to support the plant following fire damage.

themselves. Most herbivores graze selectively, suppressing favored species of plants and bolstering competitors that are less desirable as food. On the African plains, grazing ungulates follow a regular succession of species through an area, each using different types of forage. When wildebeest, the first of the successional species, were excluded from large fenced-off areas, the subsequent wave of Thompson's gazelles avoided feeding in these areas. Apparently, heavy grazing by wildebeest stimulates growth of the preferred food plants of gazelles.

Transient and Cyclic Climaxes

We view succession as a series of changes leading to a climax, determined by, and in equilibrium with, the local environment. Once established, the beech-maple forest is self-perpetuating and its general appearance does not change in spite of the constant replacement of individuals within the community. Yet not all climaxes are persistent. A simple case of a *transient climax* would be the development of animal and plant communities in seasonal ponds—small bodies of water that either dry up in the summer or freeze solid in the winter and thereby regularly destroy the communities that become established each year during the growing season. Each spring the ponds are restocked either from larger, permanent bodies of water, or from spores and resting stages left by plants and animals before the habitat disappeared the previous year.

Succession recurs whenever a new environmental opportunity appears. On African savannas, carcasses of large mammals are fed upon by a succession of vultures, beginning with large, aggressive species that devour the largest masses of flesh, followed by smaller species that glean smaller bits of meat from the bones, and finally by a kind of vulture that cracks open bones to feed on the marrow. Scavenging mammals, maggots, and microorganisms enter the sere at different points and assure that nothing edible remains. This succession has no climax, however, because all the scavengers disperse when the feast is concluded. Nonetheless, we may consider all the scavengers a part of a climax, which is the entire savanna community.

In simple communities, particular life-history characteristics in a few dominant species can create a *cyclic climax*. Suppose, for example, that species A can germinate only under species B, B can germinate only under C, and C only under A. This situation would create a regular cycle of species dominance in the order A, C, B, A, C, B, A . . . with the length of each stage determined by the life span of the dominant species. Stable cyclic climaxes, which are known from a variety of localities, usually follow the scheme presented above, often with one of the stages being bare earth. Wind or frost heaving sometimes drives the cycle. When heaths suffer extreme wind damage, shredded foliage and broken twigs create an opening for further damage and the process becomes self-accelerating. Soon a wide swath is opened in the vegetation; regeneration occurs only on the protected side of the damaged area while wind damage further encroaches upon the exposed vegetation. As a result, waves of damage and regeneration move through the community in the direction of the wind. If we watched the sequence of events at any one point, we would witness a healthy heath being reduced to bare earth by wind damage and then regenerating in repeated cycles (Figure 22-13). Similar cycles occur where hummocks of earth form in windy regions around the bases of clumps of grasses. As the hummocks grow, the soil becomes more exposed and better drained. With these changes in soil quality, shrubby lichens take over the hummock and exclude the grasses around which the hummock formed. Shrubby lichens are worn down by wind erosion and eventually are replaced by prostrate lichens, which resist wind erosion but, lacking roots, cannot hold the soil. Eventually the hummocks are completely worn down and grasses once more become established and renew the cycle.

Mosaic patterns of vegetation types are common to any climax commu-

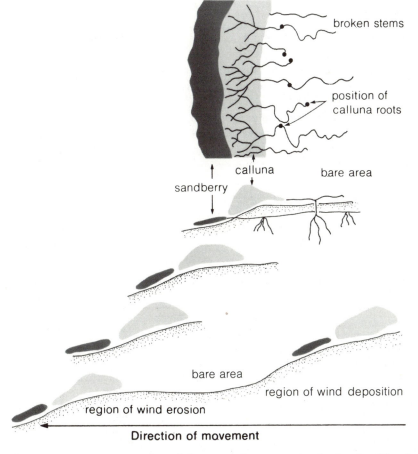

FIGURE 22-13 Sequence of wind damage and regeneration in the dwarf heaths of northern Scotland.

nity where the death of individuals alters the environment. Treefalls open the forest canopy and create patches of habitat that are dry, hot, and sunlit compared to the forest floor under unbroken canopy. These openings are often invaded by early seral forms, which persist until the canopy closes. Treefalls thus create a mosaic of successional stages within an otherwise uniform community. Indeed, adaptation by some species to grow in the particular conditions created by different-sized openings in the canopy could enhance the overall diversity of the climax community. Similar models have been developed for intertidal regions of rocky coasts, where wave damage and intense predation continually open new patches of habitat.

Cyclic patterns of changes and mosaic patterns of distribution must be incorporated into the concept of the community climax. The climax is a dynamic state, self-perpetuating in composition, even if by regular cycles of change. Persistence is the key to the climax. If a cycle persists, it is inherently as much a climax as an unchanging steady state.

Characteristics of Species in the Sere

Succession in terrestrial habitats entails a regular progression of plant forms. Plants characteristic of early stages and late stages of succession employ different strategies of growth and reproduction. Early-stage species are opportunistic and capitalize on high dispersal ability to colonize newly created or disturbed habitats rapidly. Climax species disperse and grow more slowly, but shade tolerance as seedlings and large size as mature plants gives them a competitive edge over early successional species. Plants of climax communities are adapted to grow and prosper in the environment they create, whereas early successional species are adapted to colonize unexploited environments.

Some characteristics of early and late successional stage plants are compared in Table 22-2. To enhance their colonizing ability, early seral species produce many small seeds that usually are wind dispersed (dandelion and

TABLE 22-2 General characteristics of plants during early and late stages of succession.

Character	Early stage	Late stage
Seeds	Many	Few
Seed size	Small	Large
Dispersal	Wind, stuck to animals	Gravity, eaten by animals
Seed viability	Long, latent in soil	Short
Root/shoot ratio	Low	High
Growth rate	Rapid	Slow
Mature size	Small	Large
Shade tolerance	Low	High

milkweed, for example). Their seeds are long-lived, and they can remain dormant in soils of forests and shrub habitats for years until fires or treefalls create the bare-soil conditions required for germination and growth. The seeds of most climax species, being relatively large, provide their seedlings with ample nutrients to get started in the highly competitive environment of the forest floor.

The survival of seedlings in the shaded environment of the forest floor is directly related to seed weight (Figure 22-14). The ability of seedlings to survive the shade conditions of climax habitats is inversely related to their growth rate in the direct sunlight of early successional habitats. When placed in full sunlight, early successional herbaceous species grew ten times more rapidly than shade-tolerant trees. Shade-intolerant trees, like birch and red maple, had intermediate growth rates. Shade tolerance and growth rate represent a trade-off; each species reaches a compromise between those adaptations best suited for its place in the sere.

The rapid growth of early successional species is due partly to the relatively large proportion of seedling biomass allocated to leaves. Leaves carry on photosynthesis, and their productivity determines the net accumulation of plant tissue during growth. Hence the growth rate of a plant is influenced by the allocation of tissue to the root and the aboveground parts (shoot). In the seedlings of annual herbaceous plants, the shoot typically comprises 80 to 90 per cent of the entire plant; in biennials, 70 to 80 per cent; in herbaceous perennials, 60 to 70 per cent; and in woody perennials, 20 to 60 per cent.

The allocation of a large proportion of production to shoot biomass in early successional plants leads to rapid growth and production of large crops of seeds. Because annual plants must produce seeds quickly and copiously, they never attain large size. Climax species allocate a larger proportion of

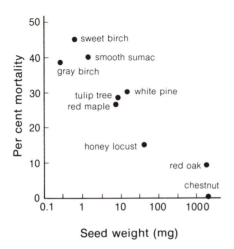

FIGURE 22-14 Relationship between seed weight and the survival of seedlings after three months under shaded conditions.

their production to root and stem tissue to increase their competitive ability; hence they grow more slowly. The progression of successional species is therefore accompanied by a shift in the compromise between adaptation for great dispersal power and adaptation for great competitive ability.

The biological properties of a developing community change as species enter and leave the sere. As a community matures, the ratio of biomass to productivity increases; the maintenance requirements of the community also increase until production no longer can meet the demand, at which point the net accumulation of biomass in the community stops. The end of biomass accumulation does not necessarily signal the attainment of climax; species may continue to invade the community and replace others whether the biomass of the community increases or not. The attainment of a steady-state biomass does mark the end of major structural change in the community; further changes are limited to the adjustment of details.

As plant size increases with succession, a greater proportion of the nutrients available to the community are tied up in organic materials. Furthermore, because the vegetation of mature communities has more supportive tissue, which is less digestible than photosynthetic tissue, a larger proportion of their productivity enters the detritus food chain rather than the consumer food chain. Other aspects of the community change as well. Soil nutrients are held more tightly in the ecosystem because they are not exposed to erosion; minerals are taken up more rapidly and stored to a greater degree by the well-developed root systems of forests; the environment near the ground is protected by the canopy of the forest; conditions in the litter are more favorable to detritus-feeding organisms.

Ecologists generally agree that communities become more diverse and complex as succession progresses, although in some seres, intermediate stages of succession may be more diverse because they contain elements of early seral stages as well as elements of the climax community. It is not known whether the increase in the diversity of a community during its early stages of succession is related to increased production, greater constancy of physical characteristics of the environment, or greater structural heterogeneity of the habitat.

The study of succession leads us to formulate four basic ecological principles. First, succession is one-directional; good colonizers with rapid growth and high tolerance of conditions on disturbed or newly exposed habitats are replaced by slowly growing species with great competitive ability. Second, successional species alter the environment by their structure and activities, often to their own detriment and to the benefit of other species. Third, the climax community is not a unit; rather it represents, at any given place, a point on a continuum of possible climax formations. The nature of the climax is influenced by climate, soil, topography, fire, and the activities of animals. Fourth, the climax may be a changing mosaic of successional stages maintained by wind, frost, or other sources of mortality acting locally within the community.

Successional changes are an inherent response of vegetation to disturbance, tending to reinstate the particular plant formation characteristic of the habitat. Succession occurs by the total replacement of populations of some species by populations of others, and thus cannot be compared to the homeostatic responses of an organism. Climax communities, nonetheless, are endowed with an inherent stability of structure and function, in part the sum of organism homeostasis and population responses, but in part the unique property of community organization. We shall examine community stability further in the next, and last, chapter of this book.

23 | Stability of the Ecosystem

STABILITY IS the inherent capacity of a system to tolerate or resist changes caused by outside influences. Suppose, for example, that precipitation fell 50 per cent below its long-term average, but plant production decreased by only 25 per cent and herbivore populations decreased by only 10 per cent. The dampening of the environmental fluctuation as it passed up the food chain is a measure of the internal stability of the system—its capacity to resist change. In this case, stability may be derived from storage of water in the soil, physiological responses of plants to drought, and if the drought lasts long enough, partial replacement of drought-sensitive herbs by drought-resistant species. The stability of a community depends upon the homeostatic responses of its constituent species.

We can visualize the concept of stability by considering a small ball placed in a bowl. If we nudge the ball, sending it a little way up the side of the bowl, the force of gravity quickly returns the ball to the bottom of the bowl. The steeper the sides of the bowl, the more powerful is the stabilizing force of gravity. This tendency of a system to be restored to a particular condition is referred to as a *stable equilibrium*. If the bowl itself represented the environmental factors acting on a population, the weight of the ball could correspond to the homeostatic capacities of the individuals in the population. A nudge of a given force would displace a steel ball bearing much less than it would a table tennis ball, just as a cold snap has less effect on a population of bears than on a population of flies.

A ball placed on a level table represents a system having no forces tending either to maintain the position of the ball or to change it. This condition is known as a *neutral equilibrium*. A nudge applied to the ball sends it across the table in one direction or the other, and its movement is unchanged until some outside force intercedes. Rarely is nature engineered like a level table. All systems have equilibrium points toward which they tend when displaced. Disturbances can, however, push a system beyond its capabilities of response like a ball sent careening out of the bowl or an insect population escaping predator controls and rising to outbreak levels.

When a system moves away from an equilibrium point, as when a ball rolls down the outside of an inverted bowl, the situation is referred to as an *unstable equilibrium*. The highest point on the outside surface of the inverted bowl is an equilibrium point because it is possible to balance the ball there. But the slightest disturbance sends the ball on its way.

The amount of movement a nudge of given force causes in a ball in a bowl depends then on the weight of the ball and the configuration of the bowl. Similarly, the amount of fluctuation in a population or a community caused by an external disturbance depends on the inherent stability of the system.

We may define the inherent stability of a system as the ratio between variation in the environment and variation in the system itself, but this definition is difficult to apply to populations or communities. Which aspect of environmental variation should we measure? Which component of the system gives the best indication of adequate function and continued persistence? Do we judge the stability of a community by the constancy of its function (production, ecological efficiency) or its structure (diversity, species turnover)? Moreover, change itself is often the best response to change. Hibernation and diapause represent a near total shutdown of biological activity in response to environmental change, yet dormancy allows the population to persist.

The biological significance of stability is even more elusive than its description and measurement. Constancy in the natural world is desirable to man because it enables him to predict conditions in advance and plan his activities accordingly. If weather and insect pests did not vary from year to year, farming would be simplified and a reasonably constant crop yield would be assured each year.

Virtually all human activity disturbs natural communities. Natural biological communities do not yield enough harvestable food to support dense human populations, and man selects the most "desirable" components of natural communities, alters their evolution through artificial selection to suit his purposes, and maintains these populations of crops, livestock, pulp trees, city parks, and backyards in a continually disturbed state, the populations constantly exposed to conditions they are not adapted to cope with. The price man pays for exploiting natural resources is the price of maintaining their stability by constant management: curbing pest infestations, maintaining soil fertility, and cleaning out weeds.

Our concern with the constancy of the natural world and with the basis for stability in natural communities is understandable. This concern is shared, unconsciously, by all species. Constancy of weather, resources, predation, and competition reduces the cost of self-maintenance (homeostasis) and increases the allocation of energy and nutrients to production. Most organisms would stand to gain from a more constant world, but because their competitors and predators would also gain, we cannot predict the net benefit of constancy to an individual or to a species.

Constancy of the environment and the community both undoubtedly enhance production and ecological efficiency because few resources are used for homeostasis, materials do not accumulate behind bottlenecks caused by population fluctuations, and adaptations of organisms become more finely tuned to the environment. But we must ask whether communities possess inherent stabilizing mechanisms in addition to the homeostatic responses of organisms and the growth responses of populations. If we are to reject the notion of community adaptations, we must also reject many contemporary ideas about community stability: that it is desirable for the community to be stable because constancy increases the efficiency of energy flow and nutrient cycling; that natural selection leads to increased complexity and diversity within the community so as to enhance the inherent stability of trophic structure and improve the ability of the community to resist perturbation; that many adaptations of organisms, such as large size, long life, low reproductive rates, and low productivity-biomass ratios, are adaptations to increase the stability of the community rather than to increase the evolutionary fitness of the organism.

Measures of community structure and function—number of species, number of trophic levels, rates of primary production, energy flow, and nutrient cycling—reflect ecological interactions among populations and between individuals and their physical environments. The ability of the community to resist ecological perturbations no doubt reflects the homeostatic mechanisms of individuals and the growth responses of populations. But beyond these separate homeostatic capacities, some properties of organization influence community stability and the efficiency of community function; that is, the homeostatic capacity of a community transcends the summed properties of its constituent populations. We might imagine, for example, that competition and ecological release help to stabilize the function of each trophic level. If the population of one species were suppressed by climate or disease, a competitor could respond by using the first population's leftover resources, and thus maintain the total production of the trophic level. We might also view predation as a destabilizing influence, magnifying population fluctuations in lower trophic levels.

In this chapter, we shall examine the inherent stability of ecological systems, and how stability of community function might be influenced by the structure of the community itself.

Some Definitions

Ecologists use stability and related terms in so many different ways that it will be necessary to provide explicit definitions here:

stability is the intrinsic ability of a system to withstand or to recover from externally caused change

constancy is a measure of the degree of variation of some attribute of a
system

predictability is a measure of regularity in patterns of change

Seasonal fluctuation in the environment usually is predictable; day-to-day
variation often is not.

The degree of fluctuation in the ecosystem is determined by three
factors, each of which will be considered separately below: (a) the constancy
and predictability of the physical environment; (b) the homeostatic mechanisms of organisms and growth responses of populations, and (c) that component of community stability uniquely contributed by the feeding and
competitive relationships of populations within the community—in other
words, the trophic organization of the community.

Variability in the Physical Environment

Temperature and rainfall have been measured for long periods at weather
monitoring stations throughout the world. Although local climatological data
do not adequately measure the environment of any particular population,
they do indicate the overall constancy of the environment. Three components of the environment are important to organisms: regular seasonal
fluctuations, variation about seasonal norms, and the predictability of short-
term variations. All habitats change daily and seasonally; coastal marine
environments vary, in addition, with a lunar rhythm. Daily, lunar, and
seasonal fluctuations reflect regular cycles in the physical world. Many years
of measurement would reveal the average conditions for each day of the
year and each time of day. But the environment rarely exhibits average
conditions. Irregularities in climate related to changing wind patterns and
random meteorological events cause the environment to vary around its
norm. For example, one hundred years of weather records in Philadelphia,
Pennsylvania, show that while the average July rainfall was 4.2 inches,
precipitation was less than half the average in nine years and more than
twice the average in four years (Figure 23-1).

Diurnal and seasonal patterns are more predictable than short-term
variations because they are tied to precise physical cycles, such as the daily
light-dark cycle and the seasonal change in daylength during the year. But
the unreliability of weather forecasts, particularly for several days ahead,
attests to the lack of predictability of short-term variations in climate; the
further one is removed in time from an event of brief duration like a
rainstorm, the less predictable it becomes. Rain can be predicted a few
minutes or a few hours before a thunderstorm by the appearance of the sky.
A change in wind direction during certain seasons signals the passage of a
front, often accompanied by precipitation and temperature change. But it
is virtually impossible to know in January, or even in May, whether June
will bring drought or deluge.

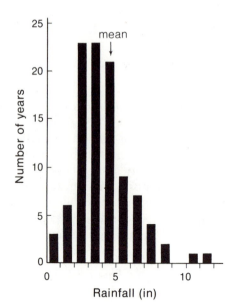

FIGURE 23-1   July precipitation for one hundred years at Philadelphia, Pennsylvania.

Rainfall is generally most variable from year to year where it is least abundant. At a given place, dry-season precipitation is more variable and less predictable than wet-season precipitation. Similarly, temperature variations are greatest when the average temperatures are lowest—geographically in polar regions and seasonally during winter. These patterns of variability indicate that in the tropics, the physical environment is more constant than in temperate and arctic regions; the tropics are warm the year around, and the climate of most tropical areas is relatively wet. The generally observed pattern could be misleading. Dry-season rainfall in some tropical areas is less than that in the driest months in many temperate localities; rainfall during the tropical dry season can be correspondingly variable and unpredictable.

Every region, no matter how constant its environment, is subject to infrequent extremes. In a wet environment, an extreme condition encountered only once in hundreds of years may be one that differs from the normal by a factor of two. In a dry locality, an extreme condition that is encountered equally infrequently may differ from the normal by a factor of ten. The homeostatic capabilities of organisms are adapted to the range of conditions normally encountered. Regardless of the degree of fluctuation in the environment, infrequent "extreme" conditions impose a stress on organisms in any region. The homeostatic mechanisms of tropical populations might, in fact, be much more poorly developed than those of temperate and arctic populations because the environment usually varies within a narrower range.

Some types of environmental variation are so drastic that they cannot be accommodated by the homeostatic mechanisms of organisms. Such events—environmental catastrophes—include hurricanes, tornadoes, fires, and hard freezes (Figures 23-2 and 23-3). Although many species are adapted to prosper in the aftermath of such disasters—the weedy species that colonize disturbed habitats, and others adapted to a regular cycle of minor fires—natural catastrophes completely destroy the fabric of most communities. Their structure must be rebuilt gradually over long periods by succession. Many human disturbances create equally catastrophic conditions —beyond the limits of variation normally encountered by organisms— completely destroying communities and sometimes preventing the re-establishment of community structure.

FIGURE 23-2   An intense, uncontrolled fire in the Willamette National Forest, Oregon. Most of the community was destroyed. A long period of succession will follow before the community again attains its mature characteristics.

FIGURE 23-3   A severe ice storm in New York damages a hardwood forest. Community function will not be completely restored for many years, during which many successional species will appear before they are crowded out by regenerating trees.

Stability at the Level of the Individual

The constancy of population size or organism activity reflects the interplay between environmental fluctuation and intrinsic stability. The outcome of this interaction can be seen by examining the growth rings of trees. A core of wood contains a record of a tree's growth rate from the sapling stage on. Because trees produce one ring each year, the rings can be dated easily, and the variation in their width compared to variation in temperature and rainfall. The sensitivity of a tree to climate fluctuation depends on where it grows. In moist habitats, water may never become sufficiently limiting to affect tree growth adversely even in the driest years. Where moisture levels are marginal for a species, drought can exert a profound effect on growth. This point is illustrated by growth ring chronologies in two populations of bristlecone pine in the White Mountains of California (Figure 23-4). Trees in a moist grove with abundant accumulation of winter snow exhibited little year-to-year variation in growth rate compared to stunted individuals growing on a dry, windswept rocky ridge. Although variation in ring width at each site paralleled variation at the other, moisture deficits severely depressed growth in the moist site only three times during the 104-year chronology—1899, 1929, and 1959.

Most of the ring-width variation in the bristlecone pine is related to the moisture level of the habitat during June, a hot month with little or no rainfall and, therefore, with a large evapotranspiration deficit. Autumn temperatures and winter moisture also influence growth during the subsequent season. The climates of different areas vary in their effect on tree growth (Figure 23-5). Moisture stress at the beginning of the growing season is more important to the bristlecone pine in California; Douglas fir responds equally to winter and spring moisture; growth of the piñon pine is determined primarily by winter moisture, indicating that it relies heavily on accumulated ground water for growth during the arid summer months. In dry habitats in Illinois, the growth of white oak responds to conditions of moisture and temperature during the growing season but, in addition, late summer drought depresses growth during the following year, perhaps by reducing storage of food in the roots or by interfering with the formation of leaf buds.

As one might expect, much of what we know about variability in communities comes from economically important species. Variation in annual production of wheat and other crops in the Prairie Provinces of Canada is related to variation in weather. In one study, variability from year to year

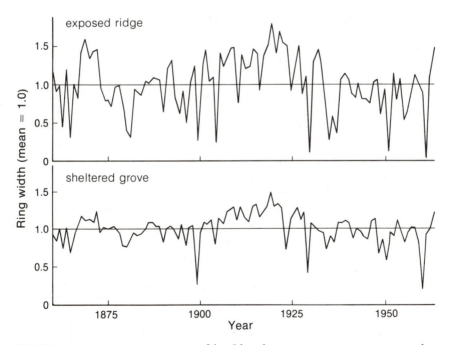

FIGURE 23-4 Variation in ring width of bristlecone pines growing on a rocky ridge (top) and in a protected grove with abundant moisture (below). The ring-width axis is scaled in such a way that the mean width at each locality equals 1.

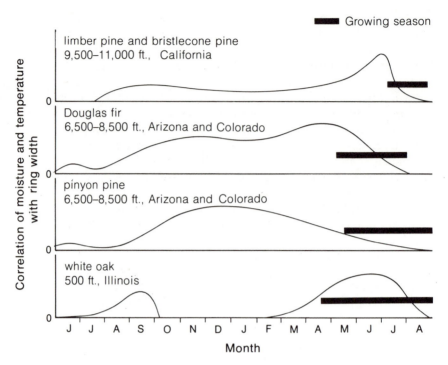

FIGURE 23-5   Correlation of tree-ring width with temperature and moisture during each of fifteen months prior to and including the growing season.

was expressed as the coefficient of variation (CV), which is the standard deviation (SD) expressed as a percentage of the mean value. Crops were studied on three soil types: brown, dark brown, and black (Table 23-1). Variation in wheat production was greatest on brown soils (CV = 35 per cent) and least on black soils (CV = 24 per cent). Furthermore, most of the variation in crop production was related to soil moisture deficits during the growing season. These deficits were higher on brown soils (142 mm ± 71 mm SD for the season) than on black soils (43 ± 82 mm). And as we would expect, variation had more influence on crop growth where moisture was least abundant. On the brown soils, 86 per cent of the variation in crop production could be related statistically to weather and soil moisture. On black soils, only 57 per cent of the variation in crop production was so related. This example shows the potential for objectively measuring stability but, defined as such, stability includes a component of the environment itself. Clearly we are dealing with a complex phenomenon.

Tree-ring and cereal crop data show that organisms do not completely compensate for variation in the environment, but we are left guessing about the inherent stability of a plant's growth processes. Some general principles concerning stability are apparent from our earlier study of homeostasis and

TABLE 23-1 Variation in annual wheat production and moisture properties on three soil types in the Prairie Provinces of Canada.

	Soil type					
	BROWN		**DARK BROWN**		**BLACK**	
	Mean	CV	Mean	CV	Mean	CV
Wheat crop (kg ha^{-1})	1,301	35	1,587	28	1,640	24
Moisture reserves in soil May 1 (mm)	113	19	117	16	147	17
Potential evapotranspiration for June (mm)	138	11	134	13	134	10
Moisture deficit (mm)						
June	69	62	64	72	65	71
July	111	32	88	44	65	53
Season	142	50	106	70	43	190
Variation in crop explained by climate and soil moisture (per cent)	86		76		57	

population growth. Large organisms have small surface-to-volume ratios; their internal environments are, therefore, more independent of the external environment than are the internal environments of small organisms. Large accumulated biomass and low turnover rate of individuals in populations also increase buffering against environmental variation. Immature organisms are generally more susceptible to environmental change than adults because they are smaller, less experienced in the case of animals, and physiologically less mature. Populations with many immature individuals tend to be less stable than populations of predominantly mature individuals.

Population Stability

The response time of the population to fluctuations in the environment is an important part of stability. Small organisms reproduce more rapidly than large organisms and their populations can respond more rapidly to change in the environment. Furthermore, small organisms may choose among several avenues of response: developmental and evolutionary responses are practical ways of coping with short-term environmental change only for small organisms with short life spans. Size and biomass-to-productivity ratios have two opposing influences on stability. The number of individuals in populations of large organisms change little but respond slowly to change;

the opposite is true of populations of small organisms. Which is more stable? Which is better attuned to fluctuation in the physical environment? The answers are not clear.

Some populations of birds have been studied in detail for long periods, and it is possible to examine the degree of variation in different parts of the life cycle (Table 23-2). Coefficients of variation in local population density are mostly between 30 and 50 per cent when populations are counted once each year. The stage of the life cycle most sensitive to change in the environment is the survival of young from independence to maturity, although in some species annual adult survival rates may be quite variable. In studies of this kind, only local populations are dealt with. In fact, local variation in population size may be considerably reduced by the movement of individuals between localities; we might expect to see larger fluctuations in populations that are more isolated. Variation in a single population also may overestimate variation in the community where populations fluctuate independently or inversely to one another. For example, in a study of deciduous forest birds in the eastern United States, the average of the coefficients of variation of populations within a habitat varied between 23 and 37 per cent, whereas the coefficients of variation in the total number of birds in each habitat varied between 8 and 18 per cent.

In the long run, populations that persist are stable, regardless of their fluctuations. The ultimate measure of instability is extinction. Populations with few individuals are disadvantaged compared to larger populations, and they are more likely to pass into oblivion following perturbations in their environment. What factors determine the size of a population? The study of taxon cycles indicates that the relative success of the adaptations of a species, compared to the adaptations of its predators, parasites, competitors, and prey, determines the degree of resource specialization and influences the size of the species' population. In an evolutionary struggle to achieve superior adaptations, species beset by few kinds of exploiters and competitors fare better than species that must confront the adaptations of many

TABLE 23-2 Relative degree of variation in annual measures of population size, reproduction, and survival of various species of birds.

Property	Range of coefficients of variation (per cent) in most studies
Population size	30–50
Clutch size	5–15
Nesting success per clutch	15–22
Young fledged per pair	25–55
Survival of fledged young	20–80
Survival of adults	10–50

antagonists. For example, a grasshopper attacked by a shrew can escape by flight. If grasshoppers are preyed upon by both shrews and sparrow hawks, each predator compromises the grasshopper's escape from the other.

The number of species in communities is greatest in the humid tropics and decreases toward the poles. Populations are surely confronted by a greater diversity of competing populations in the tropics than elsewhere, and average population size tends to be smaller. Whether each tropical species is eaten by a greater variety of predators is not known. Many ecologists have asserted that species of predators are more numerous relative to species of prey in the tropics than in temperate and arctic communities. Insect communities appear to bear this out (Table 23-3). The ratio of predators and parasites to herbivores and detritivores (the predator ratio) is greater in tropical samples than in temperate zone samples. Furthermore, within the tropics both diversity and the predator ratio increase along a moisture gradient from dry hillsides (ratio = 0.38) to wet lowland and river-bottom forest (ratio = 0.77). The greater predator ratio of moist tropical habitats is due primarily to an increase in the number of parasitic species, which tend to specialize their attack upon a few host species.

If high diversity and complexity of community organization reduced the average population size of tropical species and increased the probability of their extinction, the composition of tropical communities could be intrinsically less stable over long periods than temperate or arctic communities. These factors could be balanced, of course, by low variability of the physical environment and a preponderance of mutualistic associations in the tropics. Direct measurements of the life spans of species in tropical and temperate populations would have to be based on the fossil record, which is meager

TABLE 23-3 Percentages of trophic groups in samples of insects from various localities and habitats, arranged by decreasing latitude. In general, the tropical samples contain the most parasites and the fewest detritivores.

Locality	Habitat	Herbivores	Detritivores	Predators	Parasites	Predator ratio*
Arctic coast	Whole fauna	47	27	14	10†	0.32
Connecticut	Whole fauna	49	19	16	12	0.41
New Jersey	Whole fauna	52	19	16	10	0.37
South Carolina	Old field	68	1	19	9	0.41
Great Smoky Mountains	Average for 15 habitats	41	31	12	15	0.38
Florida Keys	Mangrove islands	41	28	22	7	0.42
Costa Rica	Average for 4 habitats	59	4	9	26	0.56

* Proportion of predators plus parasites divided by proportion of herbivores plus detritivores.
† Totals do not add to 100 per cent because some species are classified as "miscellaneous."

at best. But some groups of marine organisms are well enough represented in the fossil record that the persistence of taxa can be compared between regions. For example, genera of planktonic foraminifers (small protozoans with calcareous shells) that lived during the Cretaceous Period persisted longer in oceans north of 50° latitude than in the more diverse warm-water regions closer to the equator (Figure 23-6).

Diversity, Complexity, and Community Stability

Although high diversity may speed individual populations to extinction, thereby reducing the stability of community composition, diversity and complexity of trophic organization are widely thought to enhance the stability of community function. This principle may be stated simply: Where predators eat many kinds of prey organisms, they can specialize momentarily on whichever prey species are most abundant. This *switching* behavior makes predators less sensitive to variation in the abundance of any one prey species. In simpler communities, predators are restricted to eating few kinds of prey—perhaps only one—so their populations follow variations in their prey populations more closely (Figure 23-7). Studies of population cycles of fur-bearing mammals in arctic North America, where communities are relatively simple, are consistent with a direct relationship between diversity and stability. Populations of tropical rodents fluctuate much less than populations of temperate and arctic species.

One may, however, take a different view of the relationship between diversity, complexity, and stability. In rigorous climates, either too dry or too cold to support diverse communities, the physical environment affects most species directly and at the same time. A drought or cold snap depresses biological activity throughout the community. Milder conditions return the community as a whole to its original state. With physical conditions exerting dominant control over fluctuations in the community, all species are linked

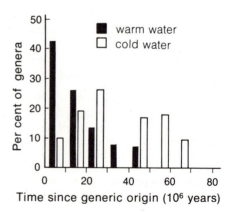

FIGURE 23-6 Life span of Cretaceous genera of planktonic foraminifera in warm-water (0 to 50° N) and cold-water (>50° N) regions.

directly to the dominant cause of the fluctuation. In the less harsh tropics, population trends are determined more by interactions with other populations than by the physical environment. Like a ripple moving across the surface of a pond, perturbations can be passed through many links of species interaction. A population in a diverse and complex tropical community can be removed by several links from an external source of disturbance. In such circumstances, time lags in population response and time lags between the links may increase rather than dampen the effects of perturbation.

An example will emphasize the point. Many species of trees in tropical forests flower during the dry season when flying insect pollinators are most abundant. Occasionally, heavy rains fall after the beginning of the dry season. Trees caught with their flowers out may not be pollinated because rain delays the season of activity of the insects. When flowers are not pollinated, trees do not produce fruit during their normal fruiting period several months later; mammals and birds that rely on fruit when other food sources are less abundant starve, curtail reproduction, or turn to novel sources of food; rodents and other gnawing mammals severely damage trees

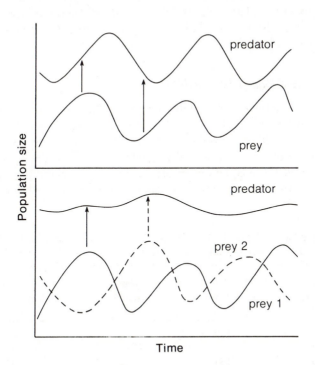

FIGURE 23-7   A diagram of population cycles of predator species feeding on either one (top) or two (bottom) kinds of prey. The predator in the two-prey system switches its feeding from one species to the other depending on which is more abundant.

and kill saplings by chewing off bark to get at the nutritious growing layers underneath; herbivorous insects that normally would have eaten the tree seedlings, and predators and parasites of these insects, are affected in turn. In this way, the initial physical perturbation is maintained in the tropical community. In simpler communities, response to physical perturbation tends to be rapid, direct, and short-lived.

Biological complexity may either dampen fluctuations, or exaggerate perturbations to communities, depending on which theory one heeds. But where theory is equivocal, we must turn to direct observation and experimentation. A few studies are pertinent to the diversity-stability problem. The extensive literature on crop pests tends not to support the hypothesis that diversity leads to stability. In addition, some species of tropical insects, members of diverse communities, exhibit extensive year-to-year variation in population size or infrequent large-scale eruptions. Kenneth Watt has examined fluctuations in populations of Canadian forest lepidoptera (moths and butterflies), comparing species that fed on few species of trees with those that had broader diets. Watt's analysis (Table 23-4) showed that population size and the relative degree of fluctuation both increase as the number of tree species eaten increases. Hence diversity of food resources does not appear to enhance the stability of butterfly and moth populations. To the contrary, the broader the diet the greater the fluctuation in population size from year to year.

Disturbance and Community Stability

If we wished to determine the ability of a community to resist disturbance, a good approach would be to disturb the community and watch its response. We would have difficulty predicting the results of our experiment beforehand, owing to the great variety of effects dependent on the nature of the disturbance and the adaptations of organisms in the community. A fire in a longleaf pine plantation in Mississippi depresses tree growth for only one

TABLE 23-4   Relationship of number of tree species eaten to the abundance and constancy of Canadian forest moth and butterfly populations.

Number of tree species eaten*	Number of lepidoptera species	Relative population size	Index of variation in population size†
1.5	179	5	0.23
4.2	173	47	0.30
10.9	134	52	0.34
24.7	34	442	0.37

* Means of four groups, gregarious species excluded.
† Variation relative to the size of the population: mean of the standard error of the logarithms of counts for each species.

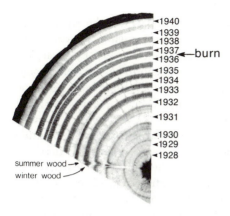

FIGURE 23-8 Cross section of a longleaf pine tree burned in January 1937 and cut in the fall of 1940 in Harrison County, Mississippi. Narrow growth rings were produced during the summer, 1937—light wood— and the winter, 1937–1938—dark wood.

year; then tree production returns to normal (Figure 23-8). When New Mexican grasslands are overgrazed, shrubs replace blue gramma and other grasses completely, and permanently alter the character of the community. Whereas grasses have extensive shallow, fibrous root systems, which hold the topsoil firmly in place, the root systems of many shrubs do not prevent soil erosion (Figure 23-9).

FIGURE 23-9 Root systems of blue gramma grass (left) and snakeweed (*Gutierrezia*, right). Overgrazing on grasslands can kill grasses and allow shrubs with little forage value and poor soil-holding qualities to take over.

One of the most consistent effects of disturbance on the structure of communities is to reduce the total number of species while allowing some of the survivors to reach abnormally high population levels. For example, polluted streams, compared to natural streams, tend to have fewer species, but some of these are extremely abundant (Figure 23-10). Strong pollution creates conditions that are lethal to most species in the community, but which are extremely favorable for a few.

When the number of plant species in a habitat is reduced by planting crops, the number of species on all trophic levels decreases. In these simplified communities, the abundance of some herbivore species increases to outbreak levels in the absence of effective control by predators. Herbivorous insects and disease organisms parasitic on plants are usually specialized to occur on one or a few host plants. Some species, adapted to feed on a particular crop, do spectacularly well in agricultural habitats, much to the farmer's dismay. On the other hand, predators require a more complex habitat and do poorly in single species stands of crops. Spruce budworms cause more damage to solid stands of balsam fir than to fir trees well spaced among other species. Infestations are worse when all the trees in the forest are the same age. Susceptibility to budworm infestation increases with age; if a forest consists entirely of old trees—often the case in managed woodlands—most of the trees become infected, easing the spread of the pest and leaving fewer unaffected trees to replace those killed by the budworm.

The relationship between community simplification and population outbreaks has been explored by David Pimentel. Insect populations on two groups of collard plants were compared. One group of collards was planted in a field that had been uncultivated for fifteen years. In the old field, collards were planted nine feet apart among vegetation that contained about

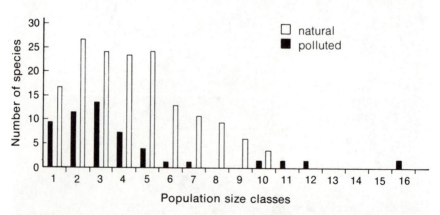

FIGURE 23-10 Distribution of abundances of diatoms in samples taken from a natural stream community and from a polluted river. Population-size classes increase from left to right by a factor of two, *e.g.*, 1-2, 2-4, 4-8, 8-16, and so on. The largest population in the polluted sample falls into the range 32,768-65,536.

300 species of plants, including five in the same family as collards (Cruciferae, the mustard family). A second group of collard plants was planted in a dense stand having no other species. The abundances of several species of insects—particularly aphids and flea beetles—reached outbreak proportions in the single species plantings. The same species of insects were kept under control by predators in the old field. The number of predators and parasitic insects collected from collard plants was also higher in the single species planting than in the old field, but they appeared too late in the season to exercise effective control over herbivore populations. Also, because Pimentel collected samples only from the collard plants in both plantings, he may, in the old field, have missed the many species on other types of plants. Such species could also have preyed on herbivorous insects on the collards. This often happens. Predators on the diamondback moth, a cabbage pest, survive the winter by eating other prey species that attack hawthorn. Many of the alternate hosts of gypsy moth parasites live in the forest understory, not in the canopy; the presence of forest undergrowth is, therefore, indirectly important to the growth of gypsy moth populations. Unfortunately, one could point to as many examples of weed or hedgerow plants maintaining populations of crop pests as populations of their predators and parasites. The role of diversity in maintaining stability is thus cast into doubt, perhaps because crop systems are so simple that a few populations, perhaps pests, perhaps their predators, can dominate a community just by chance.

In an experiment similar to Pimentel's, arthropod populations were followed in a field that was planted with millet one year and then left to develop a natural successional community during the second year. Populations were sampled ten times each summer at regular intervals. Each trophic group of insects and other arthropods was represented by more species in the second year than in the first. Populations of predatory and parasitic species were also greater in the second year, but numbers of herbivores were reduced, presumably by the predators and parasites (Table 23-5). An interesting result of the study was that variation in the numbers of predatory insects was greater during the second year in spite of the increased diversity of prey species. Once more, observations do not support a direct relationship between diversity and constancy.

In another study near Syracuse, New York, fertilizer was applied to abandoned hay fields of different ages. The experiment was designed to measure the effect of a perturbation—nutrient enrichment—on the composition of the community and to learn how this perturbation was transmitted through the trophic structure. The seventeen-year-old field was more diverse and had greater biomass of vegetation than the six-year-old field. Presumably the older field should have resisted perturbation to a greater degree, if increased diversity and biomass enhance stability. In fact, the older field was more sensitive to enrichment, and the effects were felt more strongly at the herbivore and carnivore levels than among the primary

TABLE 23-5 Diversity, density, and variation of arthropod populations in unharvested grain crop compared to the natural successional community which replaced it during the following year.

		Herbivorous insects	Predatory insects	Spiders	Parasitic insects	Total arthropods
Diversity*	1966	7.2	2.8	1.6	6.3	15.6
	1967	10.6	6.0	7.2	12.4	30.9
Density	1966	482	64	18	24	624
(ind m^{-2})	1967	156	79	38	51	355
Variation						
(per cent of	1966	24	44	33	57	19
mean density)†	1967	28	49	67	40	32

* Calculated by the equation, diversity $= (S - 1)/\log N$, where S is the number of species in the sample and N is the number of individuals.
† Coefficient of variation (per cent) = (standard deviation/mean) $\times$ 100.

producers, which were affected directly by the fertilizer (Table 23-6). It is arguable whether the increased diversity and production reflected failure of internal stabilizing mechanisms or a favorable response to enrichment, further increasing stability. The point is well made, however, that in a successionally more mature community, perturbations were transferred more efficiently throughout the trophic structure.

As we have seen, ecologists are terribly ignorant of stability in natural systems—the pertinent internal mechanisms of communities and how they work, the relative stability of different communities. Even recognizing stability where it occurs is difficult. Why has this single property of nature been so elusive? Stability is the culmination of all ecological interrelationships; it is the sum of all the components and interactions that make up the

TABLE 23-6 Percentage change in diversity and productivity of producer, herbivore, and carnivore trophic levels as a result of fertilization of six-year-old and seventeen-year-old abandoned hay fields.

Trophic level	DIVERSITY		PRODUCTIVITY*	
	6 years	17 years	6 years	17 years
Producers	−7	+3	+96	+70
Herbivores†	+24	+51	+31	+201
Carnivores†	+27	+74	−6	+108

* Mg m^{-2} day^{-1}
† Diversity and productivity measured during the earlier of two growing seasons.

community, the union of all lower-order properties of community, population, and organism. To understand stability, we must understand ecological and evolutionary responses and interrelationships at all levels. The science of ecology is not yet mature enough to mold its diverse knowledge and concepts into a unified theory of stability. It is unlikely such a theory will emerge before another decade has passed. However provocative the challenge, not every ecologist should direct his research entirely or even partly toward understanding stability. Other challenges remain, other puzzles must be solved, other speculation will engage us. Still, the significance of new knowledge and new ideas surely will be judged by their contribution to our understanding of the ecological synthesis—the stability of natural systems.

The International System of Units

Scientists throughout the world have recently adopted the International System of Units of Measurements (SI, after the French *Système International*) based primarily on the metric system. The system is described in detail in such publications as the National Bureau of Standards Special Publication 330, 1977 Ed. (U.S. Dept. Commerce, Washington, D.C.), and M. H. Green, *Metric Conversion Handbook* (Chemical Publ. Co., New York, N.Y., 1978). Briefly, there are seven basic SI units that are dimensionally independent:

Category of measurement	Name of SI unit	Symbol
length	meter	m
mass	kilogram	kg
time	second	s
electric current	ampere	A
temperature	kelvin	K
amount of substance	mole	mol
luminous intensity	candela	cd

The ampere, mole, and candela are not often used in ecology and will not be discussed here. The second is our familiar unit of time. The SI units of length and mass are metric with simple conversions to and from the English system, e.g., one meter equals 3.28 feet, and one kilogram equals 2.205 pounds. The kelvin is identical to the degree Celsius (C) except that zero on the kelvin scale is absolute zero, the coldest temperature possible (or -273 C). Most ecologists use C instead of K, because of the convenient designations for the freezing and boiling points of water (0 C and 100 C).

In the SI system, all other measurements are derived from the basic seven. For example, the SI unit of flow is the cubic meter per second, which is derived from the base units, meter and second. Some of these derived units frequently employed in ecology are:

Category of measurement	Name of SI unit	Symbol	Expression in terms of	
			Other units	Base units
area	square meter	m^2		m^2
volume	cubic meter	m^3		m^3
velocity	meter per second	$m\ s^{-1}$		$m\ s^{-1}$
flow	cubic meter per second	$m^3\ s^{-1}$		$m^3\ s^{-1}$
density	kilogram per cubic meter	$kg\ m^{-3}$		$kg\ m^{-3}$
force	newton	N		$kg\ m\ s^{-2}$
pressure	pascal	Pa	$N\ m^{-2}$	$kg\ m^{-1}s^{-2}$
energy, heat	joule	J	$N\ m$	$kg\ m^2\ s^{-2}$
power	watt	W	$J\ s^{-1}$	$kg\ m^2\ s^{-3}$

There are also a number of other units that are not part of SI but nonetheless are widely used and thus acceptable: the liter (l), a unit of volume equal to one one-thousandth of a cubic meter; the are (a), a unit of area equal to 100 square meters, and the more often used hectare (ha), equal to 100 a and 10,000 m^2; minute, hour, and day, familiar units of time. The calorie (cal), a familiar unit of heat energy, is not derived from SI base units and should be abandoned in favor of the joule (= 0.239 cal), even though ecologists continue to use the cal and kcal.

Multiples of 10 of the SI units are given special names by adding the following prefixes to the unit name:

Factor	Prefix	Symbol		Name
10^9	giga	G	1 000 000 000	billion[a]
10^6	mega	M	1 000 000	million
10^3	kilo	k	1 000	thousand
10^2	hecto	h	100	hundred
10^1	deka	da	10	ten
10^{-1}	deci	d	0.1	tenth
10^{-2}	centi	c	0.01	hundredth
10^{-3}	milli	m	0.001	thousandth
10^{-6}	micro	μ	0.000 001	millionth
10^{-9}	nano	n	0.000 000 001	billionth

[a] Milliard in the United Kingdom; the British billion is a million million (10^{12}), equivalent to the U.S. trillion, commonly used only in astronomy and the Federal budget.

Hence a thousand meters becomes a kilometer (km) and one thousandth of a meter, a millimeter (mm); a thousand joules is a kilojoule (kJ) and a million watts is a megawatt (MW).

Although the SI system is simple and eventually will become universal, at least in science, the persistence of many other systems of measurement frequently requires us to convert from one to another. The table of factors that follows applies to the most commonly needed conversions, particularly between SI, or metric, and English units.

Conversion Factors

Length

1 meter (m) = 39.4 inches (in)
1 meter = 3.28 feet (ft)
1 kilometer (km) = 3281 feet
1 kilometer = 0.621 miles (mi)
1 micron (μ) = 10^{-6} meters
1 inch = 2.54 centimeters (cm)
1 foot = 30.5 centimeters
1 mile = 1609 meters
1 Angstrom unit (Å) = 10^{-10} meters
1 millimicron (mμ) = 10^{-9} meters

Area

1 square centimeter (cm^2) = 0.155 square inches (in^2)
1 square meter (m^2) = 10.76 square feet (ft^2)
1 hectare (ha) = 2.47 acres (A)
1 hectare = 10,000 square meters
1 hectare = 0.01 square kilometer (km^2)
1 square kilometer = 0.386 square miles
1 square mile = 2.59 square kilometers
1 square inch = 6.45 square centimeters
1 square foot = 929 square centimeters
1 square yard (yd^2) = 0.836 square meters
1 acre = 0.407 hectares

Mass

1 gram (g) = 15.43 grains (gr)
1 kilogram (kg) = 35.3 ounces
1 kilogram = 2.205 pounds (lb)
1 metric ton (t) = 2204.6 pounds
1 ounce (oz) = 28.35 grams
1 pound = 453.6 grams
1 short ton = 907 kilograms

Time

1 year (yr) = 8760 hours (hr)
1 day = 86,400 seconds (s)

Volume

1 cubic centimeter (cc or cm^3) = 0.061 cubic inches (in^3)
1 cubic inch = 16.4 cubic centimeters
1 liter = 1,000 cubic centimeters
1 liter = 33.8 U.S. fluid ounces (oz)
1 liter = 1.057 U.S. quarts (qt)
1 liter = 0.264 U.S. gallons (gal)
1 U.S. gallon = 3.79 liters
1 Brit. gallon = 4.55 liters
1 cubic foot (ft^3) = 28.3 liters (l)
1 milliliter (ml) = 1 cubic centimeter
1 U.S. fluid ounce = 29.57 milliliters
1 Brit. fluid ounce = 28.4 milliliters
1 quart = 0.946 liters

Velocity

1 meter per second (m s^{-1}) = 2.24 miles per hour (mi hr^{-1})
1 foot per second (ft s^{-1}) = 1.097 kilometers per hour
1 kilometer per hour = 0.278 meters per second
1 mile per hour = 0.447 meters per second
1 mile per hour = 1.467 feet per second

Energy

1 joule = 0.239 calories (cal)
1 calorie = 4.184 joules
1 kilowatt-hour (kWh) = 860 kilocalories
1 kilowatt-hour = 3600 kilojoules
1 British thermal unit (Btu) = 252.0 calories
1 British thermal unit = 1054 joules
1 kilocalorie (kcal) = 1,000 calories

Power

1 kilowatt (kW) = 0.239 kilocalories per second
1 kilowatt = 860 kilocalories per hour
1 horsepower (hp) = 746 watts
1 horsepower = 15,397 kilocalories per day
1 horsepower = 641.5 kilocalories per hour

Energy per unit area

1 calorie per square centimeter = 3.69 British thermal units per square foot
1 British thermal unit per square foot = 0.271 calories per square centimeter
1 calorie per square centimeter = 10 kilocalories per square meter

Power per unit area

1 kilocalorie per square meter per minute = 52.56 kilocalories per hectare per year
1 footcandle (fc) = 1.30 calories per square foot per hour at 555 mμ wavelength
1 footcandle = 10.76 lux
1 lux (lx) = 1.30 calories per square meter per hour at 555 mμ wavelength

Metabolic energy equivalents

1 gram of carbohydrate = 4.2 kilocalories
1 gram of protein = 4.2 kilocalories
1 gram of fat = 9.5 kilocalories

Miscellaneous

1 gram per square meter = 0.1 kilograms per hectare
1 gram per square meter = 8.97 pounds per acre
1 kilogram per square meter = 4.485 short tons per acre
1 metric ton per hectare = 0.446 short tons per acre

Glossary

Acclimation. A reversible change in the morphology or physiology of an organism in response to environmental change.

Adaptability. Capacity for evolutionary change. Adaptability may depend on the phenotype's tolerance of environmental change as well as on the genetic variability of the population.

Adaptation. A genetically determined characteristic that enhances the ability of an organism to cope with its environment.

Aggressive mimicry. Resemblance of predators or parasites to harmless species to deceive potential prey or hosts.

Allelopathy. Direct inhibition of one species by another using noxious or toxic chemicals.

Allochthonous. Referring to materials transported into a system, particularly minerals and organic matter transported into streams and lakes.

Amino acid. One of about thirty organic acids containing the group NH_2 which are the building blocks of proteins.

Ammonification. Breakdown of proteins and amino acids with ammonia as an excretory by-product.

Anaerobic. Without oxygen.

Annual. Referring to an organism that completes its life cycle from birth or germination to death within a year.

Aposematism. Conspicuous appearance of an organism warning that it is noxious or distasteful; warning coloration.

Artificial selection. Intentional manipulation by man of the fitnesses of individuals in a population to produce a desired evolutionary response.

Aspect diversity. Variations in the outward appearance of species that live in the same habitat and are eaten by visually hunting predators.

Assimilation. Incorporation of any material into the tissues, cells, and fluids of an organism.

Assimilation efficiency. A percentage expressing the proportion of energy ingested that is absorbed into the bloodstream.

Association. A group of species living in the same place.

Autecology. The study of organisms in relation to their physical environment.

Autochthonous. Referring to materials produced within a system, particularly organic matter produced, and minerals cycled, within streams and lakes.

Autotroph. An organism that assimilates energy from either sunlight (green plants) or inorganic compounds (sulfur bacteria). *See also* Heterotroph.

471

Barren.   An area with sparse vegetation owing to some physical or chemical property of the soil.

Basal metabolism.   The energy expenditures of an organism that is at rest, fasting, and in a thermally neutral environment.

Batesian mimicry.   Resemblance of an edible species (mimic) to an unpalatable species (model) to deceive predators.

Benthic.   Bottom dwelling in rivers, lakes, and oceans.

Biomass.   Weight of living material, usually expressed as a dry weight, in all or part of an organism, population, or community. Commonly expressed as weight per unit area, a biomass density.

Biomass accumulation ratio.   The ratio of weight to annual production, usually applied to vegetation.

Biota.   Fauna and flora together.

Biotic environment.   Biological components of an organism's surroundings that interact with it, including competitors, predators, parasites, and prey.

Boreal.   Northern. Often refers to the coniferous forest regions that stretch across Canada, northern Europe, and Asia.

Calcification.   Deposition of calcium and other soluble salts in soils where evaporation greatly exceeds precipitation.

Caliche.   An alkaline salt deposit on the soil surface, usually occurring in arid regions with ground water close to the surface.

Carnivore.   An organism that consumes mostly flesh.

Carrying capacity.   Number of individuals that the resources of a habitat can support.

Cation.   A part of a dissociated molecule carrying a positive electrical charge, usually in an aqueous solution ($e.g.$ Ca^{++}, Na^+, NH_4^+, H^+).

Cation-exchange capacity.   The ability of soil particles to absorb positively charged ions, such as hydrogen (H^+) and calcium (Ca^{++}).

Character displacement.   Divergence in the characteristics of two otherwise similar species where their ranges overlap, caused by the selective effects of competition between the species in the area of overlap.

Character divergence.   Evolution of differences between similar species occurring in the same areas, caused by the selective effects of competition.

Chemoautotroph.   An organism that oxidizes inorganic compounds (often hydrogen sulfide) to obtain energy for synthesis of organic compounds: $e.g.$ sulfur bacteria.

Climatic climax.   The steady-state community characteristic of a particular climate.

Climax.   The end point of a successional sequence; a community that has reached a steady state under a particular set of environmental conditions.

Climograph.   A diagram on which localities are represented by the annual cycle of their temperature and rainfall.

Cline.   Change in population characteristics over a geographic area, usually related to a corresponding environmental change.

Closed-community concept.   The idea, popularized by F. C. Clements, that communities are distinctive associations of highly interdependent species.

Coadaptation.   Evolution of characteristics of two or more species to their mutual advantage.

Coarse-grained.   Referring to qualities of the environment that occur in large patches, with respect to the activity patterns of an organism and, therefore, among which the organism can select.

Coevolution.   Development of genetically determined traits in two species to facilitate some interaction, usually mutually beneficial. *See* Counterevolution.

Coexistence. Occurrence of two or more species in the same habitat; usually applied to potentially competing species.

Community. An association of interacting populations, usually defined by the nature of their interaction or the place in which they live.

Compensation point. Depth of water at which respiration and photosynthesis balance each other; the lower limit of the euphotic zone.

Competition. Use or defense of a resource by one individual that reduces the availability of that resource to other individuals.

Competitive exclusion principle. The hypothesis that two or more species cannot coexist on a single resource that is scarce relative to the demand for it.

Continuum. A gradient of environmental characteristics or of change in the composition of communities.

Continuum index. An artificial scale of an environmental gradient based on changes in community composition.

Convergent evolution. Development of characteristics with similar functions in unrelated species that live in the same kind of environment but in different places.

Counteradaptation. Evolution of characteristics of two or more species to their mutual disadvantage.

Counterevolution. Development of traits in a population in response to exploitation, competition, or other detrimental interaction with another population.

Crypsis. An aspect of the appearance of organisms whereby they avoid detection by others; usually applied to the prey of visually hunting predators.

Cyclic climax. A steady-state, cyclic sequence of communities, none of which by itself is stable.

Death rate. The percentage of newborn dying during a specified interval. *See* Mortality.

Decomposition. Metabolic breakdown of organic materials; the by-products are released energy and simple organic and inorganic compounds. *See* Respiration.

Denitrification. The reduction by microorganisms of nitrate and nitrite to nitrogen.

Density-dependent. Having influence on individuals in a population that varies with the degree of crowding in the population.

Density-independent. Having influence on individuals in a population that does not vary with the degree of crowding in the population.

Detritivore. An organism that feeds on freshly dead or partially decomposed organic matter.

Detritus. Freshly dead or partially decomposed organic matter.

Developmental response. Acquisition of one of several alternative forms by an organism, depending on the environmental conditions under which it grows.

Diapause. Temporary interruption in the development of insect eggs or larvae, usually associated with a dormant period.

Dimorphism. Occurrences of two forms of individuals in a population.

Direct competition. Exclusion of individuals from resources by aggressive behavior or use of toxins.

Dispersal. Movement of organisms away from the place of birth or from centers of population density.

Dispersion. Pattern of spacing of individuals in a population.

Diversity. The number of species in a community or region. Alpha diversity refers to the diversity of a particular habitat, beta diversity to the species added by pooling habitats within a region. Also, a measure of the variety of species in a community that takes into account the relative abundance of each species.

Ecocline. A geographical gradient of vegetation structure associated with one or more environmental variables.

Ecological efficiency. Percentage of energy in the biomass produced by one trophic level that is incorporated into biomass produced by the next highest trophic level.

Ecological isolation. Avoidance of competition between two species by differences in food, habitat, activity period, or geographical range.

Ecological release. Expansion of habitat and resource utilization by populations in regions of low species diversity, resulting from reduced interspecific competition.

Ecosystem. All the interacting parts of the physical and biological worlds.

Ecotone. A habitat created by the juxtaposition of distinctly different habitats; an edge habitat; a zone of transition between habitat types.

Ecotype. A genetically differentiated subpopulation that is restricted to a specific habitat.

Edaphic. Pertaining to, or influenced by, soil conditions.

Edge species. Species preferring the habitat created by the abutment of distinctive vegetation types.

Egestion. Elimination of undigested food material.

Eluviation. The downward movement of dissolved soil materials carried by percolating water.

Endemic. Confined to a certain region.

Environment. Surroundings of an organism, including the plants and animals with which it interacts.

Environmental gradient. A continuum of conditions ranging between extremes, as the gradation from hot to cold environments.

Enzymes. Organic compounds in living cells that accelerate specific biochemical transformations without themselves being affected.

Epilimnion. The warm, oxygen-rich surface layers of a lake or other body of water.

Equitability. Uniformity of abundance in an assemblage of species. Equitability is greatest when all species are equally numerous.

Euphotic zone. Surface layer of water to the depth of light penetration at which photosynthesis balances respiration. *See also* Compensation point.

Eutrophic. Referring to a body of water with abundant nutrients and high productivity.

Eutrophication. Enrichment of bodies of water, often caused by sewage and runoff from fertilized agricultural land.

Evapotranspiration. The sum of transpiration by plants and evaporation from the soil. Potential evapotranspiration is the amount of evapotranspiration that would occur, given the local temperature and humidity, if water were superabundant.

Exploitation. Removal of individuals or biomass from a population by predators or parasites.

Exploitation efficiency. The percentage of potential prey or food plants that are consumed by predators and herbivores.

Exponential rate of increase (r). Rate at which a population is growing at a particular instant, expressed as a proportional increase per unit of time.

Fall bloom. The rapid growth of algae in temperate lakes following the autumnal breakdown of thermal stratification and mixing of water layers.

Fall overturn. Vertical mixing of water layers in temperate lakes in autumn following breakdown of thermal stratification.

Fecundity. Rate at which an individual produces offspring, usually expressed only for females.

Field capacity. The amount of water that soil can hold against the pull of gravity.

Fine-grained. Referring to qualities of the environment that occur in small patches with respect to the activity patterns of an organism, and among which the organism cannot usefully distinguish.

Fitness. Genetic contribution by an individual's descendants to future generations of a population.

Floristic. Referring to studies of the species composition of plant associations.

Food chain. An abstract representation of the passage of energy through populations in the community.

Food chain efficiency. *See* Ecological efficiency.

Food web. An abstract representation of the various paths of energy flow through populations in the community.

Functional response. Change in the rate of exploitation of prey by an individual predator as a result of a change in prey density. *See also* Numerical response.

General adaptive syndrome. The set of abnormal physiological responses to the stress of social interaction in dense populations.

Generalist. A species with broad food or habitat preferences, or both.

Generation time. Average age (T_c) at which a female gives birth to her offspring, or average time (T) for a population to increase by a factor equal to the net reproductive rate.

Genetic feedback. Evolutionary response of a population to the adaptations of competitors, predators, or prey.

Geometric rate of increase (λ). Factor by which the size of a population changes over a specified period. *See* Exponential rate of increase.

Gross production. The total energy or nutrients assimilated by an organism, a population, or an entire community. *See also* Net production.

Gross production efficiency. The percentage of ingested food utilized for growth and reproduction by an organism.

Habitat. Place where an animal or plant normally lives, often characterized by a dominant plant form or physical characteristic (*i.e.*, the stream habitat, the forest habitat).

Habitat expansion. Increase in average breadth of habitat distribution of species in depauperate biotas, especially on islands, compared with species in more diverse biotas.

Habitat selection. Preference for certain habitats.

Herbivore. An organism that consumes living plants or their parts.

Heterogeneity. The variety of qualities found in an environment (habitat patches) or a population (genotypic variation).

Heterotroph. An organism that utilizes organic materials as a source of energy and nutrients. *See also* Autotroph.

Homeostasis. Maintenance of constant internal conditions in the face of a varying external environment.

Homeothermic. Able to maintain constant body temperature in the face of fluctuating environmental temperature; warm-blooded.

Homology. The condition of having similar evolutionary or developmental origin.

Hydrarch succession. Progression of terrestrial plant communities developing in an aquatic habitat such as a bog or swamp.

Hyperosmotic. Having a salt concentration greater than that of the surrounding medium.

Hypolimnion. The cold, oxygen-poor part of a lake or other body of water that lies below the zone of rapid change in water temperature. *See also* Epilimnion.

Hypo-osmotic. Having a salt concentration less than that of the surrounding medium.

Illuviation. The accumulation of dissolved substances within a soil layer.

Indirect competition. Exploitation of a resource by one individual that reduces the availability of that resource to others. *See also* Direct competition.

Innate capacity for increase (r_0). The intrinsic growth rate of a population under ideal conditions without the restraining effects of competition.

Interference. Direct antagonism between individuals whether by behavioral or chemical means.

Interspecific competiton. Competition between individuals of different species.

Intertidal zone. The region between the high and low tide marks on the shore, which is alternately covered by water and exposed to the air with each tidal cycle.

Intraspecific competition. Competition between individuals of the same species.

Intrinsic rate of increase (r_m). Exponential growth rate of a population with a stable age distribution.

Ion. The dissociated parts of a molecule, each of which carries an electrical charge.

Key factor analysis. A statistical treatment of population data designed to identify factors most responsible for change in population size.

Laterite. A hard substance rich in oxides of iron and aluminum; frequently formed when tropical soils weather under alkaline conditions.

Laterization. Leaching of silica from soil, usually in warm, moist regions with an alkaline soil reaction.

Leaching. Removal of soluble compounds from leaf litter or soil by water.

Life form. Characteristic structure of a plant or animal.

Life table. A summary by age of the survivorship and fecundity of female individuals in a population.

Life zone. A more or less distinct belt of vegetation occurring within, and characteristic of, a particular range of latitude or elevation.

Limiting resource. A resource that is scarce relative to demand for it. *See* Resource.

Loam. Soil that is a mixture of coarse sand particles, fine silt, clay particles, and organic matter.

Logistic equation. Mathematical expression for a particular sigmoid growth curve in which the percentage rate of increase decreases in linear fashion as population size increases.

Lower critical ambient temperature. Surrounding temperature below which warm-blooded animals must generate heat to maintain their body temperature.

Melanism. Occurrence of black pigment, usually melanin.

Mesic. Referring to habitats with plentiful rainfall and well-drained soils.

Metamorphosis. An abrupt change in form during development that fundamentally alters the function of the organism; often called complete metamorphosis. Incomplete metamorphosis refers to more gradual change.

Micelle. A complex soil particle resulting from the association of humus and clay particles, with negative electric charges at its surface.

Microhabitat. The particular parts of the habitat that an individual encounters in the course of its activities.

Mimicry. Resemblance of an organism to some other organism or object in the environment, evolved to deceive predators or prey into confusing the organism and that which it mimics. *See* Batesian mimicry, Müllerian mimicry, Aggressive mimicry.

Monoclimax theory. The idea that all successional sequences lead ultimately to one of a few distinctive community types, depending on the climate of the region.

Morph. A specific form, shape, or structure.

Mortality (m_x). Ratio of the number of deaths of individuals to the population, often described as a function of age (x). *See* Death rate.

Müllerian mimicry. Mutual resemblance of two or more conspicuously marked distasteful species to enhance predator avoidance.

Mutualism. Relationship between two species that benefits both parties.

Mycorrhizae. Close association of fungi and tree roots in the soil that facilitates the uptake of minerals by trees.

Natural selection. Change in the frequency of genetic traits in a population through differential survival and reproduction of individuals bearing those traits.

Negative feedback. Tendency of a system to counteract externally imposed change and return to a stable state.

Net aboveground productivity (NAP). Accumulation of biomass in aboveground parts of plants (trunks, branches, leaves, flowers, and fruits), over a specified period; usually expressed on an annual basis.

Net production. The total energy or nutrients accumulated by the organism by growth and reproduction; gross production minus respiration.

Net production efficiency. The percentage of assimilated food utilized for growth and reproduction by an organism.

Net reproductive rate. Number of offspring that females are expected to bear on average during their lifetimes.

Niche. All the components of the environment with which the organism or population interacts.

Nitrification. Breakdown of nitrogen-containing organic compounds by microorganisms, yielding nitrates and nitrites.

Nitrogen fixation. Biological assimilation of atmospheric nitrogen to form organic nitrogen-containing compounds.

Nonstabilizing factors. Influences on population growth that are independent of the size of the population.

Numerical response. Change in the population size of a predatory species as a result of a change in the density of its prey. *See also* Functional response.

Nutrient. Any substance required by organisms for normal growth and maintenance. (Mineral nutrients usually refer to inorganic substances taken from soil or water.)

Nutrient cycle. The path of an element through the ecosystem including its assimilation by organisms and its release in a reusable inorganic form.

Oligotrophic. Referring to a body of water with low nutrient content and productivity.

Omnivore. An organism whose diet is broad, including both plant and animal foods.

Open-community concept. The idea, advocated by H. A. Gleason and R. H. Whittaker, that communities are the local expression of the independent, geographic distributions of species.

Opportunistic species. A species that takes advantage of temporary or local conditions. (Populations of opportunistic species usually fluctuate markedly.)

Oscillation. Regular fluctuation through a fixed cycle above and below some mean value.

Osmoregulation. Regulation of the salt concentration in cells and body fluids.

Osmosis. Diffusion of substances in aqueous solution across the membrane of a cell.

Overturn. Vertical mixing of layers of large bodies of water caused by seasonal changes in temperature.

Parasite.   An organism that consumes part of the blood or tissues of its host, usually without killing the host. Parasites may live entirely within the host (endoparasites) or on its surface (ectoparasites).

Parasitoid.   Any of a number of so-called parasitic insects whose larvae live within and consume their host, usually another insect.

Pattern-climax theory.   The hypothesis that succession reaches a wide variety of nondiscrete climax communities depending on local climate, soil, slope, grazing pressure, and so on.

Perennial.   Referring to an organism that lives for more than one year.

pH.   A scale of acidity or alkalinity.

Photoautotroph.   An organism that utilizes sunlight as its primary energy source for the synthesis of organic compounds.

Photosynthesis.   Utilization of the energy of light to combine carbon dioxide and water into simple sugars.

Photosynthetic efficiency.   Percentage of light energy assimilated by plants; based either on net production (net photosynthetic efficiency) or on gross production (gross photosynthetic efficiency).

Phytoplankton.   Microscopic floating aquatic plants.

Plankton.   Microscopic floating aquatic plants (phytoplankton) and animals (zooplankton).

Podsolization.   Breakdown and removal of clay particles from the acidic soils of cold, moist regions.

Poikilothermic.   Unable to regulate body temperature; cold-blooded.

Polyclimax theory.   The hypothesis that succession leads to one of a variety of distinct climax communities, depending on local environmental conditions.

Potential evapotranspiration.   The amount of transpiration by plants and evaporation from the soil that would occur, given the local temperature and humidity, if water were not limited.

Predator.   An animal (rarely a plant) that kills and eats animals.

Primary consumer.   An herbivore, the lowermost eater on the food chain.

Primary producer.   A green plant that assimilates the energy of light to synthesize organic compounds.

Primary productivity.   Rate of assimilation (gross primary productivity) or accumulation (net primary productivity) of energy and nutrients by green plants and other autotrophs.

Primary succession.   Sequence of communities developing in a newly exposed habitat devoid of life.

Rain shadow.   Dry area on the leeward side of a mountain range.

Recruitment.   Addition of new individuals to a population by reproduction.

Regulatory response.   A rapid, reversible physiological or behavioral response by an organism to change in its environment.

Relative abundance.   Proportional representation of a species in a sample or a community.

Replacement series diagram.   A diagram showing the outcome of competition between two species in experiments in which the initial ratio of the two species was varied.

Resource.   A subtance or object required by an organism for normal maintenance, growth, and reproduction. If the resource is scarce relative to demand, it is referred to as a limiting resource. Nonrenewable resources (such as space) occur in fixed amounts and can be fully utilized; renewable resources (such as food) are produced at a rate that may be partly determined by their utilization.

Respiration.   Use of oxygen to break down organic compounds metabolically for the purpose of releasing chemical energy.

Riparian.   Along the bank of a river or lake.

Salinization. Accumulation of salts, such as sodium chloride and calcium sulfate, in surface layers of soil, usually in arid climates.

Saprophage. An organism that consumes detritus.

Saturation point. With respect to primary production, the amount of light that causes photosynthesis to attain its maximum rate.

Secondary succession. Progression of communities in habitats where the climax community has been disturbed or removed entirely.

Selection. *See* Artificial selection, Natural selection.

Self-thinning curve. In populations of plants limited by space or other resources, the characteristic logarithmic relationship between number and biomass.

Sere. A series of stages of community change in a particular area leading toward a stable state. *See also* Succession.

Social pathology. A syndrome of physiological and behavioral disturbances, caused by crowding, that lead to reduced fecundity and increased mortality.

Soil. The solid substrate of terrestrial communities resulting from the interaction of weather and biological activities with the underlying geological formation.

Soil horizon. A distinctive zone of soil formed at a characteristic depth by weathering and organic contributions to the soil.

Soil skeleton. Physical structure of mineral soil, referring principally to sand grains and silt particles.

Specialization. Restriction of an organism's or a population's activities to a portion of the environment; a trait that enables an organism (or one of its organs) to modify (or differentiate) in order to adapt to a particular function or environment.

Speciation. Separation of one population into two or more reproductively isolated, independent evolutionary units.

Species. A group of actually or potentially interbreeding populations that are reproductively isolated from all other kinds of organisms.

Specific heat. Amount of energy that must be added or removed to change the temperature of a substance by a specific amount. By definition, one calorie of energy is required to raise the temperature of one gram of water by one degree Celsius.

Spring overturn. Vertical mixing of water layers in temperate lakes in spring as surface ice disappears.

Stability. Inherent capacity of any system to resist change.

Stabilizing factors. Factors that tend to restore a system to its equilibrium state; specifically, the class of density-dependent factors that act to restore populations to equilibrium size.

Stable age distribution. Proportion of individuals in various age classes in a population that has been growing at a constant rate.

Subclimax. A stage of succession along a sere prevented from progressing to the climatic climax by fire, soil deficiencies, grazing, and similar factors.

Succession. Replacement of populations in a habitat through a regular progression to a stable state.

Survival (l_x). Proportion of newborn individuals alive at a given age.

Symbiosis. Intimate, and often obligatory, association of two species, usually involving coevolution. Symbiotic relationships can be parasitic or mutualistic.

Synecology. The relationship of organisms and populations to biotic factors in the environment.

Taxon cycle. Cycle of expansion and contraction of the geographical range and population density of a species or higher taxonomic category.

Temperature profile. The relationship of temperature to depth below the surface of water or the soil, or the height above the ground.

Thermal conductance. Rate at which heat flows through a substance.

Thermocline. The zone of water depth within which temperature changes rapidly between the upper warm water layer (epilimnion) and lower cold water layer (hypolimnion).

Time lag. Delay in response to a change.

Torpid. Having lost the power of motion and feeling, usually accompanied by greatly reduced rate of respiration.

Transient climax. The end of a successional sequence terminated because the appropriate habitat ceases to exist. Examples are the plant communities in seasonal pools and the series of scavengers that feed on carcasses.

Transit time. Average time that a substance or quantum of energy remains in the biological realm; ratio of biomass to productivity.

Transpiration. Evaporation of water from leaves and other plant parts.

Transpiration efficiency. The ratio of net primary production to transpiration of water by a plant, usually expressed by grams per kilogram of water.

Trophic. Pertaining to food or nutrition.

Trophic level. Position in the food chain determined by the number of energy-transfer steps to that level.

Trophic structure. Organization of the community based on feeding relationships of populations.

Ultimate factors. Aspects of the environment that are directly important to the well-being of an organism (for example, food).

Upwelling. Vertical movement of water, usually near coasts and driven by onshore winds, that brings nutrients from the depths of the ocean to surface layers.

Wilting capacity. The minimum water content of the soil at which plants can obtain water.

Xerarch succession. Progression of terrestrial plant communities developing in habitats with well-drained soil.

Xeric. Referring to habitats in which plant production is limited by availability of water.

Zooplankton. Tiny floating aquatic animals.

Selected Readings

Chapter 1. Introduction

Colinvaux, P. A. *Introduction to Ecology*. Wiley, New York (1973).

Collier, B. D., G. W. Cox, A. W. Johnson and P. C. Miller. *Dynamic Ecology*. Prentice-Hall, Englewood Cliffs, New Jersey (1973).

Kormondy, E. J. *Concepts of Ecology*. Prentice-Hall, Englewood Cliffs, New Jersey (1969).

Krebs, C. J. *Ecology: The Experimental Analysis of Distribution and Abundance* (2nd ed.). Harper & Row, New York (1978).

May, R. M. (Ed.). *Theoretical Ecology*. Saunders, Philadelphia (1976).

McNaughton, S. J. and L. L. Wolf. *General Ecology* (2nd ed.). Holt, Rinehart, & Winston, New York (1979).

Odum, E. P. 1971. *Fundamentals of Ecology* (3rd ed.). Saunders, Philadelphia (1971).

Oosting, H. J. *The Study of Plant Communities* (2nd ed.). W. H. Freeman, San Francisco (1956).

Smith, R. L. *Ecology and Field Biology* (2nd ed.). Harper & Row, New York (1974).

Ricklefs, R. E. *Ecology* (2nd ed.). Chiron, New York (1979).

Whittaker, R. H. *Communities and Ecosystems* (2nd ed.). Macmillan, New York (1975).

Young, G. L. Human ecology as an interdisciplinary concept: a critical inquiry. *Advances in Ecological Research* 8: 1–105 (1974).

Chapter 2. Life and the Physical Environment

Billings, W. D. Physiological ecology. *Annual Review of Plant Physiology* 8: 375–392 (1957).

Hadley, N. F. Desert species and adaptation. *American Scientist* 60: 338–347 (1972).

Hutchinson, G. E. The biosphere. *Scientific American* 223: 44–58 (1970).

Redfield, A. C. The biological control of chemical factors in the environment. *American Scientist* 46: 205–221 (1958).

Smith, N. G. The advantage of being parasitized. *Nature* 219: 690–694 (1968).

Tansley, A. G. The use and abuse of vegetational concepts and terms. *Ecology* 16: 284–307 (1935).

Tevis, L., Jr. and I. M. Newell. Studies on the biology and seasonal cycle of the giant red velvet mite, *Dinothrombium pandorae* (Acari, Thrombidiidae). *Ecology* 43: 497–505 (1962).

Chapter 3. Natural Selection

Antonovics, J. The effects of a heterogeneous environment on the genetics of natural populations. *American Scientist* 59: 593–599 (1971).

Darwin, C. R. *On the Origin of Species.* Murray, London (1859).

DeBach, P. *Biological Control by Natural Enemies.* Cambridge University Press, London (1974).

Futuyma, D. J. *Evolutionary Biology.* Sinauer, Sunderland, Massachusetts (1979).

Kettlewell, H. B. D. Darwin's missing evidence. *Scientific American* 200: 48–53 (1959).

Mather, K., and B. J. Harrison. The manifold effect of selection. *Heredity* 3: 1–52 (1949).

Quayle, H. J. The development of resistance in certain insects to hydrocyanic gas. *Hilgardia* 11: 183–225 (1938).

Schmidt-Nielsen, K. *Animal Physiology. Adaptation and Environment* (2nd. ed.). Cambridge University Press, London and New York (1980).

Chapter 4. Properties and Requirements of Life

Brett, J. R. Some principles in the thermal requirements of fishes. *Quarterly Review of Biology* 31: 75–87 (1956).

Gordon, M. S. *Animal Function: Principles and Adaptations.* Macmillan, New York (1968).

Henderson, L. J. *The Fitness of the Environment.* Macmillan, New York (1913).

Hochachka, P. W., and G. N. Somero. *Strategies of Biochemical Adaptation.* Saunders, Philadelphia (1973).

Potts, W. T. W., and G. Parry. *Osmotic and Ionic Regulation in Animals.* Pergamon, Oxford (1964).

Schmidt-Nielsen, K. *Animal Physiology. Adaptation and Environment* (2nd ed.). Cambridge University Press, London and New York (1980).

Treshow, M. *Environment and Plant Response.* McGraw-Hill, New York (1970).

Chapter 5. Aquatic and Terrestrial Environments

Berkner, L. V. and L. C. Marshall. History of major atmospheric components. *Proceedings of the National Academy of Sciences* 53: 1215–1226 (1965).

Deevey, E. S. A re-examination of Thoreau's "Walden." *Quarterly Review of Biology* 17: 1–11 (1942).

Gates, D. M. The energy environment in which we live. *American Scientist* 51: 327–348 (1963).

Gates, D. M. Energy, plants and ecology. *Ecology* 46: 1–13 (1965).

Hadley, N. F. Desert species and adaptation. *American Scientist* 60: 338–347 (1970).

Isaacs, J. D. The nature of oceanic life. *Scientific American* 221: 146–162 (1969).

Macan, T. T. *Freshwater Ecology.* Longmans, London (1963).

Moore, H. B. *Marine Ecology.* Wiley, New York (1958).

Ruttner, F. *Fundamentals of Limnology* (3rd ed.). University of Toronto Press, Toronto (1963).

Schmidt-Nielson, K. *How Animals Work.* Cambridge University Press, London (1972).

Schmidt-Nielson, K. and B. Schmidt-Nielson. The desert rat. *Scientific American* 189: 73–78 (1953).

Tait, R. V. *Elements of Marine Ecology.* Plenum, New York (1968).

Weisskopf, V. F. How light interacts with matter. *Scientific American* 219: 60–71 (1968).

Chapter 6. Soil Development

Brady, N. C. *Nature and Properties of Soils* (8th ed.). Macmillan, New York (1974).

Bunting, B. T. *The Geography of Soil* (rev. ed.). Aldine, Chicago (1967).

Buol, S. W., F. D. Hole, and R. J. McCracken. *Soil Genesis and Classification.* Iowa State University Press, Ames (1973).

Crocker, R. L. Soil genesis and the pedogenic factors. *Quarterly Review of Biology* 27: 139–168 (1952).

Crocker, R. L. and J. Major. Soil development in relation to vegetation and surface age at Glacier Bay, Alaska. *Journal of Ecology* 43: 427–448 (1955).

Eyre, S. R. *Vegetation and Soils* (2nd ed.). Aldine, Chicago (1968).

Harley, J. L. Fungi in ecosystems. *Journal of Animal Ecology* 41: 1–16 (1972).

Olson, J. S. Rates of succession and soil changes on southern Lake Michigan sand dunes. *Botanical Gazette* 119: 125–170 (1958).

Sanchez, P. A. (Ed.). *A Review of Soils Research in Tropical Latin America.* North Carolina Agricultural Experiment Station, Technical Bulletin No. 219 (1973).

Chapter 7. Variation in the Environment

Geiger, R. *The Climate Near the Ground.* Harvard University Press, Cambridge, Massachusetts (1966).

Flohn, H. *Climate and Weather.* World University Library, McGraw-Hill, New York (1969).

Harper, J. L., J. T. Williams, and G. R. Sagar. The behavior of seeds in soil. *Journal of Ecology* 51: 273–286 (1965).

Hutchinson, G. E. *A Treatise on Limnology, Vol. 1: Geography, Physics, and Chemistry.* Wiley, New York (1957).

Lowry, W. P. *Weather and Life.* Academic Press, New York (1969).

Merriam, C. H. Laws of temperature control of the geographic distribution of terrestrial animals and plants. *National Geographic Magazine* 6: 229–238 (1894).

Thornthwaite, C. W. An approach to a rational classification of climate. *Geographical Review* 38: 55–94 (1948).

Chapter 8. The Diversity of Biological Communities

Bourlière, F. and M. Hadley. The ecology of tropical savannas. *Annual Review of Ecology and Systematics* 1: 125–152 (1970).

Braun, E. L. *Deciduous Forests of Eastern North America.* McGraw-Hill — Blakiston, New York (1950).

Carpenter, J. R. The biome. *American Midland Naturalist.* 21: 75–91 (1939).

Dansereau, P. *Biogeography — An Ecological Perspective.* Ronald Press, New York (1957).

Eltringham, S. K. *Life in Mud and Sand.* Crane Russak, New York (1971).

Friedlander, C. P. *Heathland Ecology.* Harvard University Press, Cambridge, Massachusetts (1961).

Hardy, A. *The Open Sea: Its Natural History.* Houghton Mifflin, Boston (1971).

Holdridge, L. *Life Zone Ecology.* Tropical Science Center, San Jose, Costa Rica (1967).

Jaeger, E. C. *The North American Deserts.* Stanford University Press, Stanford, California (1957).

Odum, E. P. *Fundamentals of Ecology* (3rd ed.). Saunders, Philadelphia (1971).

Pearsall, W. H. *Mountains and Moorlands.* Collins, London (1960).

Popham, E. J. *Life in Fresh Water.* Harvard University Press, Cambridge, Massachusetts (1961).

Raunkiaer, C. *The Life Forms of Plants and Statistical Plant Geography.* Clarendon Press, Oxford (1934).

Richards, P. W. *The Tropical Rainforest.* Cambridge University Press, London (1952).

Shimwell, D. W. *Description and Classification of Vegetation.* University of Washington Press, Seattle (1971).

Shelford, V. E. *The Ecology of North America.* University of Illinois Press, Urbana (1963).

Stephenson, T. A. and A. Stephenson. *Life Between Tidemarks on Rocky Shores.* W. H. Freeman, San Francisco (1972).

Teal, J. and M. Teal. *Life and Death of a Salt Marsh.* Little, Brown, Boston (1969).

Weaver, J. E. *Grasslands of the Great Plains.* Johnsen, Lincoln, Nebraska (1956).

Whittaker, R. H. *Communities and Ecosystems* (2nd ed.). Macmillan, New York (1975).

Chapter 9. Primary Production

Berry, J. A. Adaptation of photosynthetic processes to stress. *Science* 188: 644–650 (1975).

Bray, J. R. and E. Gorham. Litter production in forests of the world. *Advances in Ecological Research* 2: 101–157 (1964).

Goldman, C. R. Aquatic primary production. *American Zoologist* 8: 31–42 (1968).

Jordan, C. F. A world pattern in plant energetics. *American Scientist* 59: 425–433 (1971).

Mann, K. H. Seaweeds: their productivity and strategy for growth. *Science* 182: 975–981 (1973).

McLaren, I. A. Primary production and nutrients in Ogac Lake, a landlocked fiord on Baffin Island. *Journal of the Fisheries Research Board of Canada* 26: 1562–1576 (1969).

Mooney, H. A. The carbon balance in plants. *Annual Review of Ecology and Systematics* 3: 315–346 (1972).

Norman, A. G. Soil-plant relationships and plant nutrition. *American Journal of Botany* 44: 67–73 (1957).

Odum, H. T. Primary production in flowing waters. *Limnology and Oceanography* 1: 102–117 (1956).

Ovington, J. D., D. Heitkamp, and D. B. Lawrence. Plant biomass and productivity of prairie, savanna, oakwood, and maize field ecosystems in central Minnesota. *Ecology* 44: 52–63 (1963).

Teeri, J., and L. Stowe. Climatic patterns and the distribution of C_4 grasses in North America. *Oecologia* 23: 1–12 (1976).

Transeau, E. N. The accumulation of energy by plants. *Ohio Journal of Science* 26: 1–10 (1926).

Whittaker, R. H. and G. Likens. The primary production of the biosphere. *Human Ecology* 1: 299–369 (1973).

Wilde, S. A. Mycorrhizae and tree nutrition. *BioScience* 18: 482–484 (1968).

Wittwer, S. H. Maximum production capacity of food crops. *BioScience* 24: 216 (1974).

Woodwell, G. M. The energy cycle of the biosphere. *Scientific American* 223: 64–74 (1970).

Chapter 10. Energy Flow in the Community

Englemann, M. D. Energetics, terrestrial field studies, and animal productivity. *Advances in Ecological Research* 3: 73–115 (1966).

Harley, J. L. Fungi in ecosystems. *Journal of Animal Ecology* 41: 1–16 (1972).

Kozlovsky, D. G. A critical evaluation of the trophic level concept. I. Ecological efficiencies. *Ecology* 49: 48–60 (1968).

Lindeman, R. L. The trophic-dynamic aspect of ecology. *Ecology* 23: 399–418 (1942).

Mann, K. H. The dynamics of aquatic ecosystems. *Advances in Ecological Research* 6: 1–81 (1969).

Odum, E. P. Relationships between structure and function in the ecosystem. *Japanese Journal of Ecology* 12: 108–118 (1962).

Odum, H. T. Trophic structure and productivity of Silver Springs, Florida. *Ecological Monographs* 27: 55–112 (1957).

Phillipson, J. *Ecological Energetics.* Edward Arnold, London (1966).

Ryther, J. H. Photosynthesis and fish production in the sea. *Science* 166: 72–76 (1969).

Stark, N. Nutrient cycling pathways and litter fungi. *BioScience* 22: 355–360 (1972).

Teal, J. M. Energy flow in the salt marsh ecosystem of Georgia. *Ecology* 43: 614–624 (1962).

Wiegert, R. G. and D. F. Owen. Trophic structure, available resources and population density in terrestrial vs. aquatic ecosystems. *Journal of Theoretical Biology* 30: 69–81 (1971).

Chapter 11. Nutrient Cycling

Bormann, F. H., and G. E. Likens. Nutrient cycling. *Science* 155: 424–429 (1969).

Brady, N. C. *Nature and Properties of Soils* (8th ed.). Macmillan, New York (1974).

Chapin, F. S., III. The mineral nutrition of wild plants. *Annual Review of Ecology and Systematics* 11: 233–260 (1980).

Cloud, P. and A. Givor. The oxygen cycle. *Scientific American* 223: 111–123 (1970).

Delwiche, C. C. The nitrogen cycle. *Scientific American* 223: 137–146 (1970).

Hardy, R. W., and U. D. Havelka. Nitrogen fixation research: A key to world food? *Science* 188: 633–642 (1975).

Likens, G. E. 1972. Eutrophication and aquatic ecosystems. *American Society of Limnology and Oceanography, Special Symposium* 1: 3–13 (1972).

Likens, G. E., F. H. Bormann, R. S. Pierce, J. S. Eaton, and N. M. Johnson. *Biogeochemistry of a Forested Ecosystem.* Springer-Verlag, New York (1977).

Ovington, J. D. Organic production, turnover, and mineral cycling in woodlands. *Biological Reviews* 40: 295–336 (1965).

Quispel, A. (Ed.). *The Biology of Nitrogen Fixation.* North-Holland, Amsterdam (1974).

Schindler, D. W. Eutrophication and recovery in experimental lakes: implications for lake management. *Science* 184: 897–899 (1974).

Vitousek, P. M., J. R. Gosz, C. C. Grier, J. M. Mellilo, and W. A. Reiners. A comparative analysis of potential nitrification and nitrate mobility in forest ecosystems. *Ecological Monographs* 52: 155–177 (1982).

Chapter 12. Regulation and Homeostasis

Daubenmire, R. *Plants and Environment* (2nd ed.). Wiley, New York (1959).

Hochachka, P. W., and G. N. Somero. *Strategies of Biochemical Adaptation.* Saunders, Philadelphia (1973).

Langley, T. L. *Homeostasis.* Reinhold, New York (1965).

Irving, L. Adaptations to cold. *Scientific American* 214: 94–101 (1966).

Potts, W. T. W., and G. Parry. *Osmotic and Ionic Regulation in Animals.* Pergamon Press, Oxford (1964).

Ricklefs, R. E., and F. R. Hainsworth. Temperature dependent behavior of the cactus wren. *Ecology* 49: 227–233 (1968).

Schmidt-Nielson, K. *How Animals Work*. Cambridge University Press, London (1972).

Schmidt-Nielson, K. *Animal Physiology. Adaptation and Environment*. (2nd. ed.). Cambridge University Press, London and New York (1980).

Chapter 13. Organisms in Heterogeneous Environments

Brett, J. R. Some principles in the thermal requirements of fishes. *Quarterly Review of Biology* 31: 75–87 (1956).

Cogger, H. G. Thermal relations of the mallee dragon *Amphibolurus fordi* (Lacertilia: Agamidae). *Australian Journal of Zoology* 33: 319–339 (1974).

Dingle, H. Migration strategies of insects. *Science* 175: 1327–1335 (1972).

Hiesey, W. M. and H. W. Milner. Physiology of ecological races and species. *Annual Review of Plant Physiology* 16: 203–216 (1965).

Levins, R. *Evolution in Changing Environments. Some Theoretical Explorations*. Princeton University Press, Princeton (1968).

Niering, W. A., R. H. Whittaker, and C. H. Lowe. The saguaro: a population in relation to environment. *Science* 142: 15–23 (1963).

Stearns, S. C. The evolution of life history traits. *Annual Review of Ecology and Systematics* 8: 145–171 (1977).

Vepsalainen, K. The life cycles and wing lengths of Finnish *Gerris* Fabr. species (Heteroptera, Gerridae). *Acta Zoologica Fennica* 141: 1–73 (1974).

Wecker, S. C. Habitat selection. *Scientific American* 211: 109–116 (1964).

Wilbur, H. M. Complex life cycles. *Annual Review of Ecology and Systematics* 11: 67–93 (1980).

Chapter 14. Environment, Adaptation, and the Distribution of Organisms

Andrewartha, H. G. and L. C. Birch. *The Distribution and Abundance of Animals*. University of Chicago Press, Chicago (1954).

Billings, W. D. The environmental complex in relation to plant growth and distribution. *Quarterly Review of Biology* 27: 251–265 (1952).

Cain, S. A. Life-forms and phytoclimate. *Botanical Reviews* 16: 1–32 (1950).

Clausen, J., D. D. Keck, and W. M. Hiesey. Experimental studies on the nature of species. III. Environmental responses of climatic races of *Achillea*. *Carnegie Institution of Washington Publications* 581: 1–129 (1948).

Good, R. *The Geography of the Flowering Plants*. Longmans, London (1964).

Harrison, A. T., E. Small, and H. A. Mooney. Drought relationships and distribution of two Mediterranean-climate California plant communities. *Ecology* 52: 869–875 (1971).

Raunkaier, C. *The Life Form of Plants and Statistical Plant Geography*. Clarendon Press, Oxford (1934).

Turesson, G. The genotypic response of the plant species to the habitat. *Hereditas* 3: 211–350 (1922).

Waring, R. H. and J. Major. Some vegetation of the California coastal region in relation to gradients of moisture, nutrients, light, and temperature. *Ecological Monographs* 34: 167–215 (1964).

Chapter 15. Population Growth and Regulation

Caughley, G. *Analysis of Vertebrate Populations*. Wiley, New York (1977).

Davidson, J., and H. G. Andrewartha. The influence of rainfall, evaporation, and the atmospheric temperature on fluctuations in the size of a natural population of *Thrips imaginis* (Thysanoptera). *Journal of Animal Ecology* 17: 193–199 (1948).

Deevey, E. S., Jr. Life tables for natural populations of animals. *Quarterly Review of Biology* 22: 283–314 (1947).

Dunham, A. E. An experimental study of interspecific competition between the iguanid lizards *Sceloporus merriami* and *Urosaurus ornatus*. *Ecological Monographs* 50: 309–330 (1980).

Elton, C. *Voles, Mice and Lemmings. Problems in Population Dynamics.* Clarendon Press, Oxford (1942).

Frank, P. W., C. D. Boll, and R. W. Kelly. Vital statistics of laboratory cultures of *Daphnia pulex* De Geer as related to density. *Physiological Zoology* 30: 287–305 (1957).

Harcourt, D. G., and E. J. Leroux. Population regulation in insects and man. *American Scientist* 55: 400–415 (1967).

Krebs, C. J., and J. Myers. Population cycles in small mammals. *Advances in Ecological Research* 8: 267–399 (1974).

Lack, D. *The Natural Regulation of Animal Numbers.* Oxford University Press, London (1954).

Morris, R. F. The dynamics of epidemic spruce budworm populations. *Memoirs of the Entomological Society of Canada.* 31: 1–332 (1963).

Neilson, M. M., and R. F. Morris. The regulation of European spruce sawfly numbers in the maritime provinces of Canada from 1937 to 1963. *Canadian Entomologist* 96: 773–784 (1964).

Nicholson, A. J. An outline of the dynamics of animal populations. *Australian Journal of Zoology* 2: 9–65 (1954).

Nicholson, A. J. The self-adjustment of populations to change. *Cold Spring Harbor Symposium on Quantitative Biology* 22: 153–173 (1958).

Solomon, M. E. Analysis of processes involved in the natural control of insects. *Advances in Ecological Research* 2: 1–58 (1964).

Southwick, C. H. The population dynamics of confined house mice supplied with unlimited food. *Ecology* 36: 212–225 (1955).

Taber, R. D., and R. F. Dasmann. The dynamics of three natural populations of the deer *Odocoileus hemionus columbianus*. *Ecology* 38: 233–246 (1957).

Wilson, E. O., and W. H. Bossert. *A Primer of Population Biology.* Sinauer, Sunderland, Massachusetts (1971).

Wynne-Edwards, V. C. Population control in animals. *Scientific American* 211: 68–74 (1964).

Chapter 16. Competition

Birch, L. C. The meanings of competition. *American Naturalist* 91: 5–18 (1957).

Connell, J. H. The influence of interspecific competition and other factors on the distribution of the barnacle *Chthamalus stellatus*. *Ecology* 42: 710–723 (1961).

DeBach, P. The competitive displacement and coexistence principles. *Annual Review of Entomology* 11: 183–212 (1966).

DeBach, P., and R. A. Sundby. Competitive displacement between ecological homologues. *Hilgardia* 34: 105–166 (1963).

Fenchel, T. Character displacement and coexistence in mud snails. *Oecologia* 20: 19–32 (1975).

Gause, G. F. *The Struggle for Existence.* Williams & Wilkins, Baltimore (1934).

Hardin, G. The competitive exclusion principle. *Science* 131: 1292–1297 (1960).

Harper, J. L. The evolution and ecology of closely related species living in the same area. *Evolution* 15: 209–227 (1961).

Harper, J. L. Approaches to the study of plant competition. *Symposium of the Society of Experimental Biologists* 15: 1–39 (1961).

Harper, J. L. A Darwinian approach to plant ecology. *Journal of Ecology* 55: 247–270 (1967).

Marshall, D. R., and S. K. Jain. Interference in pure and mixed populations of *Avena fatua* and *A. barbata. Journal of Ecology* 57: 251–270 (1969).

Miller, R. S. Pattern and process in competition. *Advances in Ecological Research* 4: 1–74 (1967).

Muller, C. H. The role of chemical inhibition (allelopathy) in vegetational composition. *Bulletin of the Torrey Botanical Club* 93: 332–351 (1966).

Paine, R. T. Trophic relationships of eight sympatric predatory gastropods. *Ecology* 44: 63–73 (1963).

Park, T. Beetles, competition, and populations. *Science* 138: 1369–1375 (1962).

Rice, E. L. *Allelopathy*. Academic Press, New York and London (1975).

Schultz, A. M., J. L. Launchbaugh, and H. H. Biswell. Relationship between grass diversity and brush seedling survival. *Ecology* 36: 226–238 (1955).

Sharitz, R. R., and J. F. McCormick. Population dynamics of two competing annual plant species. *Ecology* 54: 723–740 (1973).

Smith, W. G. Dynamics of pure and mixed populations of *Desmodium glutinosum* and *D. nudiflorum* in natural oak-forest communities. *American Midland Naturalist* 94: 99–107 (1975).

Tansley, A. G. On competition between *Galium saxatile* L. (*G. hercynicum* Weig.) and *Galium sylvestre* Poll. (*G. asperum* Schreb.) on different tpes of soil. *Journal of Ecology* 5: 173–179 (1917).

Tilman, D. Ecological competition between algae: Experimental confirmation of resource-based competition theory. *Science* 192: 463–465 (1976).

Tilman, D. Resource competition between planktonic algae: An experimental and theoretical approach. *Ecology* 58: 338–348 (1977).

Chapter 17. Predation

Batzli, G. O., and F. A. Pitelka. Influence of meadow mouse populations on California grassland. *Ecology* 51: 1027–1039 (1970).

Brooks, J. L., and S. I. Dodson. Predation, body size and composition of the plankton. *Science* 150: 28–35 (1965).

Burnet, M., and D. O. White. *Natural History of Infectious Disease* (4th ed.). Cambridge University Press, London (1972).

DeBach, P. (ed.). *Biological Control of Insect Pests and Weeds*. Chapman & Hall, London (1964).

Dodd, A. P. The biological control of prickly pear in Australia. *Monographiae Biologiae* 8: 567–577 (1959).

Errington, P. L. The phenomenon of predation. *American Scientist* 51: 180–192 (1963).

Gross, J. E. Optimum yield in deer and elk populations. *Transactions of the North American Wildlife Conference* 34: 372–386 (1969).

Harper, J. L. The role of predation in vegetational diversity. *Brookhaven Symposia on Biology* 22: 48–62 (1969).

Holling, C. S. The components of predation as revealed by a study of small mammal predation of the European pine sawfly. *Canadian Entomologist* 91: 293–320 (1959).

Holling, C. S. The functional response of predators to prey density and its role in mimicry and population regulation. *Memoirs of the Entomological Society of Canada* 48: 1–85 (1966).

Huffaker, C. B. Experimental studies on predation: dispersal factors and predator-prey oscillations. *Hilgardia* 27: 343–383 (1958).

Huffaker, C. B. Life against life—nature's pest control scheme. *Environmental Research* 3: 162–175 (1970).

Ivlev, V. S. *Experimental Ecology of the Feeding of Fishes.* Yale University Press, New Haven (1961).

Janzen, D. H. *Ecology of Plants in the Tropics.* Edward Arnold, London (1975).

Lawton, J. H., J. R. Beddington, and R. Bonser. Switching in invertebrate predators. *In* M. B. Usher and M. H. Williamson (Eds.), *Ecological Stability.* Chapman and Hall, London (1974).

Lubchenco, J., and S. F. Gaines. A unified approach to marine plant-herbivore interactions. I. Populations and communities. *Annual Review of Ecology and Systematics* 12: 405–437 (1981).

Menge, B. A., and J. Lubchenco. Community organization in temperate and tropical rocky intertidal habitats: prey refuges in relation to consumer pressure gradients. *Ecological Monographs* 51: 429–450 (1981).

Murdoch, W. W. Switching in general predators: Experiments on predator specificity and stability of prey populations. *Ecological Monographs* 39: 335–354 (1969).

Paine, R. T. 1974. Intertidal community structure. Experimental studies on the relationship between a dominant competitor and its principal predator. *Oecologia* 15: 93–120 (1974).

Paine, R. T., and S. A. Levin. Intertidal landscapes: disturbance and the dynamics of pattern. *Ecological Mongraphs* 51: 145–178 (1981).

Paine, R. T., and R. Vadas. The effects of grazing by sea urchins, *Strongylocentrotus* spp., on benthic algal populations. *Limnology and Oceanography* 14: 710–719 (1969).

Pimentel, D. Population regulation and genetic feedback. *Science* 159: 1432–1437 (1968).

Price, P. W. General concepts on the evolutionary biology of parasites. *Evolution* 31: 405–420 (1977).

Ricker, W. E. Stock and recruitment. *Journal of the Fisheries Research Board of Canada* 11: 559–623 (1954).

Slobodkin, L. B. How to be a predator. *American Zoologist* 8: 43–51 (1968).

Utida, S. Population fluctuation, an experimental and theoretical approach. *Cold Spring Harbor Symposia on Quantitative Biology* 22: 139–151 (1957).

Chapter 18. Evolutionary Responses

Antonovics, J. The effects of a heterogeneous environment on the genetics of natural populations. *American Scientist* 59: 593–599 (1971).

Ayala, F. J. Evolution of fitness. IV. Genetic evolution of interspecific competitive ability in *Drosophila. Genetics* 61: 737–747.

Brower, L. P. Ecological chemistry. *Scientific American* 220: 22–29 (1969).

Cott, H. B. *Adaptive Coloration in Animals.* Oxford University Press, London (1940).

Feeny, P. Plant apparency and chemical defense. *Recent Advances in Phytochemistry* 10: 1–10 (1976).

Fenner, F. Evolution in action: myxomatosis in the Australian wild rabbit. Pp. 463–471 in A. Kramer (ed.). *Topics in the Study of Life. The Bio Source Book.* Harper & Row, New York (1971).

Gilbert, L. E., and P. H. Raven. (Eds.). *Coevolution of Animals and Plants.* University of Texas Press, Austin (1975).

Green, G. J. Virulence changes in *Puccinia graminis* f. sp. *tritici* in Canada. *Canadian Journal of Botany.* 53: 1377–1386 (1975).

Janzen, D. H. Seed-eaters versus seed size, number, toxicity and dispersal. *Evolution* 23: 1–27 (1969).

Kettlewell, H. B. D. Darwin's missing evidence. *Scientific American* 200: 48–53 (1959).

Lack, D. Evolutionary ecology. *Journal of Animal Ecology* 34: 223–231 (1965).

Levin, D. A. The chemical defenses of plants to pathogens and herbivores. *Annual Review of Ecology and Systematics* 7: 121–159 (1976).

Maynard-Smith, J. *The Theory of Evolution.* Penguin, Baltimore (1958).

Orians, G. H. Natural selection and ecological theory. *American Naturalist* 96: 257–263 (1962).

Park, T. Beetles, competition and populations. *Science* 138: 1369–1375 (1962).

Pimentel, D., E. H. Feinberg, P. W. Wood, and J. T. Hayes. Selection, spatial distribution, and the coexistence of competing fly species. *American Naturalist* 99: 97–109 (1965).

Robinson, M. H. Defenses against visually hunting predators. *Evolutionary Biology* 3: 225–259 (1969).

Whittaker, R. H., and P. O. Feeny. Allelochemics: chemical interactions between species. *Science* 171: 757–770 (1971).

Wickler, W. *Mimicry in Plants and Animals.* World University Library (1968).

Chapter 19. Extinction

Ehrlich, P., and A. Ehrlich. *Extinction.* Random House, New York (1981).

MacArthur, R. H., and E. O. Wilson. An equilibrium theory of insular zoogeography. *Evolution* 17: 373–387 (1963).

MacArthur, R. H., and E. O. Wilson. *The Theory of Island Biogeography.* Princeton University Press, Princeton, New Jersey (1967).

Ricklefs, R. E., and G. W. Cox. Taxon cycles in the West Indian Avifauna. *American Naturalist* 106: 195–219 (1972).

Simpson, G. G. History of the fauna of Latin America. *American Scientist* 38: 361–389 (1950).

Simpson, G. G. *The Major Features of Evolution.* Columbia University Press, New York (1953).

Simpson, G. G. *Splendid Isolation. The Curious History of South American Mammals.* Yale University Press, New Haven (1980).

Stehli, F. G., R. G. Douglas, and N. D. Newell. Generation and maintenance of gradients in taxonomic diversity. *Science* 164: 947–949 (1969).

Webb, S. D. A history of savanna vertebrates in the New World. Part II: South America and the great interchange. *Annual Review of Ecology and Systematics* 9: 393–426 (1978).

Wilson, E. O. The nature of the taxon cycle in the Melanesian ant fauna. *American Naturalist* 95: 169–193 (1961).

Chapter 20. Community Ecology

Borchert, J. R. The climate of the central North American grassland. *Annals of the Association of American Geographers* 40: 1–39 (150).

Clements, F. E. Nature and structure of the climax. *Journal of Ecology* 24: 252–284 (1936).

Cody, M. L., and H. A. Mooney. Convergence versus non-convergence in Mediterranean-climate ecosystems. *Annual Review of Ecology and Systematics* 9: 265–321 (1978).

Darlington, P. J., Jr. *Zoogeography: The Geographical Distribution of Animals.* Wiley, New York (1957).

Daubenmire, R. Vegetation: identification of typal communities. *Science* 151: 291–298 (1966).

Gleason, H. A. The individualistic concept of the plant association. *American Midland Naturalist* 21: 92–110 (1939).

Hairston, N. G., F. E. Smith, and L. B. Slobodkin. Community structure, population control, and competition. *American Naturalist* 94: 421–425 (1960).

Hungate, R. E. The rumen microbial ecosystem. *Annual Review of Ecology and Systematics* 6: 39–66 (1975).

Janzen, D. H. Coevolution of mutualism between ants and acacias in Central America. *Evolution* 20: 249–275 (1966).

Janzen, D. H. Herbivores and the number of tree species in tropical forests. *American Naturalist* 104: 501–528 (1970).

Lawton, J. H., and D. Schroder. Effects of plant type, size of geographical range and taxonomic isolation on number of insect species associated with British plants. *Nature* 265: 37–140 (1977).

McIntosh, R. P. The continuum concept of vegetation. *Botanical Review* 33: 130–187 (1967).

McMillan, C. Ecotypes and ecosystem function. *BioScience* 19: 131–134 (1969).

Murdoch, W. W. "Community structure, population control, and competition"—A critique. *American Naturalist* 100: 219–226 (1966).

Strong, D. R. Biogeographic dynamics of insect-host plant communities. *Annual Review of Entomology* 24: 89–119 (1979).

Watt, A. S. The community and the individual. *Journal of Ecology* 52 (Supplement): 203–211 (1964).

Watts, W. A. The late Quaternary vegetation history of the southeastern United States. *Annual Review of Ecology and Systematics* 11: 387–409 (1980).

Whittaker, R. H. Dominance and diversity in land plant communities. *Science* 147: 250–260 (1965).

Whittaker, R. H. Gradient analysis of vegetation. *Biological Reviews* 42: 207–264 (1967).

Chapter 21. Community Organization

Colwell, R. K., and E. R. Fuentes. Experimental studies of the niche. *Annual Review of Ecology and Systematics* 6: 281–310 (1975).

Darnell, R. M. Evolution and the ecosystem. *American Zoologist* 10: 9–15 (1970).

Davis, M. B. Pleistocene biogeography of temperate deciduous forests. *Geoscience and Man* 13: 13–26 (1976).

Dobzhansky, Th. Evolution in the tropics. *American Scientist* 38: 209–221 (1950).

Fischer, A. G. Latitudinal variation in organic diversity. *Evolution* 14: 64–81 (1960).

Haffer, J. Speciation in Amazonian forest birds. *Science* 165: 131–137 (1967).

MacArthur, R. H. Patterns of species diversity. *Biological Reviews* 40: 510–533 (1965).

MacArthur, R. H. Patterns of communities in the tropics. *Biological Journal of the Linnean Society* 1: 19–30 (1969).

MacArthur, R. H., H. Recher, and M. Cody. On the relation between habitat selection and species diversity. *American Naturalist* 100: 319–332 (1966).

MacArthur, R. H., and E. O. Wilson. *The Theory of Island Biogeography.* Princeton University Press, Princeton (1967).

Orians, G. H. 1969. The number of bird species in some tropical forests. *Ecology* 50: 783–801 (1969).

Paine, R. T. Food web complexity and species diversity. *American Naturalist* 100: 65–75 (1966).

Pianka, E. R. Latitudinal gradients in species diversity: a review of concepts. *American Naturalist* 100: 33–46 (1966).

Roughgarden, J. Niche width: biogeographic patterns among Anolis lizard populations. *American Naturalist* 108: 429–442 (1974).

Schoener, T. W. Resource partitioning in ecological communities. *Science* 185: 27–39 (1974).

Simberloff, D. Using island biogeographic distributions to determine if colonization is stochastic. *American Naturalist* 112: 713–726 (1978).

Simpson, G. G. Species density of North American recent mammals. *Systematic Zoology* 13: 57–73 (1964).

Strong, D. R., Jr., L. A. Szyska, and D. S. Simberloff. Tests of community-wide character displacement against null hypotheses. *Evolution* 33: 897–913 (1979).

Wiens, J. A. On competition in variable environments. *American Scientist* 65: 590–597 (1977).

Chapter 22. Community Development

Bazzaz, F. A. The physiological ecology of plant succession. *Annual Review of Ecology and Systematics* 10: 351–371 (1979).

Bazzaz, F. A., and S. T. A. Pickett. Physiological ecology of tropical succession: A review. *Annual Review of Ecology and Systematics* 11: 287–310 (1980).

Clements, F. E. Plant succession: analysis of the development of vegetation. *Carnegie Institute of Washington Publications.* 242: 1–512 (1916).

Clements, F. E. Nature and structure of the climax. *Journal of Ecology* 24: 252–284 (1936).

Connell, J. H. Diversity in tropical rainforests and coral reefs. *Science* 199: 1302–1310 (1978).

Connell, J. H., and R. O. Slatyer. Mechanisms of succession in natural communities and their role in community stability and organization. *American Naturalist* 111: 1119–1144 (1977).

Cooper, C. F. The ecology of fire. *Scientific American* 204: 150–160 (1961).

Crocker, R. L., and J. Major. Soil development in relation to vegetation and surface age at Glacier Bay, Alaska. *Journal of Ecology* 43: 427–448 (1955).

Curtis, J. T., and R. P. McIntosh. An upland forest continuum in the prairie-forest border region of Wisconsin. *Ecology* 32: 476–496 (1951).

Drury, W. H., and I. C. T. Nisbet. Succession. *Journal of the Arnold Arboretum* 54: 331–368 (1973).

Gorham, E., P. M. Vitousek, and W. A. Reiners. The regulation of chemical budgets over the course of terrestrial ecosystem succession. *Annual Review of Ecology and Systematics* 10: 53–84 (1979).

Keever, C. Causes of succession on old fields of the Piedmont, North Carolina. *Ecological Monographs* 20: 230–250 (1950).

Knapp, R. (ed.). *Vegetation Dynamics.* Junk, The Hague (1974).

McIntosh, R. P. Plant ecology 1947–1972. *Annals of the Missouri Botanical Garden* 61: 132–165 (1974).

Olson, J. S. Rates of succession and soil changes on southern Lake Michigan sand dunes. *Botanical Gazette* 119: 125–170 (1958).

Phillips, J. Succession, development, the climax, and the complex organism: an analysis of concepts. *Journal of Ecology* 22: 554–571, 23: 210–246, 488–508 (1934–1935).

Watt, A. S. Pattern and process in the plant community. *Journal of Ecology* 35: 1–22 (1947).

Whittaker, R. H. A consideration of climax theory: the climax as a population and pattern. *Ecological Monographs* 23: 41–78 (1953).

Woodwell, G. M. Success, succession, and Adam Smith. *BioScience* 24: 81–87 (1974).

Chapter 23. Stability of the Ecosystem

Daubenmire, R. Ecology of fire in grasslands. *Advances in Ecological Research* 5: 209–266 (1968).

Frank, P. W. Life histories and community stability. *Ecology* 49: 355–357 (1968).

Fritts, H. C. Growth-rings of trees: their correlation with climate. *Science* 154: 973–979 (1966).

Goodman, D. Stability and diversity. *Quarterly Review of Biology* 50: 237–266 (1975).

Hurd, L. E., M. V. Mellinger, L. L. Wolf, and S. J. McNaughton. Stability and diversity at three trophic levels in terrestrial successional ecosystems. *Science* 173: 1134–1136 (1971).

Kozlowski, T. T., and C. E. Ahlgren (eds.). *Fire and Ecosystems*. Academic Press, New York (1974).

Margalef, R. Diversity and stability: a practical proposal and a model of interdependence. *Brookhaven Symposia on Biology* 22: 25–37 (1969).

Murdoch, W. W. Switching in general predators: experiments on predator specificity and stability of prey populations. *Ecological Monographs* 39: 335–354 (1969).

Pimentel, D. Species diversity and insect population outbreaks. *Annals of the Entomological Society of America* 54: 76–86 (1961).

Preston, F. W. Diversity and stability in the biological world. *Brookhaven Symposia on Biology* 22: 1–12 (1969).

Thompson, L. M. 1975. Weather variability, climatic change, and grain production. *Science* 188: 535–543 (1975).

VanEmden, H. F., and G. F. Williams. Insect stability and diversity in agro-ecosystems. *Annual Review of Entomology* 19: 455–475 (1974).

Watt, K. E. F. Comments on fluctuations of animal populations and measures of community stability. *Canadian Entomologist* 96: 1434–1442 (1964).

Watt, K. E. F. Community stability and the strategy of biological control. *Canadian Entomologist* 97: 887–895 (1965).

Woodwell, G. M., and H. H. Smith (eds.). Diversity and Stability in Ecological Systems. *Brookhaven Symposia on Biology* 22 (1969).

Illustration Credits,
Acknowledgements, and References

Drawings and Photographs

Figure 2-1 Photographs courtesy of the U. S. Soil Conservation Service.
Figure 2-3 Photographs courtesy of Philip L. Boyd, Deep Canyon Research Center.

Figure 3-1 After H. B. D. Kettlewell. *Heredity* 12: 51–72 (1958).
Figure 3-2 From the experiments of H. B. D. Kettlewell.
Figures 3-3, 4 After K. Mather and B. K. Harrison. *Heredity* 3: 1–52, 129–162 (1949).

Figure 4-1 After R. Emerson and C. M. Lewis. *J. Gen. Physiol.* 25: 579–595 (1942).
Figure 4-2 After D. J. Randall. *Amer. Zool.* 8: 179–189 (1968).
Figure 4-3 Adapted from K. Schmidt-Nielsen, *Animal Physiology. Adaptation and Environment.* Cambridge Univ. Press, London (1975).

Figure 5-1 After R. S. Wimpenny, *The Plankton of the Sea.* Faber and Faber, London (1966).
Figure 5-2 Photograph courtesy of the U. S. Bureau of Commercial Fisheries.
Figure 5-3 After H. V. Sverdrup, *Oceanography for Meteorologists.* Allen and Unwin, London (1945).
Figure 5-4 Photographs courtesy of P. Green and M. V. Parthasarathy.
Figure 5-5 Photograph courtesy of the U. S. Fish and Wildlife Service.

Figures 6-1, 2 Photographs courtesy of the U. S. Soil Conservation Service.
Figure 6-3 After B. T. Bunting, *The Geography of Soil* (rev. ed.). Aldine, Chicago (1967).
Figure 6-4 After E. W. Russell, *Soil Conditions and Plant Growth* (9th ed.). Wiley, New York (1961).
Figure 6-5 After S. R. Eyre, *Vegetation and Soils: A World Picture* (2nd ed.). Aldine, Chicago (1968).
Figure 6-6, 7 After N. C. Brady, *Nature and Property of Soils* (8th ed.). Macmillan, New York (1974).
Figures 6-8, 9 Courtesy of the U. S. Soil Conservation Service.

Figure 7-1 After data in H. H. Clayton. *Smithsonian Misc. Coll.* 79: 1–1199 (1944).
Figure 7-3 After E. B. Espenshade, Jr. (ed.), *Goode's World Atlas* (13th ed.). Rand-McNally, Chicago (1971).
Figure 7-5 After A. C. Duxbury, *The Earth and Its Oceans.* Addison-Wesley, Reading, Massachusetts (1971).

Figure 7-6 After data in H. H. Clayton and F. L. Clayton. *Smithsonian Misc. Coll.* 105: 1–646 (1947).

Figure 7-7 Photograph courtesy of the U. S. Forest Service.

Figure 7-8 Rainfall data from H. H. Clayton. *Smithsonian Misc. Coll.* 79: 1–1199 (1944).

Figure 7-10 Photographs courtesy of the U. S. Soil Conservation Service, W. J. Smith, and R. H. Whittaker, from R. H. Whittaker and W. A. Niering. *Ecology* 46: 429–452 (1965).

Figure 7-11 After data from H. H. Clayton and F. L. Clayton. *Smithsonian Misc. Coll.* 105: 1–646 (1947).

Figure 7-12 After C. W. Thornthwaite. *Geogr. Rev.* 38: 55–94 (1948).

Figure 8-1 After H. A. Fowells, *Silvics of Forest Trees of the U. S.* Agricultural Handbook No. 271, U. S. Department of Agriculture (1965).

Figure 8-2 After P. Dansereau, *Biogeography — An Ecological Perspective.* Ronald, New York (1957).

Figure 8-3 After C. Raunkiaer, *Plant Life Forms.* Clarendon, Oxford (1937).

Figure 8-4 After compilations in P. W. Richards, *The Tropical Rainforest.* Cambridge Univ. Press, London (1952); P. Dansereau, *Biogeography — An Ecological Perspective.* Ronald, New York (1957); and R. Daubenmire, *Plant Communities.* Harper and Rowe, New York (1968).

Figure 8-5 After L. Holdridge, *Life Zone Ecology.* Tropical Science Center, San Jose, Costa Rica (1967).

Figure 8-7 After R. H. Whittaker, *Communities and Ecosystems* (2nd ed.). Macmillan, New York (1975).

Figure 9-1 After I. A. McLaren. *J. Fish. Res. Bd. Canada* 26: 1562–1576 (1969).

Figure 9-2 After D. M. Gates. *Brookhaven Symp. Biol.* 22: 115–126 (1969); and W. Larcher, A. Cernusca, L. Schmidt, G. Grabherr, E. Nötzel, and N. Smeets. In T. Roswall and O. W. Heal (eds.), *Structure and Function of Tundra Ecosystems.* Swedish National Research Council, Stockholm, Ecol. Bulletin 20.

Figure 9-3 After J. A. Berry. *Science* 188: 644–650 (1975).

Figure 9-4 Photograph courtesy of the U. S. Forest Service.

Figure 9-5 After R. H. Whittaker, *Communities and Ecosystems.* Macmillan, New York (1970).

Figure 9-6 After R. H. Whittaker and G. E. Likens. *Human Ecology* 1: 357–369 (1973).

Figure 10-6 Based on data in H. Welch, *Ecology* 49: 755–759 (1968) and data summarized by R. E. Ricklefs, *Ecology* (2nd ed.). Chiron, New York (1979).

Figure 10-7 After C. A. Edward and G. W. Heath. Pp. 76–84 in J. Doeksen and J. van der Drift (eds.), *Soil Organisms.* North-Holland, Amsterdam (1963).

Figure 10-8 Photograph courtesy of the U. S. National Park Service.

Figure 10-9 After E. P. Odum. *Jap. J. Ecol.* 12: 108–118 (1962).

Figure 11-2 After G. Borgstrom, *Too Many.* Macmillan, New York (1969) and G. E. Hutchinson, *A Treatise on Limnology,* Vol. 1. Wiley, New York (1957).

Figure 11-3 After I. Waldron and R. E. Ricklefs, *Environment and Population.* Holt, Rinehart, and Winston, New York (1973).

Figure 11-5 Photograph courtesy of the U. S. Soil Conservation Service.

Figure 11-6 Photograph courtesy of D. W. Shindler, from D. W. Shindler. *Science* 184: 897–899 (1974). Copyright 1974 by the Amer. Assoc. for the Advancement of Science.

Figures 11-7, 8 Photographs courtesy of the U. S. Forest Service.

Figure 11-9 From G. E. Likens, F. H. Bormann, R. S. Pierce and D. W. Fisher. Pp. 553–563 in *Productivity of Forest Ecosystems.* UNESCO (1971).

Figure 11-11 After data in G. E. Likens, F. H. Borman, N. M. Johnson, and R. S. Pierce. *Ecology* 48: 772–785 (1967).

Figure 12-1 After F. E. J. Fry and J. S. Hart. *J. Fish. Res. Bd. Canada* 7: 169–175 (1948).

Figure 12-3 After L. Irving. *Sci. Amer.* 214: 94–101 (1966).

Figure 12-5 From data of H. Werntz in C. L. Prosser and F. A. Brown, *Comparative Animal Physiology* (2nd ed.). Saunders, Philadelphia (1961).

Figure 12-6 Photograph courtesy of the U. S. Fish and Wildlife Service.

Figure 12-8 After G. T. Austin. *Condor* 76: 216–217 (1974).

Figure 13-1 After J. L. Harper, J. T. Williams, and G. R. Sagar. *J. Ecology* 53: 273–286 (1965).

Figure 13-3 After N. F. Hadley. *Amer. Scient.* 60: 338–347 (1970).

Figure 13-4 Photographs courtesy of H. G. Cogger, from H. G. Cogger. *Austr. J. Ecol.* 22: 319–339 (1974).

Figure 13-5 After W. A. Beckman, J. W. Mitchell, and W. P. Porter. *J. Heat Transfer*, May 1973: 257–262 (1973).

Figure 13-6 After G. C. West. *Comp. Biochem. Physiol.* 42A: 867–876 (1972).

Figure 13-7 After B. Wallace and A. M. Srb, *Adaptation* (2nd ed.). Prentice-Hall, Englewood Cliffs, New Jersey (1964).

Figure 13-8 Courtesy of K. Vepsalainen.

Figure 13-10 Photograph courtesy of the U. S. Department of Agriculture.

Figure 13-11 Photograph courtesy of the U. S. Forest Service.

Figures 14-1, 2 After H. A. Fowells, *Sylvics of Forest Trees of the U. S.* Agriculture Handbook No. 271, U. S. Dept. of Agriculture (1965).

Figure 14-3 After R. O. Erickson. *Ann. Mo. Bot. Garden* 32: 413–460 (1945).

Figure 14-4 After R. H. Waring and J. Major. *Ecol. Monogr.* 34: 167–215 (1964).

Figure 14-5 After E. W. Beals and J. B. Cope. *Ecology* 45: 777–792 (1964).

Figure 14-7 After F. S. Bodenheimer. *Bull. Soc. Entomol. Egypte* 1924: 149–157 (1925), and W. C. Allee, A. E. Emerson, O. Park, T. Park, and K. P. Schmidt, *Principles of Animal Ecology.* Saunders, Philadelphia (1949).

Figure 14-8 After R. F. Dasmann, *Wildlife Biology.* Wiley, New York (1964).

Figure 14-9 After C. McMillan. *Ecol. Monogr.* 26: 177–212 (1956).

Figure 14-12 After H. Hellmers, J. S. Horton, G. Juhren, and J. O'Keefe. *Ecology* 36: 667–678 (1955).

Figure 14-13 After A. T. Harrison, E. Small, and H. A. Mooney. *Ecology* 52: 869–875 (1970).

Figure 14-14 After J. Clausen, D. D. Keck, and W. M. Hiesey. *Carn. Inst. Wash., Publ.* 581: 1–129 (1948).

Figure 15-2 Photograph courtesy of the American Museum of Natural History.

Figure 15-3 Based on data in N. Keyfitz and W. Flieger, *World Population. An Analysis of Vital Data.* Univ. of Chicago Press, Chicago (1968).

Figure 15-4 Photograph courtesy of the American Museum of Natural History.

Figure 15-8 After P. W. Frank, C. E. Bell, and R. W. Kelly. *Physiol. Zool.* 30: 287–305 (1957).

Figure 15-8 Modified from R. Laughlin. *J. Anim. Ecology* 34: 77–91 (1965).

Figure 15-9 Photograph courtesy of J. Ewing.

Figure 15-11 After J. Davidson. *Trans. Roy. Soc. South Australia* 62: 342–346 (1938).

Figure 15-12 After C. C. Davis. *Limnol. Oceanogr.* 9: 275–283 (1964).

Figure 15-13 After J. Davidson and H. G. Andrewartha. *J. Anim. Ecol.* 17: 193–199 (1948).

Figure 15-14 Photograph courtesy of the U. S. Forest Service.

Figure 15-15 After M. M. Neilson and R. F. Morris. *Canad. Entomol.* 96: 773–784 (1964).

Figure 15-16 After B. A. MacLulich. *Univ. Toronto Stud., Biol. Ser.* 43 (1937).

Figure 15-17 Photograph courtesy of the U. S. Forest Service.

Figures 15-18, 19 After A. J. Nicholson. *Cold Spring Harbor Symp. Quant. Biol.* 22: 1153–173 (1958).

Figure 15-20 Photograph courtesy of the U. S. Bureau of Sport Fisheries and Wildlife.

Figure 15-21 After C. H. Southwick. *Ecology* 36: 212–225, 627–634 (1955).

Figure 15-22 Photograph courtesy of the U. S. Bureau of Sport Fisheries and Wildlife.

Figure 16-1 Photograph courtesy of the U. S. Bureau of Sport Fisheries and Wildlife.

Figure 16-2 After G. F. Gause, *The Struggle for Existence.* Williams and Wilkens, Baltimore (1934).

Figure 16-3 After H. A. Bess, R. van den Bosch, and F. A. Haramoto. *Proc. Hawaiian Entomol. Soc.* 17: 367–378 (1961).

Figures 16-4, 5 After P. DeBach and R. A. Sundby. *Hilgardia* 34: 105–166 (1963).

Figures 16-7, 8 After J. L. Harper. *J. Ecology* 55: 247–270 (1967).

Figures 16-8, 9, 10 After E. R. Marshall and S. A. Jain. *J. Ecology* 57: 251–270 (1969).

Figure 16–11 After W. G. Smith. *Amer. Midl. Nat.* 94: 99–107 (1975).

Figure 16-12 After T. Park. *Physiol. Zool.* 27: 177–238 (1954).

Figures 16-13, 14 After R. D. Wright and H. A. Mooney. *Amer. Midl. Nat.* 73: 257–284 (1965).

Figure 16-16 After J. T. Schulz, *Ecological Studies on Rain Forests in Northern Suriname.* North-Holland, Amsterdam (1960).

Figure 16-17 Courtesy of the U. S. Forest Service.

Figures 16-18, 19 After A. M. Schultz, J. L. Launchbaugh, and H. H. Biswell. *Ecology* 36: 226–238 (1955).

Figure 16-20 After R. I. Yeaton and M. L. Cody. *Theoret. Pop. Bio.* 5: 42–58 (1974).

Figure 16-21 Photographs courtesy of C. H. Muller, from C. H. Muller. *Bull. Torrey Bot. Club* 93:332–351 (1966).

Figure 16-23 After R. H. MacArthur. *Ecology* 39: 599–619 (1958).

Figure 16-24 After R. W. Storer. *Auk* 83: 423–436 (1966).

Figure 16-26 After D. Lack, *Darwin's Finches.* Cambridge Univ. Press, London (1947).

Figure 17-1 Photograph courtesy of the U. S. Department of Agriculture.

Figure 17-2 After C. B. Huffaker and C. E. Kennett. *Hilgardia* 26: 191–222 (1956).

Figure 17-3 Photograph courtesy of W. H. Haseler, Dept. of Lands, Queensland, Australia.

Figure 17-4 After S. Utida. *Cold Spring Harbor Symp. Quant. Biol.* 22: 139–151 (1957).

Figure 17-5 Photograph courtesy of the U. S. Dept. of Agriculture.

Figure 17-6 Photographs courtesy of C. B. Huffaker. From C. B. Huffaker. *Hilgardia* 27: 343–383 (1958).

Figures 17-7, 8 After C. B. Huffaker. *Hilgardia* 27: 343–383 (1958).

Figure 17-12 After C. S. Holling. *Canad. Entomol.* 91: 293–320 (1959).

Figure 17-13 After V. S. Ivlev, *Experimental Ecology of the Feeding of Fishes.* Yale University Press, New Haven, Connecticut (1961).

Figure 17-15 After C. S. Holling. *Canad. Entomol.* 91: 293–320 (1959).

Figure 17-17 After J. A. Gulland. In B. D. LeCren and M. W. Holgate (eds.), *The Natural Exploitation of Natural Populations.* Wiley, New York (1962).

Figure 17-18 After F. O. Batzli and F. A. Pitelka. *Ecology* 51: 1027–1039 (1970).

Figure 17-19 After R. M. Belyea. *J. Forestry* 50: 729–738 (1952).

Figure 18-1 After F. Fenner and F. N. Ratcliffe, *Myxomatosis.* Cambridge University Press, London (1965).

Figure 18-2 Photograph courtesy of D. Pimentel, from D. Pimentel, W. P. Nagel, and J. L. Madden. *Amer. Natur.* 97: 141–167 (1963).

Figure 18-3 After D. Pimentel, E. H. Feinberg, P. W. Wood, and J. T. Hayes. *Amer. Natur.* 99: 97–109 (1965).

Figure 18-4 Photograph courtesy of D. Pimentel, from D. Pimentel. *Science* 159: 1432–1437 (1968). Copyright 1968 by the American Association for the Advancement of Science.

Figure 18-5 After D. Pimentel. *Science* 159: 143–1437 (1968).

Figure 18-8 Photographs courtesy of Thomas Eisner, from R. E. Silberglied and T. Eisner. *Science* 163: 486–488 (1969).

Figure 19-1 After R. E. Ricklefs and G. W. Cox. *Amer. Natur.* 106: 195–219 (1972).

Figure 19-2 After A. S. Romer, *Vertebrate Paleontology* (3rd ed.). Univ. Chicago Press, Chicago (1966).

Figure 19-3 After N. D. Newell. *Geol. Soc. Amer. Spec. Pap.* 89: 63–91 (1967).

Figure 19-4 After G. G. Simpson. *Amer. Sci.* 38: 361–389 (1950).

Figure 20-3 After C. D. White. Ph. D. Diss., University of Oregon (1971).

Figure 20-4 After H. A. Fowells, *Silvics of Forest Trees of the United States.* U. S. Department of Agriculture Handbook No. 271 (1965).

Figure 20-5 After R. H. Whittaker. *Ecol. Monogr.* 30: 279–338 (1960); and R. H. Whittaker and W. A. Niering. *Ecology* 46: 429–452 (1965).

Figure 20-6 After R. H. Whittaker. *Ecol. Monogr.* 26: 1–80 (1956).

Figure 20-7 After R. L. Dressler. *Evolution* 22: 202–210 (1968).

Figure 20-8 After J. A. Powell and R. A. Mackie. *Univ. Calif. Publ. Entomol.* 42: 1–46 (1966).

Figure 20-9 Photographs courtesy of D. Janzen, from D. H. Janzen. *Evolution* 20: 249–275 (1966).

Figure 21-1 After F. G. Stehli, in E. T. Drake (ed.), *Evolution and Environment.* Yale University Press, New Haven (1968).

Figure 21-3 From R. M. Darnell. *Amer. Zool.* 10: 9–15 (1970).

Figures 21-4, 5, 6 After R. H. MacArthur and E. O. Wilson. *Evolution* 17: 373–387 (1963).

Figure 21-7 After R. H. MacArthur. *Biol. J. Linnean Soc.* 1: 19–30 (1969).

Figure 21-9 Photograph courtesy of J. W. Porter from J. W. Porter. *Bull. Biol. Soc. Wash.* 2: 89–116 (1972).

Figure 22-1 Photographs by Z. Glowacinski, courtesy of O. Jarvinen from Z. Glowacinski and O. Jarvinen. *Ornis Scand.* 6: 33–40 (1975).

Figure 22-2 Photographs courtesy of the U. S. Soil Conservation Service.

Figure 22-3 Photograph courtesy of the U. S. Forest Service.

Figure 22-6 After H. J. Oosting, *The Study of Plant Communities* (2nd ed.). W. H. Freeman, San Francisco (1956).

Figure 22-7 After J. S. Beard. *Ecology* 36: 89–100 (1955); and R. H. Whittaker, *Communities and Ecosystems.* MacMillan, New York (1970).

Figure 22-8 After J. T. Curtis and R. E. McIntosh. *Ecology* 32: 476–496 (1951).

Figure 22-10 After R. L. Dix. *Ecology* 38: 663–665 (1957).

Figure 22-11 After W. A. Eggler. *Ecology* 19: 243–263 (1938).

Figure 22-13 After A. S. Watt. *J. Ecol.* 35: 1–22 (1947).
Figure 22-14 After J. P. Grime and D. W. Jeffrey. *J. Ecol.* 53: 621–642 (1965).

Figures 23-2, 3 Photographs courtesy of the U. S. Forest Service.
Figures 23-4, 5 After H. Fritts. Pp. 45–65 in *Ground Level Climatology*, AAAS, Washington (1967).
Figure 23-6 After F. G. Stehli, R. G. Douglas, and N. D. Newell. *Science* 164: 947–949 (1969).
Figures 23-8, 9 Photographs courtesy of the U. S. Forest Service.
Figure 23-10 After R. Patrick. *Ann. N. Y. Acad. Sci.* 108: 353–358 (1963).

Tables

Table 4-2 Compiled from G. K. Reid, *Ecology of Inland Waters and Estuaries.* Reinhold, New York (1961); and M. S. Gordon, *Animal Function: Principles and Adaptations.* MacMillan, New York (1968).

Table 9-1 From data in R. H. Whittaker and G. E. Likens. *Human Ecology* 1: 357–370 (1973).

Table 10-3 From data in R. H. Whittaker and G. E. Likens. Human Ecology 1: 357–370 (1973).
Table 10-4 After R. G. Weigert and D. F. Owen. *J. Theoret. Biol.* 30: 69–81 (1971).
Table 10-5 After R. L. Lindeman. *Ecology* 23: 399–418 (1942).
Table 10-6 After R. L. Lindeman (*loc. cit.*); and H. T. Odum. *Ecol. Monogr.* 27: 55–112 (1957).

Table 11-1 Based partly on E. G. Kormondy, *Concepts of Ecology.* Prentice-Hall, Englewood Cliffs, New Jersey (1969).
Table 11-2 After M. Witcamp. *Ecology* 47: 194–201 (1966).
Table 11-3 From A. Carlisle, A. H. F. Brown, and E. J. White. J. *Ecology* 54: 87–98 (1966); G. E. Likens, F. H. Bormann, N. M. Johnson, and R. S. Pierce. *Ecology* 48: 772–785 (1967); and P. Duvigneaud and S. Denaeyer- de Smet. Pp. 199–225 in D. E. Reichle (ed.), *Analysis of Temperate Forest Ecosystems.* Springer-Verlag, New York (1970).
Table 11-4 From J. D. Ovington. *Biological Reviews* 40: 295–336 (1965), and P. Duvigneaud and S. Denaeyer-de Smet (*loc. cit.*).
Table 11-5 From F. Jurion and J. Henry, *Can Primitive Farming Be Modernized?* National Institute for Agricultural Studies, Congo (1969).

Table 13-1 From F. H. Bormann. In K. V. Thimann (ed.), *The Physiology of Forest Trees.* Ronald Press, New York (1958).
Table 13-2 From K. Vepsalainen. *Acta Zool. Fenn.* 141:173 (1974), and *Ann. Acad. Sci. Fenn.*, Series A. IV, 202:1–18 (1974).

Table 14-1 After W. S. Johnson, A. Gigon, S. L. Gulmon, and H. A. Mooney. *Ecology* 55: 450–453 (1974).
Table 14-2 After A. T. Harrison, E. Small, and H. A. Mooney. *Ecology* 52: 869–875 (1971); and H. A. Mooney and E. L. Dunn. *Amer. Natur.* 104: 447–453 (1970).

Table 15-1 After data of O. Murie, in E. S. Deevey, Jr. *Quart. Rev. Biol.* 22: 283–314 (1947).
Table 15-3 Census data from Audubon Field Notes; production estimates from R. H. Whittaker and G. E. Likens. *Human Ecology* 1: 357–370 (1973).
Table 15-4 From E. L. Cheatum and C. W. Severinghaus. *Trans. N. Amer. Wildl. Conf.* 15: 170–189 (1956).

Table 15-5 From H. G. Andrewartha. *J. Scient. Indust. Res. Austr.* 8: 281–288 (1935).

Table 15-6 After D. A. Mullen. *Univ. Calif. Publ. Zool.* 85: 1–24 (1968).

Table 16-1 Data from G. W. Cox and R. E. Ricklefs. *Oikos* 29: 60–66 (1977).

Table 17-1 From C. H. Buckner and W. J. Turnock. *Ecology* 46: 223–236 (1965).

Table 17-2 From F. A. Pitelka, P. O. Tomich, and G. W. Treichel. *Ecol. Monogr.* 25: 85–117 (1955).

Table 17-3 From various sources given in Table 33-3 of R. E. Ricklefs, *Ecology* (2nd ed.). Chiron, New York (1979).

Table 19-1 From F. Harper, *Extinct and Vanishing Mammals of the World.* American Committee International Wildlife Protection, New York (1945).

Tables 19-2, 3 From R. E. Ricklefs and G. W. Cox. *Amer. Natur.* 106: 195–219 (1972).

Table 21-1 After E. J. Tramer. *Ecology* 50: 927–929 (1969); productivity data from R. H. Whittaker, *Communities and Ecosystems.* Macmillan, New York (1970).

Table 21-2 From H. G. Andrewartha and L. C. Birch, *The Distribution and Abundance of Animals.* University of Chicago Press, Chicago (1954).

Table 21-3 From R. H. MacArthur and E. O. Wilson, *The Theory of Island Biogeography.* Princeton University Press, Princeton, New Jersey (1967).

Table 22-1 From W. A. Eggler. *Ecology* 19: 243–263 (1938).

Table 23-1 After G. D. V. Williams, M. I. Joynt, and P. A. McCormick. *Canad. J. Soil Sci.* 55: 43–53 (1975).

Table 23-2 Based on R. E. Ricklefs. In D. S. Farner (ed.), *Breeding Biology of Birds.* National Research Council, Washington, D. C. (1973).

Table 23-3 Based on Table 43-3 in R. E. Ricklefs, *Ecology* (2nd ed.). Chiron, New York, (1979).

Index

The Author

ROBERT E. RICKLEFS was born in San Francisco in 1943. He graduated
from Stanford University in 1963 and received a Ph. D. from the University
of Pennsylvania in 1967. After a year of post-doctoral study at the Smithsonian Tropical Research Institute, he joined the faculty of the University
of Pennsylvania, where at present he is Professor of Biology. In 1974, he
received a Guggenheim Fellowship for ecological and evolutionary studies
of reproduction in birds. His other interests in research include the ecology
and evolution of island populations and the organization of biological communities. In addition to *The Economy of Nature*, he is author of *Ecology,
Second Edition* (1979) and co-author of *Environment and Population: Problems and Solutions* (1973).